PROTEIN DISCOVERY TECHNOLOGIES

RENATA PASQUALINI
WADIH ARAP

CRC Press
Taylor & Francis Group
Boca Raton London New York

CRC Press is an imprint of the
Taylor & Francis Group, an **informa** business

CRC Press
Taylor & Francis Group
6000 Broken Sound Parkway NW, Suite 300
Boca Raton, FL 33487-2742

First issued in paperback 2019

© 2009 by Taylor and Francis Group, LLC
CRC Press is an imprint of Taylor & Francis Group, an Informa business

No claim to original U.S. Government works

ISBN-13: 978-0-8247-5468-6 (hbk)
ISBN-13: 978-0-367-38538-5 (pbk)

Library of Congress Cataloging-in-Publication Data

Protein discovery technologies / [edited by] Renata Pasqualini and Wadih Arap.
    p. cm. -- (Drug discovery series ; 12)
  Includes bibliographical references and index.
  ISBN 978-0-8247-5468-6 (hard back : alk. paper)
  1. Proteins--Biotechnology. 2. Discoveries in science--Anecdotes. 3. Scientists--Anecdotes. I. Pasqualini, Renata. II. Arap, Wadih.

TP248.65.P76P728 2010
660.6'3--dc22
                                       2009021468

Visit the Taylor & Francis Web site at
http://www.taylorandfrancis.com

and the CRC Press Web site at
http://www.crcpress.com

Dedication

M. Judah Folkman, M.D. (1933–2008)

Dr. Folkman would have liked this book. His contribution was to have been a chapter entitled "Cryptic endogenous angiogenesis inhibitory proteins," but his untimely death precluded our reading yet another of his well-crafted, insightful, and thoroughly creative reviews. The editors and many of the authors concur that a book on protein discoveries was a highly appropriate tribute to a valued clinician-scientist and, to many of us, an uncommon colleague and friend.

As professor of pediatric surgery and cell biology at Harvard Medical School, and founder and director of the Vascular Biology Program at Children's Hospital Boston, Dr. Folkman left an enviable legacy of research, especially on angiogenesis and its regulation of tumor growth. Importantly, he will also be remembered for the trainees who have since made seminal contributions to biomedical science. He believed strongly in young, aspiring scientists and was the epitome of mentors—his questions and encouragement of interesting hypotheses in lab meetings are legendary (and true), and those who trained with him will not forget the enticements he offered for the solution of quirky problems, his favored mode of thinking.

Dr. Folkman's contributions to the fields of oncology and vascular biology have led to anti-angiogenic therapy for the remission of solid tumors in more than a million cancer patients. Over a thousand laboratories are devoted to understanding basic mechanisms regulating angiogenesis and to the development of angiogenesis inhibitors for the treatment of cancer and other diseases such as psoriasis, diabetic retinopathy, and macular degeneration. These milestones arose essentially from Dr. Folkman's observations as a surgeon—that many "successful" tumors were highly vascularized, and led to the over-arching hypothesis that a tumor, starved of its blood supply, would regress and undergo limited metastasis. Dr. Folkman spent nearly forty years refining, extending, redirecting, and remodeling this hypothesis, and his papers demonstrate his tenacity and vision, despite the disbelief and opposition he encountered along the way. He was a fine example of optimism for his students (teaching was a major priority for him), and the scope with which he viewed biology was both singular and admirable.

My own experience with Dr. Folkman was not as extensive as that of some of his trainees and colleagues, but it was a major highlight of my career in vascular and extracellular matrix biology. It was his scope that most intrigued me. We met 30 years ago, at an International Cell Biology symposium, where I was presenting a somewhat boring poster (my first) on procollagen biosynthesis by bovine aortic endothelial cells (the reliable workhorses in the early days of vascular biology). How could he have been interested in this sort of biochemistry, by an unknown postdoc?

But he was—and the questions were provocative: "Are these cells good models for angiogensis? (Not really.) Do they produce a basement membrane? (No.) Is type III procollagen an ECM component that regulates endothelial cell behavior? (Probably not.) Could you study other proteins that these cells produce? (I will try.) And, do you think culturing these cells reproduces some aspects of endothelial responses to injury? (Yes, but we need to prove it.) After working on these and other questions, I spent a sabbatical year (1992-1993) in the Folkman laboratory, on the angiostatin project. It was an exciting year, and a new direction emerged, as reflected on the lab blackboard, concerning the role of proteolytic degradation products in the regulation of angiogenesis, as can be seen from the title of Dr. Folkman's chapter that was intended for this volume.

After years of experimentation and creative contributions to the field of vascular biology, Dr. Folkman became the recipient of several coveted honors, including membership in the National Academy of Sciences and in the Institute of Medicine of the National Academy of Sciences. He was the Julia Dyckman Andrus Professor of Pediatric Surgery for forty years and served as surgeon-in-chief at the Children's Hospital from 1967-1981. He was devoted to his patients (to accompany him on his rounds I had to wear a skirt) but also felt that solving some of the mechanisms of angiogenesis required his full-time attention.

A perusal of chapter titles and their authors featured in this book on protein discovery reveals a statistically significant representation of the science that was influenced by Dr. Folkman's ideas, data, and enthusiasm. Many of us will recall the following, as one of his many insights: "We must all work very hard, nearly all the time, because we have so little time on this earth."

E. Helene Sage
Benaroya Research Institute at George Mason University
and University of Washington

Contents

Preface

This protein discovery "storytelling book" has actually been commissioned for a long time. We had been eagerly waiting for the pivotal contribution of Dr. Judah Folkman; having very much appreciated how incredibly busy Dr. Folkman surely was, we still remember how pleasantly surprised we were that he had kindly accepted our invitation. Of course, we were more than glad to wait for his promised chapter for as long as it would take—indeed several years. Finally, in early January, we were delighted to receive a fax with a firm deadline and his chapter outline for approval. But, ironically, it was just not meant to be ... within a few days, Dr. Folkman unexpectedly passed. While many honors were appropriately bestowed posthumously on Dr. Folkman, we asked his close friend and colleague, Dr. Helene Sage, to write a dedicatory foreword. There is a valid reason for it. Many discoveries are related to angiogenesis: Dr. Harold Dvorak tells the beginning of a fascinating saga of how vascular permeability factor was originally found; Drs. Michael Klagsburn and Yuen Shing remember the protein purification of basic fibroblast growth factor through binding to heparin; and Dr. Bruce Zetter discusses the use of interferon as an early agent against angiogenesis.

Other related developments in endothelial cell biology, extracellular matrix, and cell adhesion molecules are also represented. For example, we discuss the ligand-directed combinatorial mapping of the human vascular endothelium in patients. Moreover, Dr. Sage describes the coining of the term "matricellular protein" by using SPARC as the prototypic discovery of this class; Dr. Paul Bornstein talks of his pivotal finding of cross-linking in collagen-based matrix; and Dr. Raghu Kalluri and his colleagues have a new take on tumor matrix-derived fragments with regulatory attributes in tumor growth and neovascular formation. Finally, Dr. Dean Sheppard describes TGFβ activation as a central function of αvβ6 integrin-mediated cell adhesion; Dr. Erkki Koivunen and his colleagues discuss phage-display methodology and the finding of cell migration peptide inhibitors.

The stories of the seminal discovery of several other protein superfamilies are told here as well. Dr. Bharat Aggarwal and his colleagues reported the large protein family of tumor necrosis factor; Dr. Amy Lee introduces the first isolation and cloning of glucose-regulated proteins as stress-response chaperones; and Dr. Ricardo Brentani and his collaborators discuss their attempts to identify cellular prion protein receptors. Finally, Drs. Claudio Joazeiro and Tony Hunter tell the story behind their finding that RING finger proteins serve as E3 ubiquitin ligases; Drs. Jean-Francois Côté and Kristiina Vuori discover the identification of the evolutionarily conserved superfamily of DOCK180-related proteins with guanine nucleotide exchange activity.

Protein discovery may well be determined by specific pathophysiology settings in animal models of nervous or cardiac disease. Drs. Pierre-Olivier Couraud and Sandrine Bourdoulous discuss the blood–brain barrier in the context of central nervous systems diseases; Drs. Ma-Li Wong and Julio Licinio report the discovery of an endotoxin-induced myocardial disease.

Of course, molecular discoveries are not only necessarily related to proteins but to lipids, too; indeed, Dr. Richard Kolesnick tells the interesting story of the sphingo-myelin pathway.

In summary, rather than being an exhaustive or in any way comprehensive book, we merely asked a selected group of authors to write the (sometimes untold) personal stories behind their scientific discoveries. We hope that the younger researchers may learn from these short discovery vignettes and that more experienced investigators may perhaps enjoy reading them or even gain a new appreciation for some of the stories told.

Wadih Arap, MD, PhD
Stringer Professor of Medicine and Cancer Biology

Renata Pasqualini, PhD
Buchanan & Seeger Professor of Medicine and Cancer Biology

Editors

Renata Pasqualini, PhD is the Buchanan & Seeger Professor of Medicine and Cancer Biology at the University of Texas M. D. Anderson Cancer Center. She received her PhD in biochemistry at the Ludwig Institute for Cancer Research of the University of São Paulo, Brazil. She pursued postdoctoral training at Children's Hospital of Boston and at Dana Farber Cancer Institute, Harvard Medical School, and at the Burnham Institute in La Jolla, California.

Dr. Pasqualini is an internationally recognized expert in vascular biology, metastasis, and angiogenesis. She originally codeveloped in vivo phage display. This technology enables combinatorial mapping of tissue and disease-specific molecular addresses in vivo, allowing for the development of ligand-directed targeted delivery of therapeutic and imaging agents. Dr. Pasqualini has a close and long-standing scientific collaboration with Dr. Wadih Arap; together, they run a large research laboratory with several graduate and medical students, postdoctoral fellows, and technical and clerical staff. She serves on grant review boards for several national and international funding agencies. Drs. Arap and Pasqualini lead an active drug development program. "Safe to proceed" status for their first investigational new drug application has recently been granted by the Food and Drug Administration.

Wadih Arap, MD, PhD is the Stringer Professor of Medicine and Cancer Biology and serves as an attending physician and deputy chairman of the Department of Genitourinary Medical Oncology at the University of Texas M. D. Anderson Cancer Center. Dr. Arap received his MD from the University of São Paulo Medical School, Brazil, and his PhD in cancer biology from Stanford University Medical School with his cancer genetics thesis completed at the Ludwig Institute for Cancer Research, La Jolla Branch.

Dr. Arap formally trained at Memorial Sloan-Kettering Cancer Center as a Medical Oncology and Hematology Fellow. He received postdoctoral training and was promoted to staff scientist at the Burnham Institute in La Jolla, California, where he pioneered, developed, and optimized ligand-directed vascular-targeted therapy.

Dr. Arap serves on grant review boards for several national and international funding agencies. He is a member of the Board of Scientific Counselors of the National Cancer Institute. As a physician-scientist, his interests include the translation of biotechnology and the molecular diversity of human blood vessels into targeted clinical applications.

Contributors

Bharat B. Aggarwal, PhD
Department of Bioimmunotherapy
  Research
University of Texas M. D. Anderson
  Cancer Center
Houston, Texas

Wadih Arap, MD, PhD
Department of Genitourinary
  Medical Oncology
University of Texas M. D. Anderson
  Cancer Center
Houston, Texas

Alok C. Bharti
Department of Bioimmunotherapy
University of Texas M. D. Anderson
  Cancer Center
Houston, Texas

Mikael Björklund
Department of Biosciences
Division of Biochemistry
University of Helsinki
Helsinki, Finland

Paul Bornstein
Department of Biochemistry
University of Washington
Seattle, Washington

Sandrine Bourdoulous, PhD
Laboratoire d'Immuno-Pharmacologie
  Moléculaire
Centre National de la Recherche
  Scientifique
Institut Cochin de Génétique
  Moléculaire
Paris, France

Ricardo R. Brentani
Ludwig Institute for Cancer Research
São Paulo, Brazil

Jean-François Côté, PhD
Clinical Research Institute of Montreal
Montreal, Quebec, Canada

Pierre-Olivier Couraud, PhD
Departement de Biologie Cellulaire
Institut Cochin de Génétique
  Moléculaire
Paris, France

Sandro J. De Souza
Computational Biology Group
Ludwig Institute for Cancer Research
São Paulo, Brazil

Harold F. Dvorak, MD
Beth Israel Deaconess Medical Center
Harvard Medical School
Boston, Massachusetts

Tanjore Harikrishna
Department of Medicine
Beth Israel Deaconess Medical Center
Harvard Medical School
Boston, Massachusetts

Tony Hunter, PhD
The Salk Institute
Molecular and Cell Biology Laboratory
La Jolla, California

Claudio A. P. Joazeiro, PhD
The Scripps Research Institute
Department of Cell Biology
and
The Salk Institute for Biological Studies
La Jolla, California

Raghu Kalluri, PhD
Department of Medicine
Beth Israel Deaconess Medical Center
Harvard Medical School
Boston, Massachusetts

Aino Kangasniemi
Department of Biosciences
Division of Biochemistry
University of Helsinki
Helsinki, Finland

Michael Klagsbrun, PhD
Children's Hospital
Harvard Medical School
Boston, Massachusetts

Erkki Koivunen, PhD
Department of Genitourinary
    Medical Oncology
University of Texas M. D. Anderson
    Cancer Center
Houston, Texas

Richard Kolesnick, PhD
Molecular Pharmacology and
    Chemistry
Memorial Sloan-Kettering Cancer
    Center
New York, New York

Amy S. Lee, PhD
University of Southern California
Norris Cancer Center
Los Angeles, California

Julio Licinio
Center on Pharmacogenomics
Department of Psychiatry and
    Behavioral Sciences
University of Miami

Lingge Lu
Department of Medicine
Beth Israel Deaconess Medical Center
Harvard Medical School
Boston, Massachusetts

Vilma R. Martins
Molecular and Cellular Biology Group
Ludwig Institute for Cancer Research
São Paulo, Brazil

Leonard M. Miller
School of Medicine
Batchelor Children's Research Institute
Miami, Florida

Renata Pasqualini, PhD
Department of Genitourinary
    Medical Oncology
University of Texas M. D. Anderson
    Cancer Center
Houston, Texas

Tanja-Maria Ranta
Department of Biosciences
Division of Biochemistry
University of Helsinki
Helsinki, Finland

Terhi Ruohtula
Department of Biosciences
Division of Biochemistry
University of Helsinki
Helsinki, Finland

E. Helene Sage, PhD
Benaroya Research Institute
George Mason University
Fairfax, Virginia
and
University of Washington
Seattle, Washington

Dean Sheppard, MD
University of California San Francisco
San Francisco General Hospital
San Francisco, California

Yuen Shing
Children's Hospital
Harvard Medical School
Boston, Massachusetts

Shishir Shishodia
Department of Bioimmunotherapy
University of Texas M. D. Anderson
 Cancer Center
Houston, Texas

Michael Stefanidakis
Department of Biosciences
Division of Biochemistry
University of Helsinki
Helsinki, Finland

Kristiina Vuori, MD, PhD
Cell Adhesion-Extracellular Matrix
 Biology Program
The Burnham Institute
La Jolla, California

Ma-Li Wong
Center on Pharmacogenomics
Department of Psychiatry and
 Behavioral Sciences
University of Miami Leonard M. Miller
 School of Medicine
Batchelor Children's Research Institute
Miami, Florida

Bruce R. Zetter, PhD
Departments of Cell Biology and
 Surgery
Children's Hospital
Harvard Medical School
Boston, Massachusetts

1 Discovery of Vascular Permeability Factor (VPF, VEGF, VPF/VEGF, VEGF-A¹⁶⁵)

Harold F. Dvorak

CONTENTS

1.1 INTRODUCTION

The story goes back to the late 1970s. I was at the time an immunopathologist investigating a new form of cellular immunity that we had discovered called cutaneous
basophil hypersensitivity (CBH) (1,2). CBH is a form of lymphocyte-mediated
immune response in which basophilic leukocytes play a prominent role, so much
so that they may account for half or more of the cellular infiltrate. Originally
described with purified protein antigens in guinea pigs, we had found a similar
basophil-rich response to contact allergens (e.g., poison ivy) and other medically
relevant entities such as viral pathogens. I was therefore curious to see whether
tumors also elicited a CBH response.

I began a study of syngeneic tumors in guinea pigs and found, to my delight, that
they also elicited a response rich in basophils along with lymphocytes and monocytes
(3). However, there was a much more interesting finding, namely, that the growing

1

tumors developed within a fibrin gel cocoon (4,5). When harvested at early inter-
vals after implantation, before a mass became palpable, tumor cells were found to be
growing in small clumps that were separated from each other by apparently empty
space (Fig. 1.1a and b). However, on closer inspection and with the use of sophisticated
morphologic techniques, the empty space between tumor clumps was found to be
filled with extravasated edema fluid and thin strands of fibrillar material. This fibrillar
material was definitively identified by both immunoperoxidase (Fig. 1c) and electron
microscopy as cross-linked fibrin. Further, the fibrin took the form of a gel, i.e., it
consisted of a three-dimensional meshwork in which long chains of insoluble fibrin
strands were dispersed in and immobilized a substantially larger volume of serum
exudate. This, of course, is the structure that fibrin takes when plasma is allowed to
clot in a test tube. We quickly generalized the finding of fibrin deposition to many
other tumors, animal and human (Fig. 1d), transplantable and autochthonous (6–10).

1.2   HOW DOES FIBRIN GET DEPOSITED IN TUMORS?

I reasoned that for fibrin to be deposited outside of blood vessels in tumors the vasculature must first become leaky. Fibrin results from the clotting of fibrinogen, a large
plasma protein that is normally retained within the circulation. Therefore, for fibrin
to be deposited in the extravascular space, local blood vessels must first become
hyperpermeable so that plasma fibrinogen can extravasate. Obviously a second event
is also required, i.e., the clotting system must be activated to transform extravasated
plasma fibrinogen into fibrin.

With regard to fibrinogen leakage, there was already in the 1970s an extensive
literature on the general subject of vascular permeability. It had been known since
the days of Starling that water and other small molecules in plasma extravasate freely
from the proximal portions of normal *capillaries*, only to be reabsorbed distally by
capillaries as well as by lymphatics. Very small amounts of albumin (65 kDa) and
IgG immunoglobulin (150 kDa) also escape, but the vast majority of plasma proteins,
especially very large plasma proteins such as fibrinogen (350 kDa), remain within
the vasculature.

In inflammation, however, the situation is quite different. Here *post-capillary
venules* and *small veins* respond to permeability factors such as histamine, serotonin,
PAF, etc., and become highly leaky not only to small molecules but also to large
molecules such as plasma proteins (11). The result is an inflammatory exudate comprised of extravasated plasma proteins that clot to deposit a fibrin gel. Therefore, the
finding of edema and fibrin in tumors suggested that an analogous process was taking
place, i.e., that tumor blood vessels were permeable to plasma and plasma proteins and
that mechanisms were in place to induce clotting of extravasated plasma fibrinogen.

1.3   TUMOR BLOOD VESSELS ARE HYPERPERMEABLE TO
PLASMA AND PLASMA PROTEINS

Literature search revealed that others had already provided evidence that the
blood vessels of at least some animal tumors were hyperpermeable to circulating

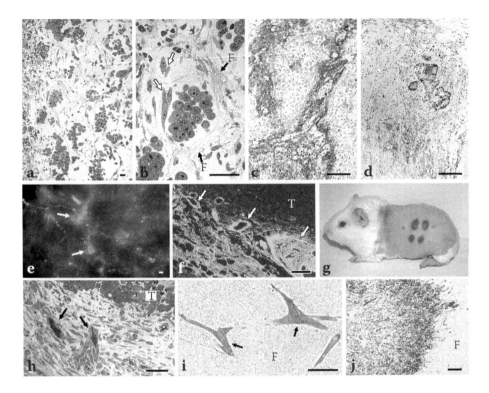

FIGURE 1.1 (See color insert following page 112.) (a) and (b) Line 10 tumor cells 48 hours after transplant into the subcutaneous space of syngeneic strain 2 guinea pigs. Fibrin (F) forms a water-trapping gel that serves as a provisional stroma that separates tumor cells into discrete islands and that provides a favorable matrix for fibroblast (*white arrows*) and endothelial cell migration. (c) and (d) Immunohistochemical demonstration of fibrin (brown staining) in guinea pig line 1 and human colorectal adenocarcinomas, respectively. (e) and (f) Blood vessels (*arrows*) supplying line 10 guinea pig tumors are hyperpermeable to circulating macromolecular fluoresceinated dextran. (g) Miles assay illustrating permeability to Evans blue-albumin complex at sites of intradermal injections of the following: (*top row, left to right*) neutralizing anti-VEGF-A antibody, line 10 guinea pig tumor ascites fluid, mix of line 10 tumor ascites fluid and control IgG, mix of tumor ascites fluid and neutralizing anti-VEGF-A antibody; (*bottom row*) line 1 tumor ascites fluid, mix of line 1 tumor ascites fluid and control IgG, and mix of line 1 tumor ascites fluid and neutralizing anti-VEGF-A antibody. (h) Fibroblasts and blood vessels (*black arrows*) invade line 1 tumor fibrin gel, replacing it with fibrous connective tissue. (i) Fibroblasts (*arrows*) migrate through fibrin gel (F) in vitro. (j) Implanted fibrin gel (F) in subcutaneous space is replaced by ingrowth of fibroblasts and new blood vessels, creating granulation-like vascular connective tissue. Scale bars, 25 μm (b, i), 50 μm (a, c, d, h), and 100 μm (e, f, j). *Sources:* Reproduced from H. F. Dvorak, J. A. Nagy, D. Feng, L. F. Brown, and A. M. Dvorak. Vascular Permeability Factor/Vascular Endothelial Growth Factor and the Significance of Microvascular Hyperpermeability in Angiogenesis. *Curr Top Microbiol Immunol* 237 (1999): 97–132; H. F. Dvorak. Rous-Whipple Award Lecture. How Tumors Make Bad Blood Vessels and Stroma. *Amer J Path* 162 (2003): 1747–57; and H. F. Dvorak, V. S. Harvey, P. Estrella, L. F. Brown, J. McDonagh, and A. M. Dvorak. Fibrin Containing Gels Induce Angiogenesis. Implications for Tumor Stroma Generation and Wound Healing. *Lab Invest* 57 (1987): 673–86, with permission.

macromolecules (reviewed in 6,7). Also, the Irish pathologist O'Meara had suggested that fibrin was present in some tumors (12). We set out to investigate these results in tumor-bearing mice and guinea pigs using a variety of macromolecular tracers that included colloidal carbon, fluoresceinated dextrans, radioactively labeled plasma proteins, horseradish peroxidase, and ferritin. We found that all of these tracers leaked selectively from tumors as compared with normal blood vessels (Fig. 1.1e and f) (13–15). Using radioactively labeled fibrinogen and albumin, we demonstrated that the permeability of tumor blood vessels was four to ten times that of the normal vasculature (16) and that clotting of extravasated plasma resulted in the deposition of extensive cross-linked fibrin. Later, using transmission electron microscopy we determined that, as in inflammation, most of this leakage came from venules (and from new, abnormal tumor blood vessels derived from venules) and took place by a trans-endothelial route in which tracer proteins passed through a subsequently recognized organelle in venular endothelium, the vesiculo-vacuolar organelle (VVO) (17–22).

Tumor blood vessel hyperpermeability was of particular interest to me because of my aforementioned interest in the basophil-rich lesions of CBH (1). Basophilic leukocytes and their first cousins, mast cells, are stuffed with cytoplasmic granules that contain, among other components, histamine. In inflammation, various agents trigger histamine release, which acts on venules to cause plasma and plasma protein leakage, edema, and fibrin deposition. I inferred (erroneously as it turned out) that tumors were making a substance that caused infiltrating basophils and tissue mast cells to degranulate, thereby triggering a histamine-induced exudate as in inflammation.

1.4 TUMOR CELLS SECRETE VASCULAR PERMEABILITY FACTOR

Because of my background in immunology, I was familiar with an assay that had been developed by Sir Ashley Miles for assessing altered vascular permeability in inflammation (23). In the Miles assay, Evans blue dye is injected intravenously into experimental animals such as guinea pigs or mice. Immediately before or after, test substances such as histamine are injected intradermally into the animal's depilated flank skin. Evans blue is a low-molecular-weight dye that would be expected to leak freely from normal capillaries along with water and other small molecules, except for the fact that it binds tightly but non-covalently to plasma proteins, particularly albumin. Therefore, because it is bound to plasma proteins that are largely retained within the circulation, Evans blue extravasates only very slowly from normal blood vessels (over a time frame measured in hours). However, when venules are rendered hyperpermeable by histamine or other vascular permeabilizing agents, the Evans blue-plasma protein complex extravasates rapidly, beginning within a minute or two and generating a clearly defined blue spot that reaches maximum intensity in about 20 minutes.

We used the Miles assay to determine whether tumor cells made and secreted a substance that caused normal vessels to become leaky (Fig. 1.1g). We cultured a variety of human and animal tumor cells overnight in serum-free medium (to avoid confounding serum factors, some of which can also render venules leaky); thereafter, cultures were centrifuged to separate tumor cells and the resulting supernatants were tested for their ability to permeabilize blood vessels in normal guinea pig skin. To my great delight, culture supernatants from nearly all of the tumor cell lines we tested generated intense blue spots, indicative of vascular permeability, whereas supernatants from a variety of normal control cells did not (5). We called this bluing activity *vascular permeability factor* or VPF.

1.5   CONSEQUENCES OF VASCULAR HYPERPERMEABILITY

An obvious immediate consequence of increased microvascular permeability is leakage of plasma and plasma proteins into the surrounding connective tissue. Extravascular fluid accumulation is initially compensated by lymphatic drainage but, when lymphatic capacity is exceeded, fluid continues to accumulate as edema, with a consequent buildup of interstitial tissue pressure. Drs. Pietro Gullino and Rakesh Jain had shown that, compared with normal tissues, solid tumors contain increased amounts of plasma and plasma proteins along with increased interstitial pressure (24,25). However, tumor-induced plasma extravasation is less inhibited by building tissue pressure when tumors grow in open tissue compartments such as the pericardial, pleural, or peritoneal cavities. When the blood vessels lining these body cavities become hyperpermeable, large amounts of plasma can escape, largely unopposed by increasing tissue pressure. The resulting large accumulation of plasma protein-rich fluid (ascites) is a characteristic feature of many common carcinomas growing in body cavities (e.g., breast, ovary).

We found that a second consequence of the vascular hyperpermeability and plasma leakage induced by VPF (or by other vascular permeabilizing agents) was activation of the clotting system (Fig. 2) (26). The vascular hyperpermeability induced by VPF is relatively nondiscriminatory with respect to molecules of varying size and shape. Therefore, plasma proteins participating at all stages in the clotting cascade extravasate readily from hyperpermeable vessels, e.g., clotting factors V, VII, X, XIII, and prothrombin in addition to fibrinogen. Following plasma protein leakage, clotting proceeded rapidly, resulting in deposition of an extravascular fibrin gel. Clotting was initiated by tissue factor, a glycoprotein expressed on the surfaces of tumor cells as well as on many normal tissue cells. Tissue factor activates the extrinsic clotting pathway, serving as a cofactor that increases the enzymatic activity of clotting factor VIIa by several orders of magnitude; this greatly enhances factor VIIa's ability to cleave factor X to Xa. Factor Xa, in turn, forms a complex with factor Va and prothrombin on an appropriate phospholipid surface. This prothrombinase complex possesses the enzymatic activity that generates thrombin, which in turn cleaves fibrinogen to fibrin. We also found that tumor cell plasma membranes, as well as the plasma membrane vesicles that tumors commonly shed, provided an effective surface for assembling the prothrombinase complex (27). The clotting process was completed when factor XIII, itself activated by thrombin, cross links and thereby stabilizes fibrin.

1.6   INITIAL CHARACTERIZATION OF VPF

We next performed what today would be regarded as extremely primitive experiments to characterize VPF (5,28). We showed that it was non-dialyzable and therefore likely to be a macromolecule or at least bound to a macromolecule. Cold profoundly depressed tumor cell secretion of VPF as did inhibition of protein synthesis by

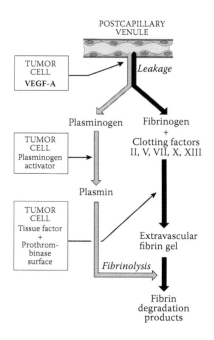

FIGURE 1.2 Schematic diagram illustrating the mechanisms by which VPF/VEGF's induction of vascular hyperpermeability leads to deposition of extravascular fibrin and its subsequent degradation.

puromycin. Heat (70°C for 30 minutes) largely inactivated the VPF activity present in tumor culture supernatants, as did common proteases. From these results we concluded that VPF was a protein.

However, a surprise came when we began to look at VPF's mechanism of action. As expected, pretreatment of guinea pigs with an antihistamine profoundly abolished the vascular permeabilizing response of mast cell degranulating agents or of histamine in the Miles assay. However, antihistamines did not inhibit the permeability effect induced by the VPF in tumor culture supernatants (5,28). Furthermore, tumor culture supernatants did not cause tissue mast cells or isolated guinea pig basophils or mast cells to degranulate. We were forced to conclude from these experiments that the VPF activity present in tumor culture supernatants was acting directly to permeabilize blood vessels, not indirectly by causing histamine release from basophils or mast cells, as I had originally hypothesized. This finding was consistent with the observation that, although tumors that made VPF had hyperpermeable blood vessels, this vascular hyperpermeability was unrelated to the presence of basophils or mast cells; in fact, subsequent experiments showed that many tumors contain few basophils or mast cells.

1.7   PURIFICATION OF VPF

At just this time (1979) I moved to the Department of Pathology at the Beth Israel Hospital in Boston and for that reason there was a short hiatus in our studies.

However, I was fortunate to recruit a young post-doctoral fellow, Dr. Donald Senger, to join the lab, and he was given the task of purifying and further characterizing VPF. Our work came to be supported for a time by the Monsanto Company, which had been interested in identifying the putative tumor angiogenesis factor first proposed by Dr. Judah Folkman. Biological factors tend to be potent substances that are present only in very small quantities. Therefore, we needed large amounts of starting material for purifying VPF. Fortunately, the Monsanto Company had large-scale tissue culture facilities, and Dr. Joseph Feder, who directed these facilities and was an enthusiastic proponent of the project, prepared many liters of serum-free tumor culture supernatant from guinea pig tumor cells that we provided. These were concentrated and shipped to us in the form of large ice cubes (measured in cubic feet), which served as the starting material for VPF purification. Using this material, and the Miles assay to assess activity, Dr. Senger was able to purify VPF to homogeneity using heparin-Sepharose and hydroxylapatite chromatography. VPF turned out to be a 34–43 kDa dimeric protein (5,28,29). On reduction, several major bands appeared and all vascular permeabilizing activity was lost. We later went on to show that VPF was an N-linked glycoprotein and that non-glycosylated forms were equally potent in the Miles assay (30).

Studies with the purified protein indicated that VPF induced vascular permeability with a potency some 50,000 times that of histamine on a molar basis (28). We showed it to be distinct from other vascular permeabilizing agents such as bradykinin, plasma kallikrein, PAF, leukokinins, prostaglandins, and Pf/dil. The effect of a single intradermal injection was rapid (permeability in the Miles assay was evident within 1–2 minutes) and transient (little permeability-inducing activity remained after 20 minutes). Dr. Senger sequenced the N-terminus of VPF, thereby facilitating the cloning of VPF by our colleagues at the Monsanto Company (31). He also made use of this sequence to prepare a rabbit antibody against a peptide corresponding to amino acids 1–24 (28). This antibody abolished all of the permeability-increasing activity present in culture medium from several guinea pig and rat tumors; also, it prevented circulating albumin from accumulating in tumor ascites fluid (28). Together these findings identified VPF as a potential target for anti-angiogenesis therapy for tumors and, as it turned out later, for many other pathologies in which VPF played a prominent role (6–8).

We went on to define other VPF activities. Tissue culture studies indicated that VPF provoked a three- to four-fold increase in endothelial cell $[Ca^{2+}]_i$ and also generated inositol triphosphate (IP_3) (32). Like other mediators of endothelial cell calcium flux, VPF induced substantial release of von Willebrand factor. Taken together, these experiments provided the first conclusive evidence that VPF acted directly on endothelial cells, i.e., its effects were not mediated by some secondary process. Of interest, and at the time somewhat disappointing, experiments (unpublished) with the late Dr. Tom Maciag failed to demonstrate that VPF had mitogenic activity for endothelial cells under standard culture conditions. Subsequently VPF was shown to be an endothelial cell mitogen in vitro and an angiogenic factor in vivo by Daniel Connolly and colleagues at Monsanto (31). Independently, Napoleone Ferrara and colleagues at Genentech showed in the same year that VPF was an endothelial cell mitogen in vitro and gave it the name, which has stuck, vascular endothelial growth factor or VEGF (33). However, we now know that VPF/VEGF is a relatively weak mitogen, active only under highly specialized (very low serum) culture conditions that are unlikely to occur in vivo. Therefore, although VPF/VEGF is a major angiogenic factor, it is likely that much of the endothelial cell proliferation associated with angiogenesis in vivo is mediated indirectly, for example, by activation of clotting, generation of thrombin, deposition of fibrin, release of platelet growth factors, etc. (reviewed in 7,8,34,35).

1.8   EXPRESSION OF VPF/VEGF IN TUMORS, NORMAL TISSUES, AND ANGIOGENESIS

Making use of in situ hybridization, Northern and Western blotting, and immuno-histochemistry (36), we demonstrated that VPF/VEGF was overexpressed by most malignant tumors (reviewed in 6,7). VPF/VEGF was also expressed, though at lower levels, in a number of normal adult tissues, including lung (by type II pneumocytes), kidney (by glomerular podocytes), adrenal (cortical cells), heart (myocytes), and liver (37). These findings made it clear that at lower levels VPF/VEGF did not induce angiogenesis and suggested that it might have other functions in normal physiology such as preserving normal endothelial cell structure and function. It was later shown that VPF/VEGF is an endothelial cell survival factor (38) and prevents endothelial cell senescence (39).

The finding that VPF/VEGF was not a unique tumor product also suggested that it might have a role in other forms of pathological angiogenesis that were not associated with tumors. In fact, this proved to be the case. We found that VPF/VEGF had a central role in wound healing, whether of the skin or of the heart as in myocardial infarcts (40,41). In healing skin wounds, VPF/VEGF was highly upregulated within a day in keratinocytes at the viable wound edge, returning gradually to normal low levels with wound closure and re-epithelialization (40); VPF/VEGF was similarly overexpressed in viable cardiac myocytes immediately adjacent to infarcted heart muscle (41). VPF/VEGF was also overexpressed in various inflammatory pathologies associated with new blood vessel formation including the keratinocytes of psoriasis (42), bullous pemphigoid (43), and delayed hypersensitivity (44); the synovial cells of rheumatoid arthritis (45); glial cells adjacent to infarcted brain, as well as in macrophages infiltrating healing wounds, delayed hypersensitivity, and other inflammatory reactions. VPF/VEGF was also highly expressed in common forms of physiological angiogenesis such as corpus luteum formation (46,47) and was found to have a central role in the vasculogenesis of embryonic development (48,49).

1.9   SIGNIFICANCE OF VASCULAR HYPERPERMEABILITY AND FIBRIN DEPOSITION FOR GENERATING ANGIOGENESIS AND STROMA

These studies raised a number of important questions. Why do tumors overexpress VPF/VEGF? What advantage do tumors gain by increasing vascular permeability? What is the significance of fibrin deposition? My answer to these questions is that by increasing vascular permeability and activating the clotting system tumors mimic the basic steps and mechanisms of wound healing and inflammation and in this manner generate the vascular supply and other stromal components they need for survival and growth.

Like normal tissues, solid tumors are comprised of two interdependent compartments, parenchyma and stroma. In tumors, the neoplastic cells are the parenchyma and the vascular connective tissue they induce is the stroma. Whether transplanted into tissues or arising autochthonously, tumors require stroma for survival, and to grow beyond minimal size they must generate new stroma (50). The vascular hyperpermeability, edema, and clotting induced by VPF/VEGF generate a provisional stroma, the fibrin gel. Fibrin deposited between clumps of tumor cells provides structural support, space, and a surface on which stromal cells (fibroblasts, endothelial cells, pericytes, etc.) can migrate (Fig. 1.1h), as well as a serum-rich environment. Stimulated by VPF/VEGF and serum growth factors, endothelial cells migrate on fibrin strands to enter tumors and form new blood vessels; fibroblasts migrate in similar fashion (51) and synthesize and secrete the matrix components of mature tumor stroma. Fibrin itself provides only a temporary or provisional stroma, as it is degraded by proteases such as plasmin that result from the action of tumor-secreted plasminogen activators on plasminogen, another plasma protein that extravasates from leaky blood vesels (7,52). Hence, the fibrin deposited in tumors at any moment in time reflects a balance between clotting and fibrin deposition on the one hand and proteolytic degradation of fibrin on the other (Fig. 1.2).

The process just described in tumors closely resembles that occurring in healing wounds, and I have for this reason suggested that tumors behave as wounds that do not heal (53). In wounds, whether to the skin or in the form of myocardial infarcts or strokes, initial bleeding is usually staunched within a few minutes by mechanisms such as vasoconstriction, platelet activation, and intravascular clotting. However, over the course of hours and days, the adjacent vasculature becomes hyperpermeable as VPF/VEGF expression is rapidly upregulated in neighboring cells (presumably because of local tissue hypoxia). As a result, plasma extravasates and the resulting clotting of extravasated fibrinogen leads to deposition of a fibrin gel that serves as a provisional stroma that favors inward migration of vascular and connective tissue cells, just as in tumors.

Considerable independent evidence supports this model of stroma generation. In vitro studies have provided direct evidence that cross-linked fibrin of the type deposited in tumors affords a matrix that supports and favors mesenchymal and inflammatory cell migration (Fig. 1.1i) (51). Furthermore, we have found that simple implantation of cross-linked fibrin gels in normal animal tissues, in the absence of tumor cells, induces the progressive ingrowth of new blood vessels and fibroblasts (Fig. 1.1j); over time, these replace the fibrin gel with mature vascularized connective tissue (34). Thus, tumor cell secretion of VPF/VEGF sets in motion a process that results in vascular hyperpermeability and fibrin deposition, events that themselves initiate angiogenesis and the formation of mature stroma.

Taken together, the connective tissue stroma of adult tissues is designed for normal homeostasis, a state in which there is little migration or turnover of vascular endothelium and other stromal cells. Insertion of fibrin gel into tissues upsets the homeostatic balance and provides a provisional matrix that is highly supportive of mesenchymal cell adhesion, migration, and division. Though compositionally very different from the stroma of normal embryonic development, fibrin shares with it the property of offering a matrix that is conducive to cell migration. However, the end result of fibrin deposition is very different from that which occurs in normal development; whether in tumors, wound healing, or inflammation, the stroma that forms is highly abnormal in structure and function, a caricature of normal tissue stroma.

1.10   SUMMARY AND PERSPECTIVE

Much has happened in the 25 years since VPF/VEGF was discovered. The nomenclature has changed and the protein we named VPF is now most widely known as the 165 amino acid isoform of VEGF-A, i.e., VEGF-A^{165} (VEGF-A^{164} in the mouse) (7,54–56). VEGF-A is encoded by a single gene that is alternatively spliced to generate several different isoforms; VEGF-A^{165} is generally the most abundantly expressed, but isoforms of 121 and 189 amino acids are also noteworthy. Further, VEGF-A belongs to a family of proteins that now includes VEGFs B, C, D, E, and placenta growth factor (PlGF), each with distinct properties and activities in the development of blood and lymphatic vessels. VEGF-A^{165} is now known to be a multifunctional cytokine with central roles in vasculogenesis, in both physiological and pathological angiogenesis, in maintenance of endothelial cell differentiation and integrity, and most recently in lymphangiogenesis (57) and in the growth and maintenance of nerves (58). Although all of these functions are important, VEGF-A's most distinctive activity in pathology is that for which it was originally named, its ability to render microvessels hyperpermeable. In so doing, VEGF-A initiates a cascade of events that result in extravasation of plasma and plasma proteins; edema; and fibrin deposition. Together these events contribute to the formation of new blood vessels and connective tissue stroma. Thus, the discovery of VPF 30 years ago resulted not only in the discovery of a new molecule, but also in the elucidation of a process—one that, though not as yet completely understood, is responsible for tissue repair following injury, but that, unfortunately, also facilitates the growth and spread of malignant tumors.

ACKNOWLEDGMENTS

Supported in part by U.S. Public Health Service grants CA-50453, HL-59316, and P01 CA92644, and by a contract from the National Foundation for Cancer Research.

REFERENCES

1. H. F. Dvorak, A. M. Dvorak, B. A. Simpson, H. B. Richerson, S. Leskowitz, and M. J. Karnovsky. "Cutaneous Basophil Hypersensitivity. II. A Light and Electron Microscopic Description." *J Exp Med* 132 (1970): 558–82.
2. R. Colvin and H. Dvorak. "Role of the Clotting System in Cell-Mediated Hypersensitivity. II. Kinetics of Fibrinogen/Fibrin Accumulation and Vascular Permeability Changes in Tuberculin and Cutaneous Basophil Hypersensitivity Reactions." *J Immunol* 114 (1975): 377–87.

3. H. F. Dvorak, A. M. Dvorak, and W. H. Churchill. "Immunologic Rejection of Diethylnitrosamine-Induced Hepatomas in Strain 2 Guinea Pigs: Participation of Basophilic Leukocytes and Macrophage Aggregates." *J Exp Med* 137 (1973): 751–55.
4. H. F. Dvorak, A. M. Dvorak, E. J. Manseau, L. Wiberg, and W. H. Churchill. "Fibrin Gel Investment Associated with Line 1 and Line 10 Solid Tumor Growth, Angiogenesis, and Fibroplasia in Guinea Pigs. Role of Cellular Immunity, Myofibroblasts, Microvascular Damage, and Infarction in Line 1 Tumor Regression." *J Nat Cancer Inst* 62 (1979): 1459–72.
5. H. F. Dvorak, N. S. Orenstein, A. C. Carvalho, W. H. Churchill, A. M. Dvorak, S. J. Galli, J. Feder, A. M. Bitzer, J. Rypysc, and P. Giovinco. "Induction of a Fibrin-Gel Investment: An Early Event in Line 10 Hepatocarcinoma Growth Mediated by Tumor-Secreted Products." *J Immunol* 122 (1979): 166–74.
6. L. Brown, M. Detmar, K. Claffey, J. Nagy, D. Feng, A. Dvorak, and H. Dvorak. "Vascular Permeability Factor/Vascular Endothelial Growth Factor: A Multifunctional Angiogenic Cytokine," in *Regulation of Angiogenesis,* eds. I. Goldberg and E. Rosen Basel, Switzerland: Birkhauser Verlag, 1997.
7. H. F. Dvorak, J. A. Nagy, D. Feng, L. F. Brown, and A. M. Dvorak. "Vascular Permeability Factor/Vascular Endothelial Growth Factor and the Significance of Microvascular Hyperpermeability in Angiogenesis." *Curr Top Microbiol Immunol* 237 (1999): 97–132.
8. H. F. Dvorak. Rous-Whipple Award Lecture. "How Tumors Make Bad Blood Vessels and Stroma." *Amer J Path* 162 (2003): 1747–57.
9. N. L. Harris, A. M. Dvorak, J. Smith, and H. F. Dvorak. "Fibrin Deposits in Hodgkin's Disease." *Am J Pathol* 108 (1982): 119–29.
10. H. F. Dvorak, G. R. Dickersin, A. M. Dvorak, E. J. Manseau, and K. Pyne. "Human Breast Carcinoma: Fibrin Deposits and Desmoplasia. Inflammatory Cell Type and Distribution. Microvasculature and Infarction." *J Nat Cancer Institute* 67 (1981): 335–345.
11. G. Majno, V. Gilmore, and M. Leventhal. "On the Mechanism of Vascular Leakage Caused by Histamine Type Mediators. A Microscopic Study In Vivo." *Circulation Research* 21 (1967): 833–47.
12. R. A. O'Meara. "Coagulative Properties of Cancers." *Ir J Med Sci* 394 (1958): 474–79.
13. H. F. Dvorak, J. A. Nagy, J. T. Dvorak, and A. M. Dvorak. "Identification and Characterization of the Blood Vessels of Solid Tumors that Are Leaky to Circulating Macromolecules." *Am J Path* 133 (1988): 95–109.
14. L. F. Brown, B. Asch, V. S. Harvey, B. Buchinski, and H. F. Dvorak. "Fibrinogen Influx and Accumulation of Cross-Linked Fibrin in Mouse Carcinomas." *Cancer Res* 48 (1988): 1920–25.
15. L. F. Brown, L. Van De Water, V. S. Harvey, and H. F. Dvorak. "Fibrinogen Influx and Accumulation of Cross-Linked Fibrin in Healing Wounds and in Tumor Stroma." *Amer J Pathol* 130 (1988): 455–65.
16. H. F. Dvorak, V. S. Harvey, and J. McDonagh. "Quantitation of Fibrinogen Influx and Fibrin Deposition and Turnover in Line 1 and Line 10 Guinea Pig Carcinomas." *Cancer Res* 44 (1984): 3348–54.
17. S. Kohn, J. A. Nagy, H. F. Dvorak, and A. M. Dvorak AM. "Pathways of Macromolecular Tracer Transport Across Venules and Small Veins. Structural Basis for the Hyperpermeability of Tumor Blood Vessels." *Lab Invest* 67 (1992): 596–607.
18. D. Feng, J. Nagy, J. Hipp, H. Dvorak, and A. Dvorak. "Vesiculo-vacuolar Organelles and the Regulation of Venule Permeability to Macromolecules by Vascular Permeability Factor, Histamine, and Serotonin." *J Exp Med* 183 (1996): 1981–86.
19. D. Feng, J. Nagy, J. Hipp, K. Pyne, H. Dvorak, and A. Dvorak. "Reinterpretation of Endothelial Cell Gaps Induced by Vasoactive Mediators in Guinea-Pig, Mouse and Rat: Many Are Transcellular Pores." *J Physiol (London)* 504 (1997): 747–61.

20. D. Feng, J. Nagy, K. Pyne, I. Hamel, H. Dvorak, and A. Dvorak. "Pathways of Macromolecular Extravasation Across Microvascular Endothelium in Response to VPF/VEGF and Other Vasoactive Mediators." *Microcirculation* 6 (1999): 23–44.
21. D. Feng, J. Nagy, A. Dvorak, and H. Dvorak. "Different Pathways of Macromolecule Extravasation from Hyperpermeable Tumor Vessels." *Microvascular Research* 59 (2000): 24–37.
22. A. M. Dvorak, S. Kohn, E. S. Morgan, P. Fox, J. A. Nagy, and H. F. Dvorak. "The Vesiculo-vacuolar Organelle (VVO): A Distinct Endothelial Cell Structure that Provides a Transcellular Pathway for Macromolecular Extravasation." *J Leukocyte Biol* 59 (1996): 100–15.
23. A. A. Miles and E. M. Miles. "Vascular Reaction to Histamine, Histamine-Liberator, and Leukotaxine in the Skin of Guinea Pigs." *J Physiol (London)* 118 (1952): 228–57.
24. P. M. Gullino. "Extracellular Compartments of Solid Tumors," in *Cancer: A Comprehensive Treatise*, ed. F. F. Becker (New York: Plenum Press, 1975), 327–54.
25. R. K. Jain. "Vascular and Interstitial Barriers to Delivery of Therapeutic Agents in Tumors." *Cancer Metastasis Rev* 9 (1990): 253–66.
26. H. F. Dvorak, D. S. Senger, A. M. Dvorak, V. S. Harvey, and J. McDonagh. "Regulation of Extravascular Coagulation by Microvascular Permeability." *Science* 227 (1985): 1059–61.
27. L. VanDeWater, P. B. Tracy, D. Aronson, K. G. Mann, and H. F. Dvorak. "Tumor Cell Generation of Thrombin via Functional Prothrombinase Assembly." *Cancer Res* 45 (1985): 5521–25.
28. D. R. Senger, S. J. Galli, A. M. Dvorak, C. A. Perruzzi, V. S. Harvey, and H. F. Dvorak. "Tumor Cells Secrete a Vascular Permeability Factor that Promotes Accumulation of Ascites Fluid." *Science* 219 (1983): 983–85.
29. D. R. Senger, D. T. Connolly, L. Van De Water, J. Feder, and H. F. Dvorak. "Purification and NH_2-Terminal Amino Acid Sequence of Guinea Pig Tumor-Secreted Vascular Permeability Factor." *Cancer Res* 50 (1990): 1774–78.
30. T.-K. Yeo, D. R. Senger, H. F. Dvorak, L. Freter, and K.-T. Yeo. "Glycosylation Is Essential for Efficient Secretion but Not for Permeability-Enhancing Activity of Vascular Permeability Factor (Vascular Endothelial Growth Factor)." *BBRC* 179 (1991): 1568–75.
31. P. J. Keck, S. D. Hauser, G. Krivi, K. Sanzo, T. Warren, J. Feder, and D. T. Connolly. "Vascular Permeability Factor, an Endothelial Cell Mitogen Related to PDGF." *Science* 246 (1989): 1309–12.
32. T. A. Brock, H. F. Dvorak, and D. R. Senger. "Tumor-Secreted Vascular Permeability Factor Increases Cytosolic Ca^{2+} and Von Willebrand Factor Release in Human Endothelial Cells." *Amer J Path* 138 (1991): 213–21.
33. D. W. Leung, G. Cachianes, W.-J. Kuang, D. V. Goeddel, and N. Ferrara. "Vascular Endothelial Growth Factor Is a Secreted Angiogenic Mitogen." *Science* 246 (1989): 1306–9.
34. H. F. Dvorak, V. S. Harvey, P. Estrella, L. F. Brown, J. McDonagh, and A. M. Dvorak. "Fibrin Containing Gels Induce Angiogenesis. Implications for Tumor Stroma Generation and Wound Healing." *Lab Invest* 57 (1987): 673–86.
35. H. F. Dvorak and F. R. Rickles. Malignancy and Hemostatis In: Colman R. W., Marder, V. J., Clowes, A., et al. editors. *Hemostasis and thrombosis. basic principles and clinical practice*. 5th ed. Philadelphia: Lippincott Williams & Wilkins; 2006. p. 851–73.
36. H. F. Dvorak, T. M. Sioussat, L. F. Brown, J. A. Nagy, A. Sotrel, E. Manseau, L. Van De Water, and D. R. Senger. "Distribution of Vascular Permeability Factor (Vascular Endothelial Growth Factor) in Tumors: Concentration in Tumor Blood Vessels." *J Exp Med* 174 (1991): 1275–78.

37. B. Berse, L. F. Brown, L. Van De Water, H. F. Dvorak, and D. R. Senger. "Vascular Permeability Factor (Vascular Endothelial Growth Factor) Gene Is Expressed Differentially in Normal Tissues, Macrophages, and Tumors." *Molecular Biol Cell* 3 (1992): 211–20.

38. L. E. Benjamin and E. Keshet. "Conditional Switching of Vascular Endothelial Growth Factor (VEGF) Expression in Tumors: Induction of Endothelial Cell Shedding and Regression of Hemangioblastoma-Like Vessels by VEGF Withdrawal." *Proc Natl Acad Sci U S A* 94 (1997): 8761–66.

39. Y. Watanabe, S. W. Lee, M. Detmar, I. Ajioka, and H. F. Dvorak. "Vascular Permeability Factor/Vascular Endothelial Growth Factor (VPF/VEGF) Delays and Induces Escape from Senescence in Human Dermal Microvascular Endothelial Cells." *Oncogene* 14 (1997): 2025–32.

40. L. F. Brown, K. T. Yeo, B. Berse, T. K. Yeo, D. R. Senger, H. F. Dvorak, and L. van de Water. "Expression of Vascular Permeability Factor (Vascular Endothelial Growth Factor) by Epidermal Deratinocytes During Wound Healing." *J Exp Med* 176 (1992): 1375–79.

41. J. Li, L. Brown, M. Hibbeerd, J. Grossman, J. Morgan, and M. Simons. "VEGF, *flk*-1, and *flt*-1 Expression in a Rat Myocardial Infarction Model of Angiogenesis." *Am J Physiol* 270 (1996): H1803–11.

42. M. Detmar, L. F. Brown, K. P. Claffey, K. T. Yeo, O. Kocher, R. W. Jackman, B. Berse, and H. F. Dvorak. Overexpression of Vascular Permeability Factor/Vascular Endothelial Growth Factor and Its Receptors in Psoriasis. *J Exp Med* 180 (1994): 1141–46.

43. L. F. Brown, T. J. Harrist, K. T. Yeo, B. M. Stahle, R. W. Jackman, B. Berse, K. Tognazzi, H. F. Dvorak, and M. Detmar. "Increased Expression of Vascular Permeability Factor (Vascular Endothelial Growth Factor) in Bullous Pemphigoid, Dermatitis Herpetiformis, and Erythema Multiforme." *J Invest Dermatol* 104 (1995): 744–49.

44. L. F. Brown, S. M. Olbricht, B. Berse, R. W. Jackman, G. Matsueda, K. A. Tognazzi, E. J. Manseau, H. F. Dvorak, and L. Van de Water. "Overexpression of Vascular Permeability Factor (VPF/VEGF) and Its Endothelial Cell Receptors in Delayed Hypersensitivity Skin Reactions." *J Immunol* 154 (1995): 2801–7.

45. R. A. Fava, N. J. Olsen, G. G. Spencer, K. T. Yeo, T. K. Yeo, B. Berse, R. W. Jackman, D. R. Senger, H. F. Dvorak, and L. F. Brown. "Vascular Permeability Factor/Endothelial Growth Factor (VPF/VEGF): Accumulation and Expression in Human Synovial Fluids and Rheumatoid Synovial Tissue." *J Exp Med* 180 (1994): 341–46.

46. H. S. Phillips, J. Hains, D. W. Leung, and N. Ferrara. "Vascular Endothelial Growth Factor Is Expressed in Rat Corpus Luteum." *Endocrinology* 127 (1990): 965–67.

47. B. R. Kamat, L. F. Brown, E. J. Manseau, D. R. Senger, and H. F. Dvorak. "Expression of Vascular Permeability Factor/Vascular Endothelial Growth Factor by Human Granulosa and Theca Lutein Cells: Role in Corpus Luteum Development." *Amer J Path* 146 (1995): 157–65.

48. G. Breier, U. Albrecht, S. Sterrer, and W. Risau. "Expression of Vascular Endothelial Growth Factor during Embryonic Angiogenesis and Endothelial Cell Differentiation." *Development* 114 (1992): 521–32.

49. W. Risau. "Mechanisms of Angiogenesis." *Nature* 386 (1997): 671–74.

50. J. Folkman. "The Role of Angiogenesis in Tumor Growth." *Seminars Cancer Biology* 3 (1992): 65–71.

51. L. F. Brown, N. Lanir, J. McDonagh, K. Czarnecki, P. Estrella, A. M. Dvorak, and H. F. Dvorak. "Fibroblast Migration in Fibrin Gel Matrices." *Amer J Pathol* 142 (1993): 273–83.

52. R. Clark. "Wound Repair: Overview and General Considerations," in *The Molecular and Cellular Biology of Wound Repair*, 2nd ed., ed. R. Clark (New York: Plenum Press, 1996), 3–50.

53. H. F. Dvorak. "Tumors: Wounds That Do Not Heal. Similarities Between Tumor Stroma Generation and Wound Healing." *N Engl J Med* 315 (1986): 1650–59.

54. N. Ferrara. "Vascular Endothelial Growth Factor: Molecular and Biological Aspects." *Curr Top Microbiol Immunol 237* (1999): 1–30.

55. T. Veikkola and K. Alitalo. "VEGFs, Receptors and Angiogenesis." *Semin Cancer Biol* 9 (1999): 211–20.

56. P. Carmeliet. "Angiogenesis in Health and Disease." *Nat Med* 9 (2003): 653–60.

57. J. A. Nagy, E. Vasile, D. Feng, C. Sundberg, L. F. Brown, M. J. Detmar, J. A. Lawitts, et al. "Vascular Permeability Factor/Vascular Endothelial Growth Factor Induces Lymphangiogenesis as well as Angiogenesis." *J Exp Med* 196 (2002): 1497–1506.

58. E. Storkebaum, D. Lambrechts, and P. Carmeliet. "VEGF: Once Regarded as a Specific Angiogenic Factor, Now Implicated in Neuroprotection." *Bioessays* 26 (2004): 943–54.

2 Tumor Necrosis Factor and Its Family Members

Bharat B. Aggarwal, Alok C. Bharti, and
Shishir Shishodia

CONTENTS

2.1 INTRODUCTION

Discovered 20 years ago, tumor necrosis factors (TNFs) are now known to consist of a family of 19 ligands with 15%–25% amino acid sequence homology. They bind to 29 different cell surface receptors that activate various transcription factors and protein kinases. Although initially described as antitumor agents, this family of cytokines is now known to mediate apoptosis, proliferation, survival, differentiation, viral replication, inflammation, septic shock, and other biological responses. This review provides an overview of this novel family of cytokines.

2.2 DISCOVERY OF TNF

In 1868, the German physician P. Brunes noted spontaneous regression of tumors in patients following acute bacterial infections, e.g., tuberculosis (for references see [1–3]). This observation led William B. Coley (1894) to study the effects of bacteria-free filtrate preparations from *Streptococcus* and other bacteria on tumors in humans; in many cases the tumor volume was reduced by these preparations, which have been referred to in the literature as Coley's toxins [4]. In 1931, A. Gratia et al. reported the regression of liposarcomas in guinea pigs after they were given *Escherichia coli* culture filtrates [5]. By 1944, one of the major inducers of TNF, endotoxin, had been isolated as bacterial polysaccharide from a Gram-negative bacterium by Murray Shear et al. [6]. This polysaccharide caused hemorrhagic necrosis of tumors. Glenn Algire et al. reported in 1952 that the mechanism of this necrosis was systemic hypotension leading to collapse of the tumor vasculature, resulting in tumor cell anoxia and cell death [7]. In 1962, Bert O'Malley reported the production of a factor in the serum of normal mice challenged with endotoxin; since this serum could induce necrosis when administered to tumor-bearing animals, it was named tumor necrosis serum (TNS) [8]. When E. A. Carswell's group confirmed the activity in serum of mice injected with endotoxin, they renamed it tumor necrosis factor (TNF) [9]. That same year, TNF was found to be cytotoxic to tumor cells in culture [10], but its molecular identity was not clarified until a decade later.

In 1984 and 1985, two different proteins that caused lysis of tumor cells were isolated and structurally identified, and their genes cloned [11–16]. These studies indicated that TNF is a homotrimer with a subunit molecular mass of 17 kDa that is produced by a wide variety of cell types in response to various inflammatory stimuli. We now know that it plays a major role in growth regulation, differentiation, inflammation, viral replication, tumorigenesis, autoimmune diseases, and the response to viral, bacterial, fungal, and parasitic infections (for references see [2,3,8]). TNF is also referred to as TNF-α, cachectin, and differentiation-inducing factor (DIF) [17,18]. TNF has continued to be a major topic of scientific investigation as indicated by over 46,000 citations published to date.

2.3 TNF FAMILY MEMBERS

By now, 19 different homologues of TNF with 15%–25% identity to each other have been reported (Fig. 2.1, Tables 2.1 and 2.2). Unlike TNF and lymphotoxin (LT), other members of this family were cloned based on either expression cloning or sequence homology searches of the human genome database. The receptors for most of these cytokines have been identified (Fig. 2.1), and they too belong to the TNF receptor superfamily with maximum homology in their extracellular domain [19].

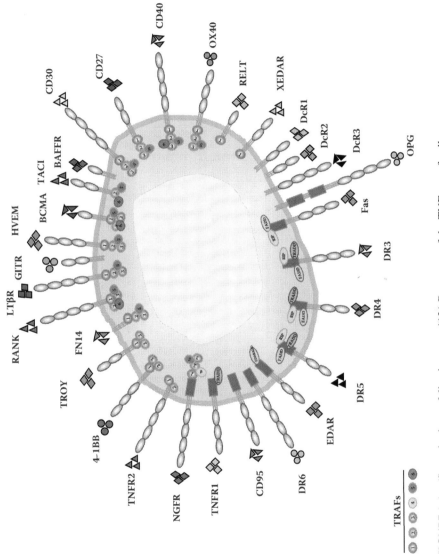

FIGURE 2.1 (See color insert following page 112.) Receptors of the TNF superfamily.

TABLE 2.1
Nomenclature of Members of TNF Superfamily

Ligand	Alternative Names	Receptor	Alternative Names
LT (TNFSF1)	LTα, TNFβ	TNFR1 (TNFRSF1A)	CD120a, p55
TNF (TNFSF2)	TNFα, DIF	TNFR2 (TNFRSF1B)	CD120b, p75
LTβ (TNFSF3)	TNFC, p33	LT-βR (TNFRSF3)	TNFR2-RP, TNFR-RP, TNFCR, TNF-R-III
OX40L (TNFSF4)	gp34, TXGP1	OX-40 (TNFRSF4)	ACT35, TXGPIL
CD40L (TNFSF5)	hCD40L, TRAP, CD154, gp39, CD40LG, IMD3, HIGM1	CD40 (TNFRSF5)	p50, Bp50
FasL (TNFSF6)	Apo-1L, APT1LG1	Fas (TNFRSF6)	CD95, APO-1, APT1
		DcR3 (TNFRSF6B)	TR6, M68
CD27L (TNFSF7)	CD70, CD27LG	CD27 (TNFSF7)	Tp55, S152
CD30L (TNFSF8)	CD30LG	CD30 (TNFRSF8)	Ki-1, D1S166E
4-1BBL (TNFSF9)		4-IBB (TNFRSF9)	CD137, ILA
TRAIL (TNFSF10)	Apo-2L, TL2	DR-4 (TNFRSF10A)	Apo2, TRAILR-1
		DR5 (TNFRSF10B)	KILLER, TRICK2A, TRAIL-R2, TRICKB
		DcR1(TNFRSF10C)	TRID, TRAILR3, LIT
		DcR2 (TNFRSF10D)	TRUNDD, TRAILR4
RANKL (TNFSF11)	TRANCE, OPGL, ODF	RANK (TNFRSF11A)	TR8
		OPG (TNFRSF11B)	OCIF, TR1, FDCR-1
TWEAK (TNFSF12)	DR3LG, APO3L	DR3 (TNFRSF12)	TRAMP, WSL-1, LARD, WSL-LR, DDR3, TR3, APO-3
APRIL (TNFSF13)	TRDL-1	TACI (TNFRSF13)	
BAFF (TNFSF13B)	THANK, BlyS, TALL-1, TALL1, zTNF4	BAFF-R (TNFRSF13B)	
LIGHT (TNFSF14)	LTγ, HVEM-L	HVEM (TNFRSF14)	TR2, ATAR, LIGHTR, HVEA
VEGI (TNFSF15)	TL1, TL1A	TR-1 (TNFRSF15)	
		BCMA (TNFRSF17)	
AITRL (TNFSF18)	TL6, hGITRL	AITR (TNFRSF18)	GITR
		TROY (TNFRSF19)	TAJ, APO4, TRAIN-R, OAF065
		DR6 (TNFRSF21)	RELT, TANGO129, T129
		EDAR	DL

The biological significance of the TNF superfamily has been explored using gene deletion experiments and studies on gene knockout animals. Deletion of ligands and receptors of the TNF superfamily have indicated a critical role of these cytokines in protection from microorganisms, in the formation of lymph nodes and the development of the immune system, and in bone metabolism (Fig. 2.2A and 2.B).

TABLE 2.2
Journey Leading to Discovery of TNF Superfamily

Discovery	Discoverer(s)	Year	Ref.
Tumor regression in humans after bacterial infection	Brunes	1868	[1]
Bacterial extracts to treat cancers (Coley's toxins)	Coley	1894	[4]
Tumor regression in guinea pigs by bacterial extracts	Gratia and Linz	1931	[5]
Isolated LPS that induced tumor necrosis	Shear and Turner	1944	[6]
LPS induces systemic hypotension, collapse of tumor vasculature, tumor cell anorexia and death	Algire et al.	1952	[7]
LPS-challenged mice produce serum TNF	O'Malley et al.	1962	[8]
Described lymphotoxin as secreted cytotoxin	Granger and Williams	1968	[30]
LPS-induced serum factor named as TNF	Carswell et al.	1975	[9]
Isolated and sequenced TNFα protein	Aggarwal et al.	1985	[14]
Isolated and sequenced TNFβ proteins	Aggarwal et al.	1984	[11]
Cloned cDNA for TNF-α	Pennica et al.	1984	[16]
Cloned cDNA for TNF-β	Gray et al.	1984	[15]
TNF is cachetin	Beutler et al.	1985	[17]
4-1BB ligand	Kwon and Weissman	1989	[110]
OX-40 ligand	Miura et al.	1991	[39]
CD27 ligand	Goodwin et al.	1993	[82]
Fas ligand	Suda et al.	1993	[65]
CD30 ligand	Smith et al.	1993	[91]
CD40 ligand	Armitage et al.	1995	[55]
TRAIL	Wiley et al.	1995	[114]
RANKL	Wong et al.	1997	[144]
	Anderson et al.	1997	[145]
TWEAK	Chicheportiche et al.	1997	[184]
APRIL	Hahne et al.	1998	[187]
LIGHT	Zhai et al.	1998	[194]
VEGI	Zhai et al.	1999	[189]
THANK/Blys/BAFF	Mukhopadhyay et al.	1999	[191]
GITR	Gurney et al.	1999	[197]

The members of the TNF superfamily have a number of similarities in their signal transduction mechanisms and the cellular responses they mediate. Most members of the TNF receptor superfamily recruit one or more TNFR-associated factors (TRAFs) for signaling. Six different TRAFs have been so far identified that are required for the signaling of the TNF superfamily. Deletion studies have shown that TRAFs are indispensable for normal development and immune function (for reference see [20]; Fig. 3). All family members of the TNF superfamily activate the transcription factor NF-κB, though not all to the same extent [21]. For instance, TNF activates NF-κB in virtually all cell types so far examined, whereas FasL and TRAIL activate NF-κB only in certain cell types. The reason for the cell type–specific signaling by different cytokines is not understood. Another characteristic feature of the members of the TNF superfamily is that all of them activate c-Jun N-terminal kinase (JNK).

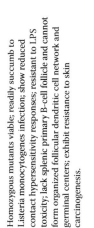

Homozygous mutants viable; readily succumb to Listeria monocytogenes infection; show reduced contact hypersensitivity responses; resistant to LPS toxicity; lack splenic primary B-cell follicle and cannot form organized follicular dendritic cell network and germinal centers; exhibit resistance to skin carcinogenesis.

Homozygous mutants viable but normal growth only till three weeks, severely retarded growth following weaning; marked osteopetrosis, lack osteoclasts, but normal hematopoietic osteoclast precursors; defect in tooth eruption; lack all lymph nodes but have normal spleni c structure and Peyer's patches; fail to form lobulo-alveolar mammary structures during pregnancy.

Homozygous mutants viable; susceptible to experimental and spontaneous tumor metastasis.

Homozygous mutants viable; reduced IgM and IgG; severe loss of B220+ cell in secondary lymphoid organs; reduced mature circulating B cells blocked B-cell development.

Homozygous mutants viable; do not develop germinal centers to thymus-dependent antigens; severely impaired virus-specific CD4 T-cell response; inability to develop memory B-cell response, impaired T-cell mediated macrophage activation.

Homozygous mutants viable; lack lymph nodes and Peyer's patches; failure of normal segregation of B and T cells in splenic white pulp; abnormal lymphocyte clusters accumulate in periportal and perivascular regions.

Homozygous mutants viable; lack Peyer's patches, peripheral lymph nodes, splenic germinal centers, and follicular dendritic cells; mesenteric lymph nodes with germinal center-like regions but with no follicular dendritic cells.

Spontaneous mutation at gld locus ; lymphadenopathy and systemic autoimmunity; enlarged spleen; increased number of T, B, and null lymphocytes; develops immune comple x glomerulonephritis.

Homozygous mutants viable; reduced proliferative response of Vb8+ CD8+ cells than Vb8+ CD4+ cells to staphylococcal enterotoxin B; reduced induction and cytokine secretion of CD8 + CTL to MHC class I-restricted peptides.

Homozygous mutants viable; impaired contact hypersensitivity response; dendritic cells are defective in costimulating T-cell cytokine production; impaired intrinsic antigen presenting cell functions.

Homozygous mutants viable; decreased CD8T cell expansion.

Homozygous mutants viable; hypoplastic hair, teeth, and eccrine sweat glands.

FIGURE 2.2 (A)　Phenotypic effects of gene deletions of the TNF superfamily ligands.

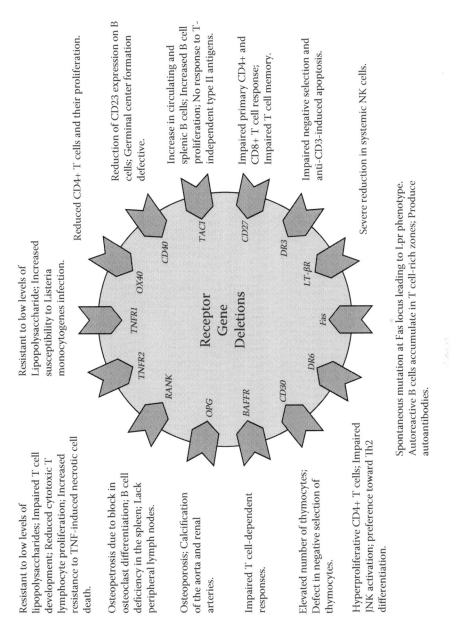

FIGURE 2.2 (B) Phenotypic effects of gene deletions of the TNF superfamily receptors.

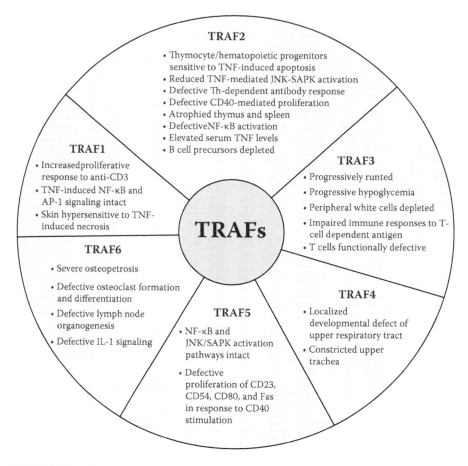

FIGURE 2.3   TRAFs: effects of gene deletion.

With respect to growth modulation, some members of this family such as TNF, Fas, and TRAIL induce apoptosis [22]; others (such as RANKL) promote cell survival; and some (such as TNF, CD28L, CD30L, and CD40L) induce cell proliferation (Table 2.3). How members of this family induce these biological responses is becoming increasingly more apparent. Interestingly, however, the enigma that still remains unresolved is the fact that the activation of NF-κB by TNF and other cytokines inhibits apoptosis induced by the same cytokine (Fig. 2.4) [22,23].The mechanism by which TNF induces apoptosis was reviewed in detail recently [24].

Research during the past two decades has shown that TNF superfamily cytokines are important and required for normal biological responses, but their inappropriate expression has been implicated in a number of diseases (Fig. 2.5). Because of the systemic toxicity of TNF, its role as potent anticancer agent has diminished; however, recent studies showed that TNF is an effective treatment for locally advanced soft-tissue sarcomas of the limbs when it is administered by isolated limb perfusion [25]. In contrast, TRAIL has been found to specifically kill tumor cells, but without harming normal cells, and so it is being explored for the treatment of cancer.

TABLE 2.3
List of Ligands, Their Receptors, and Cellular Responses of TNF Superfamily

Ligand	Receptor	Interacting Protein(s) Death Domains	Proliferation	Apoptosis	NF-κB	JNK	p42MAPK	p38MAPK	References
TNFα	TNFR1	TRADD	+	+	+	+	+	+	[201,202]
TNFα	TNFR2	—	+	+	+	+	+	+	[201–203]
LT	TNFR1	TRADD	+	+	+	+	+	+	[204]
LT	TNFR2	—	+	+	+	+	+	+	[204]
FasL	Fas	RIP, FADD	—	+	+	+	+	+	[76,205]
FasL	DcR3	—	—	(−)	+	—	—	—	[77]
VEGI	DR3	TRADD, FADD, RIP	—	+	+	—	+	+	[190]
VEGI	DcR3	—	—	(−)	—	—	—	—	[190]
TRAIL	DcR1	—	(−)	—	+	+	+	—	[206,207]
TRAIL	DcR2	—	(−)	—	+	—	—	—	[206,207]
TRAIL	DR4	TRADD, FADD, RIP	—	+	+	+	+	+	[208,209]
TRAIL	DR5	TRADD, FADD, RIP	—	+	+	+	+	+	[208,209]
TRAIL	OPG	—	(−)	+	—	—	—	—	[116]
LT-β	LT-βR	—	+	+	+	+	—	—	[210–212]
LIGHT	LT-βR	—	—	+	+	—	—	—	[210–212]
LIGHT	HVEM	—	—	+	+	—	—	—	[213]
LIGHT	DcR3	—	—	—	—	—	—	—	[214]
CD27L	CD27	—	+	+	+	+	+	—	[202,215,216]
CD30L	CD30	—	+	+	+	+	—	—	[102,217]
CD40L	CD40	—	+	—	+	+	+	—	[203,218]
OX40L	OX 40	—	+	—	—	—	—	—	[49,202]
4-1BBL	4-1BB	—	—	+	+	+	—	—	[202,219]
RANKL	RANK	—	+	—	+	+	+	+	[220,221]

continued

TABLE 2.3 (continued)
List of Ligands, Their Receptors, and Cellular Responses of TNF Superfamily

Ligand	Receptor	Interacting Protein(s) Death Domains	Proliferation	Apoptosis	NF-κB	JNK	p42MAPK	p38MAPK	References
RANKL	OPG	—	—	—	—	—	—	—	[222]
APRIL	TACI	—	+	—	+	+	—	—	[223,224]
APRIL	BCMA	—	—	—	+	+	+	+	[225,226]
BAFF	TACI	—	+	—	+	+	—	—	[223,224]
BAFF	BCMA	—	—	—	+	+	+	+	[225,226]
BAFF	BAFF-R	—	+	—	+	—	—	—	[227,228]
GITRL	GITR	—	+	—	+	—	—	—	[197]
ND	TROY	—	+	—	+	+	—	—	[199]
ND	DR6	—	—	—	+	+	—	—	[200]
TWEAK	Fn 14	—	+	+	+	+	—	—	[229,230]
EDA-A1	EDAR	EDARADD	—	+	+	+	—	—	[231,232]
EDA-A2	XEDAR		—	—	+	—	—	—	[233]

Note: ND, not determined; +, activation; (−), no induction/activation; —, not reported.

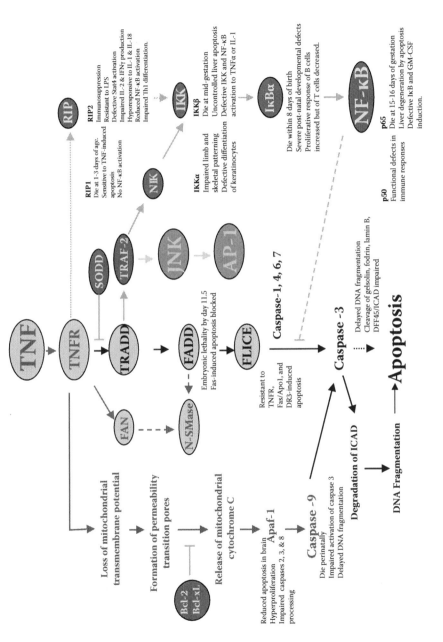

FIGURE 2.4 Pathway leading to activation of NF-κB, JNK, AP-1, and apoptosis.

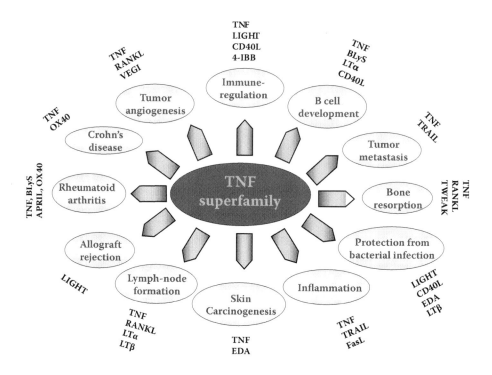

FIGURE 2.5 The main physiological and pathological effects linked to members of the TNF superfamily.

Ligands of the TNF superfamily are required for immune regulation at several levels. TNF, LT, and RANKL provide crucial signals for the morphogenesis of secondary lymphoid organs. Pro-apoptotic members, such as TNF, CD95L, and TRAIL, contribute to the function of cytotoxic effector cells in the recognition and destruction of virus-infected cells. Ligands of the TNF superfamily are produced by hematopoietic cells and have a crucial role in regulating their proliferation. It has been shown that deletion of either TNF or TNFR1 in mice leads to rapid death from infection and resistance to LPS-mediated septic shock, indicating a crucial role of this cytokine in protection from infection [26]. Antibodies specific for TNF (such as infliximab) and soluble TNFRs (such as Enbrel) have been approved for use in the treatment of Crohn's disease and rheumatoid arthritis, respectively.

Contrary to the original belief, it is now clear that TNF can contribute to tumorigenesis by mediating the proliferation, invasion, and metastasis of tumor cells. For example, it has an important role in the pathogenesis of type II diabetes mellitus. It interferes with an insulin-signaling mechanism by inhibiting the tyrosine kinase activities of the insulin receptor and serine phosphorylation of the insulin receptor substrate 1. It has also been shown to be an autocrine growth factor for a wide variety of tumors. Several TNF ligands have been implicated in the development of autoimmunity. Besides these, ligands of the TNF superfamily have been linked with chronic heart failure, bone resorption, AIDS, Alzheimer's disease, transplant rejection, atherosclerosis, and hepatotoxicity (for reference see [19]).

2.3.1 TUMOR NECROSIS FACTOR α

Tumor necrosis factor α (TNFα, originally called TNF) was isolated as a cytotoxic factor from the human promyelomonocytic HL-60 cell line in 1985 by Aggarwal et al. [14]. Although TNF was initially identified because of its antitumor activity, it turned out later that the activities of TNF were not limited only to tumor regression. A number of other activities were assigned to TNF in the murine system in due course of time. Beutler et al. isolated a protein that was responsible for endotoxin-induced cachexia in mice and named it cachectin [17]. In 1986, Takeda et al. discovered a protein and called it T-cell differentiation-inducing factor (DIF) [18]. The amino acid sequencing of both these proteins showed that they had a structure that was similar to TNFα.

Human TNF is a type II transmembrane protein consisting of 233 amino acid residues. Mature soluble human TNF is a 157 amino acid long nonglycoprotein containing a single disulfide bridge and known to exist as a homotrimer in aqueous solution [14]. The gene for TNF was localized on human chromosome 6 in a region between p23 and q12. It was found to be encoded by a single gene of 3.6 kb that is split by three introns [27].

TNF binds to TNF receptors I and II and either mediates apoptosis or survival, differentiation or proliferation through the activation of pathways involving NF-κB, JNK, ERK, or p38 MAPK. The normal physiological role of TNF in vivo is unclear. It is believed that TNF is required for protection against bacterial, fungal, parasitic, and viral infections [19]. Whether TNF is able to block tumorigenesis and metastasis in vivo is still not fully understood; however, a report by Orosz et al. shows that TNF is involved in tumorigenesis and metastasis [28] and mice deficient in TNF are resistant to skin carcinogenesis [29]. Besides this, TNF is involved with viral replication, septic shock, fever, inflammation and autoimmune diseases including Crohn's disease, rheumatoid arthritis, and graft-versus-host disease.

2.3.2 TUMOR NECROSIS FACTOR β

Tumor necrosis factor β (TNFβ) was earlier known as a soluble factor produced in supernatants of mitogen-activated lymphocytes and was found to be lytic to certain tumor cells in culture [30]. It was also described as a product of antigen-activated lymphocytes that killed fibroblasts [31]. This soluble factor was isolated by our laboratory in 1984 as a cytotoxic factor from the human B lymphoblastoid cell line RPMI-1788 and called lymphotoxin [11]. Later it was renamed TNFβ after the discovery of TNFα in 1985 [14].

The amino acid sequences of TNFα and TNFβ and the isolation of their cDNA revealed that TNFα and TNFβ are structurally homologous factors [15,16]. TNFβ is expressed by a large number of cell lines such as natural killer cells, T lymphocytes, B lymphoblastoid cells, and myeloma cells, to name a few. The production of TNFβ can be induced by phorbol esters, viruses, interleukin 2, interferons, and mitogens. TNFβ binds to both the TNF receptors and is involved in inflammation and cachexia, and mediates endotoxin-induced shock, autoimmunity, wound healing angiogenesis, bone resorption, and cancer.

2.3.3 LYMPHOTOXIN-β

Lymphotoxin (LT) was initially described as a secreted cytotoxin [30]. Later, Ware et al. identified a nonsecreted form of a cell surface protein expressed by activated T cells [32]. It was originally called LTα-binding protein. LTα is present on the surfaces of activated T, B, and LAK cells as a complex with a 33-kDa glycoprotein, and cloning of the cDNA encoding the associated protein, called LTβ, revealed it to be a type II membrane protein with significant homology to TNF, LTα, and the ligand for the CD40 receptor [33]. LTβ is produced by activated lymphocytes, including T and B cells, and NK cells. The human CD4+ T-cell hybridoma II-23.D7 produces abundant LTβ upon stimulation with phorbol ester and was used as a source for the purification of LTβ [33]. The secreted forms of LT revealed significant sequence and functional similarities to TNF [15,16]. LT, now designated as LTα, exists as a heterotrimer, whereas surface LT is formed as a heterocomplex composed of LTα and LTβ subunits. The two related LT proteins, along with TNF, are now recognized as regulators of lymphoid organ development and immune function. LTα was previously referred to as TNFβ, whereas TNFα is now simply designated TNF.

Evidence indicates that the LTβ protein is not normally expressed in the absence of LTα. Two distinct LTαβ complexes have been identified on the surface of activated T cells, designated LTα1β2 and LTα2β1 heterotrimers. LTβ provides the membrane anchor for the LTαβ complexes. The cell surface LT complex binds and signals through the LTβR, which is related to the two known receptors of TNF. Neither TNFR60 nor TNFR80 interacts directly with the LTβ, indicating that LTβ provides the specificity for binding to the LTβR [34].

Upon stimulation, LTβR recruits TRAFs 3 and 5 and signals for the activation of NF-κB, as well as induction of apoptosis. However, deletion of TRAF3 abrogates cell death signaling but not activation of NF-κB [35,36].

LTs play a unique role in immune physiology and have emerged as critical developmental regulators for the peripheral lymphoid system. Mice that are homozygous mutant for LTβ are viable but lack peripheral lymph nodes, Peyer's patches, splenic germinal centers, and follicular dendritic cells [37]. Deletion of the LTβR gene dramatically reduces the number of NK cells, suggesting that signaling via LTβR on bone marrow stromal cells by membrane LT is an important pathway for early NK cell development [38]. Thus, LTβ plays an important role in controlling the formation of tissue microenvironments where immune responses are convened.

2.3.4 OX-40L

The human ligand for OX-40 was originally cloned by Miura et al. [39] and named gp34 by them. It was described as a protein that is regulated by the *tax* gene of the human retrovirus HTLV-1 [39]. Subsequently, this protein was shown to bind to the OX-40R [40,41]. OX-40L is a type II membrane protein with limited homology to TNF and is costimulatory to OX-40+ T cells in vitro [40,41]. The murine and human OX-40L cDNAs have 68% homology at the nucleotide level and 46% at the amino acid level.

The OX-40R was discovered when one of the mouse monoclonal antibodies, MRC OX-40, prepared against rat T-cell blasts recognized an antigen that differed from any previously described in that its expression was detected only on T blasts that also expressed the CD4 antigen. The MRC OX-40 antibody recognized a CD4+ rat T-cell surface "blast" antigen with molecular mass of 50 kDa [42]. This unique antigen was expressed only on terminally differentiated CD4+ T cells that had been stimulated with a mitogen.

The rat OX-40 cDNA was cloned and shown to be a member of the TNF-R superfamily with homology to CD40 and NGFR [43]. The OX-40 receptor is a transmembrane protein found on the surface of activated CD4+ T cells. The murine and human OX-40R cDNAs have also been cloned [44,45]. The murine cDNA is 93% homologous to the rat, and all the N-terminal cysteines are conserved. The human OX-40 cDNA sequence is 65% and 68% homologous with the rat and mouse sequences, respectively.

The human OX-40 gene is located on chromosome 1 and is clustered with the TNFR family members CD30, 4-1BB, TNFR-II, and DR3 at the distal band 1p36 [45]. Human and murine OX-40R are expressed on both CD4+ and CD8+ T cells after stimulation with Con A or PHA [40]. The OX-40R can be induced on naïve or effector T cells solely by the engagement of the T-cell receptor, in the apparent absence of other signals.

Human OX-40L stimulates human T cells exclusively, while murine OX-40L stimulates both human and mouse T cells [40]. Cell surface OX-40L expression is quite restricted, and this may limit the number of T cells stimulated through the OX-40R in vivo. Activated APC express OX-40L and can transmit the OX-40L:OX-40R signal during presentation of antigen to CD4+ T cells. Activation through the antigen receptor and CD40 on B cells or CD40 on dendritic cells upregulates OX-40L expression, and engagement of OX-40L transmits a potent signal to the APC [46]. The signal appears to promote differentiation of B cells to Ig secreting plasma cells but has no role in B cell memory development [46,47]. OX-40L signaling is important for differentiation of human dendritic cells and leads to increased production of IL-12, TNF-α, IL-1β, and IL-6 [48]. Activation of OX40 signal transduction pathways leads to TNFR-associated factor (TRAF)-mediated NF-κB activation. The NF-κB activation resulting from OX-40 stimulation is mediated by both TRAF2 and TRAF5 and is likely to be negatively modulated by TRAF3 [49]. In addition, OX-40L is expressed on freshly isolated vascular endothelial cells. Interestingly, adhesion of activated T cells to endothelial cells can be blocked by antibodies to OX-40R and OX-40L [50].

Deletion studies in mice showed that, though the homozygous mutants for OX-40L were viable, they had impaired contact hypersensitivity response and their dendritic cells were defective in costimulating T-cell cytokine production [51]. The mutation also resulted in impaired intrinsic antigen-presenting cell functions [52]. Deletion of OX-40 in mice reduced the number and proliferation of CD4+ T cells [53].

2.3.5 CD40L

In humans, anti-CD40 mAb induces B-cell proliferation, thereby implicating CD40 as an important triggering molecule for B cells. In the absence of a suitable antibody to mouse CD40, a soluble fusion protein of the extracellular domains of human CD40 and the Fc domain of hIgG (CD40-Ig) was prepared; it caused a dose-dependent inhibition of the B-cell RNA synthesis that was induced by the plasma membrane preparation from anti-CD3 or Con A-activated T_H cells [54]. Biochemical studies revealed that CD40-Ig was bound to a 33- to 39-kDa membrane protein on T_H cells, named gp39. Additional support for the role of gp39 in B-cell activation came from studies showing that a high affinity hamster mAb specific to gp39 blocked the activation of resting B cells by an active plasma membrane preparation [54]. Armitage et al. reported the molecular cloning of the ligand for CD40, which they called CD40L [55]. Through expression cloning of a library constructed from EL4 thymoma cells, a 33-kDa molecule was identified. Upon transfection of CD40L into a COS-like cell line, the expressing cell line induced the proliferation of heterogeneous splenic B cells and, in the presence of IL-4, induced the production of IgE [55]. CD40L and gp39 appeared to be identical. To identify the possible human homologue of CD40L, a variant Jurkat cell line that induced B-cell growth and differentiation was isolated [56]. An mAb, 5c8, specific to the variant cell line identified a membrane protein of 30 kDa. The mAb blocked the ability of Jurkat cells and activated T cells to induce B-cell activation. Later studies suggested that the antigen recognized by 5c8 was the ligand for CD40 [57].

CD40 was first identified serologically as a membrane protein expressed on immature and mature B cells [58], some carcinomas, interdigitating cells present in the T-cell zones of secondary lymphoid organs, follicular dendritic cells (FDCs), and thymic epithelium [54,59]. CD40 is a membrane protein of 45–50 kDa. The CD40 ligand is a type II transmembrane glycoprotein of the TNF family and exists as a trimer on the cell surface. The human protein consists of a 22 amino acid cytoplasmic tail, a single 24 amino acid transmembrane domain, and a 215 amino acid extracellular region (214 amino acids in the mouse); human CD40L is 83% identical to murine CD40L at the nucleotide level [54,55].

Upon stimulation, CD40 induces phosphorylation of multiple substrates and activates tyrosine protein kinases, including Lyn and Syk, and the serine/threonine protein kinases JNK and MAP [60]. Phosphorylation leads to activation of various transcription factors, including NF-κB, c-Jun, and NF-AT. CD40 mediates its effects through receptor-associated proteins, in particular, several members of the TRAF family [61].

In mice with a genetic deletion of CD40L, homozygous mutants are viable but do not develop germinal centers to thymus-dependent antigens. They have severely impaired virus-specific CD4 T-cell response, are unable to develop memory B cell response, and have an impaired T cell–mediated macrophage activation response [62,63]. The deletion of the CD40 gene in mice also reduces CD23 expression on B cells [64].

2.3.6 FAS L

The Fas ligand was identified by Suda et al. in 1993 using a soluble form of mouse Fas prepared by fusion with human immunoglobulin Fc [65]. Fas ligand was detected on the cell surface of a cytotoxic T-cell hybridoma, PC60-d10S. A cell population that highly expresses Fas ligand could be isolated using FACS, and its cDNA was identified in the sorted cells by expression cloning. The amino acid sequence indicated that the Fas ligand is a type II transmembrane protein that belongs to the TNF superfamily. The purified Fas ligand has a molecular mass of 40 kDa. Its cytolytic activity specifically targets cells expressing Fas.

However, using murine models of systemic lupus erythematosus (SLE), Mountz et al. (1985) found selective abnormal proto-oncogene expression associated with the characteristic abnormal cell growth or differentiation of lymphocytes of autoimmune mice. The lymph nodes of MRL-lpr/lpr mice were packed with unusual T cells. The gld/gld mouse has a very similar unusual T cell in the lymph nodes [66]. Later it was determined that functional differences between lpr and gld become apparent after bone marrow transfer experiments. The molecule altered by the gld mutation is expressed only in the bone marrow cells, whereas the molecules altered by the lpr mutation are expressed by both bone marrow-derived cells and by one or more peripheral radioresistant cell populations [67]. This study also predicted that the Fas ligand was mutated in gld mice and was a cytokine that interacted with the mutated gene product present in lpr mice. The study established a connection between the Fas defect in lpr mice and Fas ligand defect in in gld mice.

Fas ligand is produced by Th0 and Th2 cells, B cells, macrophages, natural killer (NK) cells, and certain nonhematopoietic cells including testes, ovary, and the salivary gland. It can be induced by anti-TCR antibody and IFN-γ in T cells [68,69] and lipopolysaccharide in B cells and macrophages [70,71]. The Fas ligand binds to the Fas receptor (the name Fas is derived from the FS-7 fibroblast surface activation antigen) [72].

The Fas signaling pathway is initiated by trimerization of Fas by Fas ligand, resulting in conformational changes within the cytoplasmic domain. This results in the binding of Fas-associated death domain (FADD) to the Fas death domain and also in binding of the Fas receptor interacting protein (RIP). Interaction with membrane-bound FasL reorganizes these complexes and allows the formation of a death-inducing signaling complex (DISC). The Fas DISC contains the adapter protein FADD and caspases 8 and 10, which can initiate the process of apoptosis. Fas-L induced clustering of Fas, FADD, and caspase-8 or -10 within the DISC results in autoproteolytic processing of these caspases by induced proximity and in release of the processed active proteases [73]. Fas-mediated apoptosis can occur in two distinct ways. Type I apoptosis is characterized by rapid formation of DISC and robust production of caspase 8 that is followed by the release of caspase 3, whereas type II apoptosis exhibits a slow DISC assembly and activation of the caspase cascade. Interestingly, the caspase-8-activating capacity of the Fas DISC is mainly regulated by FADD-like interleukin-1β-converting enzyme (FLICE)-like inhibitory protein (FLIP) [74] in both the cases described above. In addition, Fas-mediated apoptosis is controlled by a number of regulators including SMAC inhibitor of apoptosis proteins, which are Bcl-2 family members [75]. Fas-mediated cell death may also occur by necrosis, which requires the adaptor protein FADD and Fas-interacting RIP, whereas caspase 8 seems to be dispensable [76]. Stimulation of Fas by membrane-bound Fasl can be antagonized by the soluble decoy receptor DcR3 [77], by various Fas isoforms lacking the transmembrane and/or death domains, and by soluble Fasl generated by proteolytic processing or alternative splicing.

Defects in the expression of or function of either the receptor or the ligand lead to lymphoproliferative diseases, which include autoimmune diseases and cancer [78,79].

2.3.7   CD27L

CD27L was first detected on the surface of non-Hodgkin's lymphoma cells and Reed-Sternberg cells diagnostic of Hodgkin's lymphoma. Originally, it was described as the Ki-24 antigen, a ligand for a monoclonal antibody raised against a Hodgkin's disease cell line [80], and later as CD70, based on antibody clustering [81]. Expression cloning of the gene by binding to a soluble CD27 construct gave it its third name, the CD27 ligand [82]. Later, expression cloning by binding to anti-CD70 monoclonal antibodies confirmed the identity of CD70 as the CD27 ligand.

CD27, a 193 amino acid long protein expressed on a wide variety of cells of lymphoid origin, is activated by T or B cells [83]. It is also expressed by T and B cells transformed with EBV or HTLV-I and nasopharyngeal carcinoma cells associated with EBV expression [84]. Resting lymphoid cells do not express detectable levels of CD27 ligand.

CD27 ligand binds to CD27, which is a surface antigen found on T and B cells that has homology to a family of molecules, including the receptors for tumor necrosis factor (TNF) and nerve growth factor [85]. A cDNA encoding a ligand for CD27 was isolated by a direct-expression cloning strategy using a fusion protein composed of the extracellular domain of CD27 linked to the constant domain of a human immunoglobulin G1 molecule as a probe. The predicted protein product is a type II transmembrane protein whose gene maps to 19p13 and that shows homology to TNF and the ligand for CD40 [82]. Biological characterization indicates that the cloned ligand induces the proliferation of costimulated T cells and enhances the generation of cytolytic T cells.

In addition, CD27 ligand has been seen associated with such pathologies as dermatopathic lymphadenopathy, nasopharyngeal carcinoma, and chronic lymphocytic leukemia. Sequence analysis has identified this molecule as a member of the ligands for the TNF receptor superfamily. CD27 ligand has been shown to act as a costimulator of T-cell activation, enhancing both helper and cytolytic activities. It has also been found to enhance IgM and IgG synthesis by mitogen-stimulated B cells.

2.3.8   CD30L

Like CD27, CD30 was initially described as an antigen expressed on Reed-Sternberg cells of Hodgkin's disease [86]. Screening for CD30 expression in normal and neoplastic cells using monoclonal antibodies led to the identification of a new category of non-Hodgkin's lymphoma: CD30/Ki-1 positive anaplastic large cell lymphoma [87], and to recognition of CD30 as an activation-induced antigen expressed mostly on lymphoid cells [87]. Subsequently, a soluble form of CD30 (sCD30) was identified. Increased serum levels are seen in patients with CD30-positive neoplasms, some viral infections, and Th2-type immune response [88,89]. In 1992, the cDNA encoding human CD30 was cloned, thus establishing CD30 as a member of the TNF receptor TNFR superfamily [90]. Cloning of the cDNA for CD30L in 1993 in turn revealed that it belonged to the TNF superfamily [91].

CD30 is a type I transmembrane glycosylated protein of 120/105 kDa derived from a 90-kDa non-glycosylated precursor [92]. Two mRNA species of 3.8 kb and a minor 2.6 kb form differ in the 3′ noncoding region. The human CD30 comprises 595 amino acid residues, whereas murine and rat CD30 have 498 and 493 amino acids, respectively [93,94]. The 362-residue extracellular domain has six cysteine-rich regions in a duplicated structure, whereas murine and rat CD30 lack the second cluster. This region shows significant homology to those of other TNFR superfamily members, whereas none was found in the intracellular domain. Like other members of the TNFR superfamily, catalytic domains were not found in the cytoplasmic domain, although there are several potential phosphorylation sites. The human CD30 gene locates to chromosome 1p36, [95], like other members of this superfamily, such as the human TNFR2 and OX40.

Using a CD30-Fc fusion protein, the ligand for CD30 was identified on the surface of anti-CD3 stimulated human PBMC and the murine T-cell line 7B9. Hence, human CD30 cross-reacts with murine CD30L. A murine cDNA encoding the latter was isolated by expression screening of the cDNA library, using a CD30-Fc fusion protein, and that of the human counterpart by cross-hybridization [91]. CD30L is a type II membrane glycoprotein with a molecular mass of 40 kDa encoded by a 1.8-kb mRNA. The human CD30L gene locates to 9q33. Human CD30L protein has an extracellular domain comprising carboxy terminal 172 amino acids and a cytoplasmic domain of N-terminal 40 residues. The extracellular domain shows significant homology to TNFα, TNFβ, and the CD40L. Like other TNF family proteins, CD30L forms a trimer, which is considered to be the functional form. It is presently unclear whether CD30L also exists in a soluble form like TNFα and FasL. Recombinant CD30L appears to be functional only when immobilized [91,96]. Expression of CD30L was evident in activated T cells, neutrophils, eosinophils, B cells in the resting state, the medulla of the thymus epithelial cells, and Hassal's corpuscles, as well as in various leukemic cells [91,97–99].

Stimulation of CD30 by agonistic monoclonal antibodies M44 and M67 or CD30L in lymphoid cells induces pleiotropic biological effects, including proliferation, activation, differentiation, and cell death, depending on cell type, stage of differentiation, and presence of other stimuli [96]. Cross-linking of CD30 activates NF-κB mediated by TNFR-associated factor TRAF proteins that interact with the cytoplasmic tail of the CD30 protein [100–102]. In addition to activating NF-κB, CD-30 can activate SAPK/JNK.

Elevated levels of CD30 have been observed in viral infections, infectious mononucleosis caused by EBV, hepatitis virus B infection, and human immunodeficiency virus type 1 infection. Increased sCD30 levels were noted in the majority of HIV-1-infected individuals [103]. Increased levels of sCD30 have also been found in autoimmune diseases, such as systemic lupus erythematosus, rheumatoid arthritis, systemic sclerosis, atopic dermatitis, Wegener's granulomatosis, Graves' disease, and Hashimoto's thyroiditis [104–107]. The serum concentration of sCD30 increases in some neoplastic diseases as the result of release by neoplastic or reactive cells expressing CD30. Increased levels have been found in patients with Hodgkin's disease, anaplastic large cell lymphoma, adult T-cell leukemia and angioimmuno-blastic lymphadenopathy-like T-cell lymphoma [108,109].

2.3.9   4-1BBL

The 4-1BBL protein was identified as the cognate for the receptor for 4-1BB, which was originally characterized as an inducible murine T-cell cDNA [110]. Subseqently, 4-1BB was found to belong to a family of molecules referred to as the TNF receptor superfamily [111]. The ligand is a type II membrane glycoprotein that has homology to members of the TNF superfamily.

The murine 4-1BB ligand was expression-cloned using a library prepared with RNA from the thymoma cell line EL4. Likewise, the human ligand was expression-cloned from an activated CD4+ T-cell clone. The expression of 4-1BBL has been primarily characterized by analysis of its mRNA. The gene maps to mouse chromosome 17, but at a considerable distance from the TNF and LT genes.

The 4-1BB ligand is expressed in bone marrow, brain, kidney, spleen, thymus, placenta, skeletal muscle, and activated T and B cells. It is also expressed in D11 (murine bone marrow stromal), F4 (murine thymic stromal), Abl 1.1 (murine pre B), A20 (murine B-cell lymphoma), RAW 264.7 (murine macrophage), THP-1 (human monocytic), and SK-N-SH (human neuroblastoma) cell lines.

Resting T lymphocytes express little or no 4-1BB ligand, but stimulation of these cells with immobilized anti-CD3 leads to expression of 4-1BBL transcripts and surface protein [112]. Cross-linking of the 4-1BB molecule by agonistic antibody transmits a distinct and potent costimulatory signal leading to the activation and differentiation of CD4+ and CD8+ cells. 4-1BB transmits signals through the TRAF2-NIK pathway and activates NF-κB [113]. Signals relayed through 4-1BB inhibit activation-induced cell death and rescue the immune system during the post-CD28 phase.

Antibodies to 4-1BB can increase severity of graft-versus-host disease (GVHD), accelerate the rejection of cardiac allograft and skin transplants, and eradicate established tumors. 4-1BB–deficient mice have dysregulated immune responses and mount elevated Ig responses to T-dependent antigens. Blocking the 4-1BB/4-1BB ligand interaction, with either a soluble form of the receptor or with blocking antibodies to either protein perhaps in combination with inhibition of the CD28/B7 interaction, should lead to an immunosuppressive response in vivo.

2.3.10   TRAIL (TNF-RELATED APOPTOSIS-INDUCING LIGAND)

One of the first TNF-related ligands that was identified independently by two groups using expressed sequence tags was named TRAIL [114], also known as Apo2L [115]. TRAIL is a ubiquitous type II transmembrane protein of 281 amino acids in the human form and 291 amino acids in the murine form. It can be cleaved from the membrane by a protease to yield a soluble protein. TRAIL specifically interacts with four membrane-bound receptors known as TRAIL R1 through TRAIL R4 (also called DR4, DR5, DcR1, and DcR2, respectively) and one soluble receptor, osteoprotegrin (OPG), which can inhibit TRAIL-induced apoptosis [116]. TRAIL-R1(DR4) and TRAIL-R2(DR5) can initiate intracellular signaling that leads to death of the target cells. However, DR4 and DR5 have different capabilities for stimulating the JNK pathway and differ also in their cross-linking requirements for activation by recombinant ligands. DR4 responds to either cross-linked or non-cross-linked soluble TRAIL and signals NF-κB activation and apoptosis, whereas DR5 signals NF-κB activation, apoptosis, and JNK activation only in response to cross-linked TRAIL [117].

Investigations of the intracellular signaling pathways responsible for TRAIL receptor-induced apoptosis have produced controversial results, but most recent studies suggest DR5 signals through a FADD- and caspase 8–dependent pathway [118]. FADD must be essential for DR4- and DR5-mediated apoptosis since TRAIL failed to induce apoptosis in FADD-deficient mouse embryonic fibroblasts [119]. TRAIL induces apoptosis through caspase 8 and 3 in human melanoma cells that are resistant to FasL-induced cell killing [120]. Bax is required for TRAIL-induced apoptosis of certain cancer cell lines, possibly by allowing release of second mitochondria-derived activator of caspases (Smac)/direct inhibitor of apoptosis protein (IAP)-binding protein with low pH (DIABLO) and antagonizing the IAP family [121]. Bax gene ablation led to resistance to TRAIL [122,123], and reintroduction of Bax into Bax-deficient cells restored TRAIL sensitivity [121].

The physiological role of TRAIL in mice was determined by three different methods: (a) a neutralizing anti-mouse (m)TRAIL mAb (N2B2) [124]; (b) soluble recombinant human DR5 [125]; and (c) TRAIL gene-targeting [126]. TRAIL mRNA was found in a variety of tissues and cells [114]. Freshly isolated T cells, NKT cells, B cells, dendritic cells (DC), monocytes, or NK cells did not express a detectable level of TRAIL on their surface [124,127–130], whereas mouse liver NK cells expressed TRAIL constitutively in a type II IFN-dependent manner [129,130]. TRAIL is highly expressed on most NK cells, T cells, CD11c+ DC, and monocytes after stimulation, which then acquire the ability to kill tumor cells [124,127,128,131–134]. These expression patterns reflect the broad role that TRAIL likely plays in innate immune responses involving NK cells, monocytes, and DC.

Cretney et al. demonstrated the key role of TRAIL in suppressing tumor initiation and metastasis using TRAIL gene-targeted mice [126]. Liver and spleen mononuclear cells from TRAIL gene-targeted mice showed no TRAIL expression or TRAIL-mediated cytotoxicity. TRAIL gene-targeted mice were more susceptible to experimental and spontaneous tumor metastasis, and the immunotherapeutic value of alpha-galactosylceramide was diminished in these mice. TRAIL gene-targeted mice were also more sensitive to the chemical carcinogen methylcholanthrene.

TRAIL possesses immense potential in anticancer therapeutics because of its ability to induce apoptosis in a variety of transformed cell types without affecting normal (nontransformed) cells [114,135]. Most remarkably, systemic TRAIL is tumoricidal, causing significant tumor regression in mice without causing toxic side effects [136,137]. However, not all cancer cells are sensitive to the cytotoxic effects of TRAIL. Resistance has been ascribed to DR5 mutations in some cancers of the head and neck, breast and lung, and Hodgkin's lymphoma [138], and recent studies in the mouse suggest TRAIL immunoselects tumors for increased TRAIL resistance [126]. DR5 has been implicated in the cellular response to DNA-damaging radiation or chemotherapy as a target of p53 [139]. One attractive feature of TRAIL is its ability to kill cancers with p53 mutations; however, its combination with chemotherapeutic agents has also been found to be particularly effective against cancers with wild-type p53, presumably through induction of DR5 expression [140]. In addition, Bax mutation in mismatch repair-deficient tumors can cause resistance to TRAIL therapy, but pre-exposure to chemotherapy rescues tumor sensitivity [123].

Manipulation of TRAIL function in humans has not yet been tried. We now await the administration of recombinant human TRAIL to cancer patients, but additional trials comparing the efficacy of TRAIL-expressing NK cells and T cells in humans following adoptive transfer in either solid tumor or allogeneic bone marrow transplant patients may also be of interest. Recombinant forms of TRAIL have been shown to kill tumor cells in either a type I (not protected by bcl-2 family proteins)– or a type II (protected by bcl-2 family proteins)–dependent manner. These different forms of target cell sensitivity are not well understood molecularly, and which of these pathways is used by natural TRAIL expressed on lymphocytes remains unclear. Defining the contexts in which TRAIL may eliminate or spare normal tissues is also of great importance when considering manipulating the TRAIL pathway. The ability of some soluble forms of TRAIL to kill normal cells such as hepatocytes [141] and the possible role of TRAIL in other immune responses [125,142] raise some concerns about proposed simplistic approaches. The demonstrated killing activity of antihuman DR5 mAbs against liver cancer cells, but not normal hepatocytes [143], at least offers hope that rational engineering of recombinant molecules will provide new strategies to exploit the TRAIL/TRAIL-R pathway in cancer therapeutics.

2.3.11 RECEPTOR ACTIVATOR OF NF-κB LIGAND

Receptor activator of NF-κB ligand (RANKL) was cloned independently by four groups as an apoptosis-regulatory gene (TRANCE, for TNF-related activation-induced cytokine) [144], as a factor required for T-cell growth and dendritic cell function [145], as an osteoclast differentiation factor [146], and as a ligand for the soluble TNFR family member OPG [147]. It is named RANKL because over-expression of its receptor RANK could activate NF-κB independent of receptor-ligand cross-linking [145]. RANKL also stimulates JNK activity in mouse thymocytes and T-cell hybridomas, but not B cells [148]. The human and mouse RANKL share 85% identity [145]. Human RANKL has approximately 30% homology to TRAIL and CD40 and approximately 20% homology to Fas ligand [145,147–149]. It has now been shown to exist in two forms: a 40- to 45-kDa cellular and a 31-kDa type II transmembrane soluble form derived by cleavage of the full-length form at position 140 or 145 [147,150]. RANKL mRNA is expressed at highest levels in bone and bone marrow, and lymphoid tissues (lymph node, thymus, spleen, fetal liver, and Peyer's patches) [145,147–149]. Its major role in bone is the stimulation of osteoclast differentiation [147,151] and activity [147] and the inhibition of osteoclast apoptosis [151]. In the presence of low levels of macrophage colony-stimulating factor (M-CSF), RANKL appears to be both necessary and sufficient for the complete differentiation of osteoclast precursor cells into mature osteoclasts [146,147,151]. In addition, it is clear that RANKL has a number of effects on immune cells, including activation of c-Jun N-terminal kinase (JNK) in T cells [148], inhibition of apoptosis of dendritic cells by upregulation of Bcl-X$_L$ [144], induction of cluster formation by dendritic cells, and effects on cytokine-activated T-cell proliferation [145]. Consistent with these findings, RANKL knockout mice have severe osteopetrosis with defects in tooth eruption [152], and a complete absence of osteoclasts. They exhibit defects in early differentiation of T and B cells and in thymic differentiation and lack lymph nodes, but have a normal splenic structure and Peyer's patches [152]. A somewhat unexpected finding in these mice is that they also have defects in mammary gland development [153]. In particular, they fail to form lobulo-alveolar structures during pregnancy, so the newborns starve [153].

RANKL specifically interacts with transmembrane RANK and with a soluble OPG receptor. RANK initiates downstream signaling events, whereas OPG works as a receptor decoy to RANK to reduce effective RANKL concentration [146,154]. RANK interacts with various TRAF family members to activate JNK and NF-κB [155,156].

Deletion analysis of the cytoplasmic region of RANK showed that RANK has three independent TRAF-interacting motifs (TIMs), for binding TRAF2, -5, and -6 at its C-terminal 85-amino acid tail [155]. The membrane-proximal TRAF6 binding site appears to be highly specific. The other two sites each bind TRAF2 and -5. However, the carboxyl-terminal site is most specific for TRAF5 and the more amino-proximal site is most specific for TRAF2 [157]. The amino-terminal RZF domain of TRAF2 and -6 activates downstream signals to NF-κB-inducing kinase (NIK), I-κB kinase (IKK), JNK, and p38 kinase [157,158]. Osteoclast differentiation is blocked in mice deficient in the p50 and p52 forms of NF-κB, demonstrating the critical role of this factor [159,160]. However, RANK activation of both the JNK and p38 kinase pathways has also been demonstrated to be important for osteoclast differentiation and function [161,162]. TRAF2 and -5 appear to have similar activities. Interestingly, TRAF2-/- mice do not exhibit osteopetrosis [163], an observation indicating that this TRAF is either not important for osteoclastogenesis or, more likely, is complemented by other TRAFs.

The carboxy-terminal receptor-binding/trimerization domain of TRAF6 is distinct from that of other TRAFs in that it contains a short, proline-rich loop capable of binding to the SH3 domain of the Src tyrosine kinase [156]. This loop provides a means for the activation of Src, which, though constitutively membrane-associated via amino-terminal myristylation, is inhibited by intramolecular SH2 and SH3 interactions [164]. The activation of Src by TRAF6-mediated SH3 competition provides a mechanism for the reported activation of phosphatidylinositol 3-kinase (PI3 kinase) by RANK [156]. This activation may not require involvement of the Src kinase function and may depend only upon interaction between the SH3 domain of Src and the proline-rich sequence of the PI3 kinase p85 regulatory subunit [165]. This could explain how src-/- mice, which exhibit a severe osteopetrotic phenotype [165], can be rescued by a kinase-defective Src [166]. The functional association between TRAF6 and Src is also supported by the observation that both src-/- and TRAF6-/- mice exhibit a similar phenotype of osteopetrosis in which there are abundant osteoclasts but a defect in osteoclastic bone resorption [166,167]. This is in contrast to both the RANK-/- and RANKL/ODF-/- mice, which lack osteoclasts [168,169].

The disorders most clearly related to alterations in the RANKL/RANK/OPG system are familial expansile osteolysis, a rare autosomal dominant disorder characterized by focal areas of enhanced bone resorption, and familial Paget's disease, both of which are due to mutations in the signal peptide region of the RANK protein [170]. These mutations may lead to an accumulation of defective RANK translation products in the secretion pathway, resulting perhaps in receptor self-association and increased constitutive RANK signal transduction. In addition, bone marrow stromal cells in Paget's disease have been shown to have enhanced RANKL expression, and preosteoclastic cells from affected lesions have increased sensitivity to RANKL [171]. This combination of abnormalities may explain, at least in part, the increased numbers of osteoclasts in pagetic bone.

The role of the RANKL/RANK/OPG system in the pathogenesis of more common disorders, such as postmenopausal or age-related osteoporosis, remains controversial. Estrogen increases OPG production by osteoblastic [172] and marrow stromal cells [173]. However, serum OPG levels are, if anything, higher in postmenopausal women with osteoporosis and increased bone turnover [174], perhaps as a homeostatic mechanism limiting their bone loss. In addition, although OPG production by marrow stromal cells appears to decline with age [175], serum OPG levels have consistently been found to increase with age in women and in men [174]. RANKL also plays an important role in glucocorticoid-induced and parathyroid hormone-induced osteoporosis [176–178]. Activated T cells (as in rheumatoid arthritis) have increased levels of RANKL expression [152], and in the presence of M-CSF, RANKL can induce synovial macrophages to differentiate into osteoclastic cells [179], thus potentially leading to the periarticular bone loss commonly seen in various forms of inflammatory arthritis.

Blocking of RANKL activity by OPG or inhibitors of RANK signaling may have therapeutic utility in conditions associated with accelerated bone resorption, including skeletal metastases from multiple myeloma or other tumors and postmenopausal osteoporosis. Indeed, OPG blocked skeletal destruction and pain in a mouse model of sarcoma-induced bone destruction [180]. In addition, a RANK-Fc fusion protein was effective in suppressing bone resorption and hypercalcemia in a murine model of humoral hypercalcemia of malignancy due to xenografts of human lung cancer [181]. Finally, a single dose of an OPG-Fc fusion protein resulted in a profound (by up to 80%) and sustained (for up to three weeks) suppression of bone resorption in postmenopausal women [182]. We have attempted to block RANKL signaling and found that curcumin, a blocker of RANKL-induced NF-κB activation, could inhibit the process of RANKL-induced osteoclastogenesis [183]. However, whether these approaches will translate into viable new therapies for these disorders remains to be seen.

2.3.12   TNF-Related WEAK Inducer of Apoptosis

TNF-related WEAK inducer of apoptosis (TWEAK) was first identified as a clone that weakly hybridized to an erythropoietin probe whose primary sequence was similar to ligands of the TNF family [184]. An identical molecule was identified through a screen of an EST database by its homology to TNF family members and was named Apo3L [185]. TWEAK is a 249-amino-acid type II transmembrane protein whose mRNA is expressed in essentially all tissues examined. Soluble recombinant TWEAK caused IL-8 secretion in HT29, A375, WI-38, and A549 cells [184].

Additionally, TWEAK caused weak induction of apoptosis in HT29 cells when cultured with IFN [184]. In contrast, others have shown that TWEAK activates apoptosis strongly in MCF-7 cells, the activation being dependent on FADD and caspase activation [185]. TWEAK specifically interacts with the death receptor, DR3 [185]. The activation of NF-κB by TWEAK was also demonstrated to be TRAF2, TRADD, RIP, and NIK dependent [185]. TWEAK induces proliferation in a variety of normal endothelial cells and in aortic smooth muscle cells and reduces requirements for serum and growth factors in culture [186]. TWEAK induces a strong angiogenic response when implanted in rat corneas, suggesting a physiological role for TWEAK in vasculature formation in vivo [186].

2.3.13   A PROLIFERATION-INDUCING LIGAND

A proliferation-inducing ligand (APRIL) was discovered by screening a public database using a profile search based on an optimal alignment of all the currently known TNF ligand family members [187]. An identical molecule was identified by a similar search and named TALL2 [188]. The cDNA clone encoded a type II transmembrane protein of 250 amino acids, which contained 28 amino acids in the cytoplasmic domain, 21 amino acids in the transmembrane domain, and 201 amino acids in the extracellular domain. The APRIL sequence showed highest similarity in its extracellular domain with FasL (21%), TNF (20%), and LT (18%). APRIL mRNA was weakly expressed and restricted to a few tissues, most notably prostate, colon, spleen, pancreas, and peripheral blood lymphocytes. Interestingly, APRIL was expressed in various tumor cell lines including HL60, HeLa S3, K562, Molt-4, Raji, SW-480, A549, and G361. Remarkably, APRIL mRNA was increased in thyroid carcinoma and in lymphoma, but in the corresponding normal tissue the expression was either weak or absent.

APRIL's expression in tumor-derived but not normal tissue suggested that it might serve in tumor growth proliferation. Indeed, recombinant APRIL caused proliferation in Jurkat T lymphoma cells, in some B-cell lymphomas (that is, Raji, mouse A20), and in some cell lines of epithelial origin such as COS, HeLa, and some melanomas. Further, NIH-3T3 cells engineered to express APRIL increased tumor growth rates in nude mice as compared with NIH-3T3 cells expressing no ligand. The mechanism by which APRIL induces cellular proliferation is not known, but it does activate NF-κB and JNK. Given that APRIL may play a part in tumorigenesis, antagonistic antibodies to APRIL or its receptor may have a potential for therapeutic intervention.

2.3.14   VASCULAR ENDOTHELIAL GROWTH INHIBITOR

To identify an autocrine inhibitor of angiogenesis specific to endothelial cells, a cDNA library was constructed from RNA derived from various endothelial cells. A search for TNF homologues in this EST database showed a type II transmembrane protein of 174 amino acids with 20%–30% homology to TNF family members. As the new protein was subsequently found to be able to inhibit endothelial cell growth, it was designated vascular endothelial growth inhibitor (VEGI) [189]. Unlike other members of the TNF family, VEGI is expressed predominantly in endothelial cells. Local production of a secreted form of VEGI via gene transfer caused complete suppression of the growth of MC-38 murine colon cancers in syngeneic C57BL/6 mice. Histological examination showed marked reduction of vascularization in MC-38 tumors that expressed soluble but not membrane-bound VEGI or were transfected with control vector. The conditioned media from soluble VEGI-expressing cells showed marked inhibitory effect on in vitro proliferation of adult bovine aortic endothelial cells. Recent studies showed VEGI interacts with an orphan receptor of TNF superfamily, DR3, and the decoy receptor DcR3 and that it functions as a T-cell costimulator [190].

VEGI is a novel angiogenesis inhibitor of the TNF family and functions in part by directly inhibiting endothelial cell proliferation, suggesting that VEGI may be highly valuable in angiogenesis-based cancer therapy.

2.3.15   TNF HOMOLOGUE THAT ACTIVATES APOPTOSIS, NF-κB, AND JNK/BLYS/BAFF

By using an amino acid sequence motif of TNF and searching an EST database, a novel TNF homologue encoding 285 amino acids was identified and named TNF homologue that activates apoptosis, NF-κB, and JNK (THANK) [191]. The predicted extracellular domain of THANK is 15%, 16%, 18%, and 19% identical to LIGHT, FasL, TNF, and LTa, respectively. Northern blot analysis of its mRNA indicated expression in peripheral blood lymphocytes, spleen, thymus, lung, placenta, small intestine, and pancreas. THANK mRNA expression was highest in HL60 followed by K562, A549, and G361, but there was no expression in HeLa, Molt-4, Raji, and SW-480. Recombinant THANK protein activated NF-κB and JNK in the promyeloid cell line U937. Additionally, THANK induced activation of apoptosis in U937 cells. Identical molecules to THANK were identified and named TALL1 [188] and BAFF [192]. This member can bind specifically to BAFF-R and TACI and has some affinity to BCMA [193]. Gene targeting of the receptor did not affect the viability of the mice, but the humoral immune response was severely compromised with reduced levels of IgM, IgG, and mature B cells because of blocked B-cell development. These observations indicate that THANK is indispensable in B-cell maturation.

2.3.16   LIGAND FOR HERPESVIRUS ENTRY MEDIATOR

An additional member of the TNF family, named ligand for herpesvirus entry mediator (LIGHT), was identified by searching an EST database for sequence similarity to TNF family members [194,195]. An identical molecule was identified by its interaction with HVEM and designated HVEM-L [196]. LIGHT mRNA is highly expressed in splenocytes, activated peripheral blood lymphocytes, CD8+ tumor-infiltrating lymphocytes, granulocytes, and monocytes, but it is not expressed in the thymus or in tumor cells. Additionally, LIGHT is upregulated in CD4+ and CD8+ T cells when exposed to PMA. LIGHT encodes a type II transmembrane protein of 240 amino acids. It binds not only to herpesvirus entry mediator (HVEM), but also to the LT receptor. A soluble, secreted form stimulates proliferation of T lymphocytes during allogeneic responses, inhibits HT-29 cell growth, and weakly stimulates NF-κB-dependent transcription [196].

The MDA-MB-231 human breast carcinoma transected with LIGHT caused complete tumor suppression in mice. Histological examination showed marked neutrophil infiltration and necrosis [194]. IFN dramatically increases LIGHT-mediated apoptosis, and LIGHT induces apoptosis of various tumor cells that express both LT and HVEM receptors. However, LIGHT was not cytolytic to the tumor cells that express only the LTβR or HVEM or hematopoietic cells that express only HVEM, such as peripheral blood lymphocytes, Jurkat cells, or CD8+ TIL cells. In contrast, treatment of the activated PBLs with LIGHT resulted in release of IFN. Targeted disruption of *LIGHT* gene resulted in viable mice with apparently normal immune systems except some impairment of cytotoxic T-cell functions.

2.3.17   GLUCOCORTICOID-INDUCED TNFR FAMILY–RELATED LIGAND

The glucocorticoid-induced TNFR family–related ligand (GITRL) was identified by a yeast-based signal sequence trap method from a human umbilical vein endothelial cell cDNA library [197]. This ligand was also identified in an EST database search for TNF-related ligands [198]. GITRL encodes a 177 amino acid type II transmembrane protein with a calculated mass of 20 kDa. Analysis of its mRNA revealed highest expression in small intestine, ovary, testis, and kidney and lower to no expression in other tissues. Expression of membrane-bound GITRL was detected on cultured umbilical vein cells [197]. Expression of either GITRL or its receptor or both the ligand and receptor in Jurkat cells inhibited activation-induced cell death [197]. Consistent with the inhibition of apoptosis, GITRL activated the proapoptotic transcription factor NF-κB [197,198]. Thus, GITRL may modulate peripheral T-cell interaction with blood vessels in the periphery.

2.3.18   TNFRSF EXPRESSED ON THE MOUSE EMBRYO

TNFRSF expressed on the mouse embryo (TROY) is a newly identified member of the tumor necrosis factor receptor superfamily that exhibits homology with Edar and is expressed in embryonic skin and hair follicles. During a signal sequence trap screening of the murine brain, Kojima et al. identified this new member of the tumor necrosis factor receptor superfamily and designated it as TROY [199].

TROY is a type I membrane protein of 416 amino acids with characteristic cysteine-rich motifs in the extracellular domain and a TRAF2 binding sequence in the cytoplasmic domain of 223 amino acids. Overexpression of TROY induced activation of NF-κB. The extracellular domain of TROY exhibits an extensive homology with that of Edar, a receptor that specifies hair follicle fate. TROY mRNA is strongly expressed in brain and embryo and moderately expressed in the heart, lung, and liver. In the embryo, the expression level is particularly strong in the skin. The *Troy* gene is located near the waved coat (Wc) locus, a mutant related to abnormalities in skin and hair.

2.3.19   RECEPTOR EXPRESSED IN LYMPHOID TISSUES

Receptor expressed in lymphoid tissues (RELT) was discovered by Sica et al. while looking for novel molecules homologous to TNFR superfamily members that are important for activation and/or differentiation of T lymphocytes. During a search with the cysteine-rich extracellular domain of OX40, they identified a partial EST sequence corresponding to a new member of the TNFR superfamily that was designated as RELT [200].

RELT is a type I transmembrane glycoprotein with a cysteine-rich extracellular domain, possessing significant homology to other members of the TNFR superfamily, especially TROY, DR3, OX40, and LT receptor. The mRNA of RELT is especially abundant in hematologic tissues such as spleen, lymph node, and peripheral blood leukocytes in addition to certain cancers like the leukemias and lymphomas. RELT selectively binds TRAF1 and is able to activate the NF-κB and may be a potential regulator of immune responses.

2.4   CONCLUSION

Although much has been learned about TNF-α, information about other members of this superfamily is still very limited. The ligands for death receptor 6 (DR6), receptor expressed in lymphoid tissue (RELT), and TNFRSF expressed on the mouse embryo (TROY), remain to be identified. Furthermore, the mechanisms of release of these cytokines need to be elucidated. Likewise, the mechanism of signaling and the interaction between the signaling pathways of various members of the TNF superfamily need to be examined. Gene deletion, overexpression, dominant-negative, transgene, and protein–protein interaction techniques have provided results that are inconsistent. Thus, a better understanding of the mechanism of action of the TNF superfamily members will allow for the development of more specific therapeutics against these ailments.

ABBREVIATIONS

APRIL, a proliferation-inducing ligand; BAFF, Blys, B cell–activating factor; BCMA, B-cell maturation antigen; BIR, baculovirus IAP repeat; cIAP, cellular inhibitor of apoptosis; DcR, decoy receptor, DD, death domain; DED, death effector domain; DR, death receptor; ECD, extracellular domain; FADD, Fas-associated death domain; FAN, factor associated with N sphingomyelinase activation; FLICE, FADD-like ICE; GITR; glucocorticoid-induced tumor necrosis factor receptor family receptor; HVEM, herpesvirus entry mediator; ICD, intracellular domain; ICE, interleukin-1 converting enzyme; JNK, c-Jun kinase; LIGHT, ligand for HVEM; LT, lymphotoxin; MACH, MORT-associated CED homologue; MORT, mediator of receptor-induced toxicity; NF-κB, nuclear factor kappa B; NIK, NF-κB-inducing kinase; NSD, neutral Smase-activating domain; OPG; osteoprotegrin; RANK, receptor activator of NF-κB; RIP, receptor-interacting protein; TACI, transmembrane activator and cyclophilin ligand interactor; TANK (also called I-TRAF), TRAF family member–associated NF-κB activator; TNF, tumor necrosis factor; TRAF, TNF receptor-associated factor; TRAIL, TNF-related death-inducing ligand.

REFERENCES

1. P. Brunes. "Die Heilwirkung des Erysipels auf Geschwulste." *Beitr Klin Chir* 3 (1868): 443–46.
2. L. J. Old. "Tumor Necrosis Factor (TNF)." *Science* 230, no. 4726 (1985): 630–32.
3. B. B. Aggarwal. "Tumor Necrosis Factors: TNF-Alpha and TNF-Beta Their Structure and Pleiotropic Biological Effects." *Drugs of the Future* 12, no. 9 (1987): 891–98.
4. W. B. Coley. "Contribution to the Knowledge of Sarcoma." *Ann Surg* 14 (1891): 199–220.
5. A. Gratia and R. Linz. "C. R. Seances." *Soc Biol Ses Fil* 108 (1931): 421–28.
6. M. J. Shear, and F. C. Turner. "Chemical Treatment of Tumors. V. Isolation of the Hemorrhage-Producing Fraction from *Serratia marcescens* (*Bacillus prodigiosus*) Culture Filtrate." *J Natl Cancer Inst* 4 (1943): 81–97.
7. G. H. Algire, F. Y. Legallais, and B. F. Anderson. "Vascular Reactions of Normal and Malignant Tissues In Vivo. V. The Role of Hypotension in the Action of a Bacterial Polysaccharide on Tumors." *J Natl Cancer Inst* 12, no. 6 (1952): 1279–95.
8. W. E. O'Malley, B. Achinstein, and M. J. Shear. "Action of Bacterial Polysaccharide on Tumors. II: Damage of Sarcoma 37 by Serum of Mice Treated with *Serratia marcesscens* Polysaccharide and Induced Tolerance." *J Natl Cancer Inst* 29 (1962): 1169–75.
9. E. A. Carswell, et al. "An Endotoxin-Induced Serum Factor That Causes Necrosis of Tumors." *Proc Natl Acad Sci U S A* 72, no. 9 (1975): 3666–70.
10. L. Helson, et al. "Effect of Tumour Necrosis Factor on Cultured Human Melanoma Cells." *Nature* 258, no. 5537 (1975): 731–32.
11. B. B. Aggarwal, B. Moffat, and R. N. Harkins. "Human Lymphotoxin. Production by a Lymphoblastoid Cell Line, Purification, and Initial Characterization." *J Biol Chem* 259, no. 1 (1984): 686–91.
12. B. B. Aggarwal, T. E. Eessalu, and P. E. Hass. "Characterization of Receptors for Human Tumour Necrosis Factor and Their Regulation by Gamma-Interferon." *Nature* 318, no. 6047 (1985): 665–67.
13. B. B. Aggarwal, et al. "Primary Structure of Human Lymphotoxin Derived from 1788 Lymphoblastoid Cell Line." *J Biol Chem* 260, no. 4 (1985): 2334–44.
14. B. B. Aggarwal, et al. "Human Tumor Necrosis Factor. Production, Purification, and Characterization." *J Biol Chem* 260, no. 4 (1985): 2345–54.
15. P. W. Gray, et al. "Cloning and Expression of cDNA for Human Lymphotoxin, a Lymphokine with Tumour Necrosis Activity." *Nature* 312, no. 5996 (1984): 721–24.
16. D. Pennica, et al. "Human Tumour Necrosis Factor: Precursor Structure, Expression and Homology to Lymphotoxin." *Nature* 312, no. 5996 (1984): 724–29.
17. B. Beutler, et al. "Identity of Tumour Necrosis Factor and the Macrophage-Secreted Factor Cachectin." *Nature* 316, no. 6028 (1985): 552–54.
18. K. Takeda, et al. "Identity of Differentiation Inducing Factor and Tumour Necrosis Factor." *Nature* 323, no. 6086 (1986): 338–40.
19. B. B. Aggarwal. "Signalling Pathways of the TNF Superfamily: A Double-Edged Sword." *Nat Rev Immunol* 3, no. 9 (2003): 745–56.
20. B. B. Aggarwal, et al. "The Role of TNF and Its Family Members in Inflammation and Cancer: Lessons from Gene Deletion." *Curr Drug Targets Inflamm Allergy* 1, no. 4 (2002): 327–41.

21. B. B. Aggarwal. "Tumour Necrosis Factors Receptor Associated Signalling Molecules and Their Role in Activation of Apoptosis, JNK and NF-KappaB." *Ann Rheum Dis* 59, suppl 1 (2000): i6–16.

22. B. B. Aggarwal. "Tumor Necrosis Factors: A Double-Edged Sword." *Journal of Clinical Ligand Assay* 23 (2000): 181–92.

23. B. B. Aggarwal. "Apoptosis and Nuclear Factor-Kappa B: A Tale of Association and Dissociation." *Biochem Pharmacol* 60, no. 8 (2000): 1033–39.

24. P. C. Rath and B. B. Aggarwal. "TNF-Induced Signaling in Apoptosis." *J Clin Immunol* 19, no. 6 (1999): 350–64.

25. A. M. Eggermont, J. H. de Wilt, and T. L. ten Hagen. "Current Uses of Isolated Limb Perfusion in the Clinic and a Model System for New Strategies." *Lancet Oncol* 4, no. 7 (2003): 429–37.

26. I. Garcia, et al. "Transgenic Mice Expressing High Levels of Soluble TNF-R1 Fusion Protein Are Protected from Lethal Septic Shock and Cerebral Malaria, and Are Highly Sensitive to *Listeria monocytogenes* and *Leishmania major infections.*" *Eur J Immunol* 25, no. 8 (1995): 2401–7.

27. G. E. Nedwin, et al. "Human Lymphotoxin and Tumor Necrosis Factor Genes: Structure, Homology and Chromosomal Localization." *Nucleic Acids Res* 13, no. 17 (1985): 6361–73.

28. P. Orosz, et al. "Enhancement of Experimental Metastasis by Tumor Necrosis Factor." *J Exp Med* 177, no. 5 (1993): 1391–98.

29. R. J. Moore, et al. "Mice Deficient in Tumor Necrosis Factor-Alpha Are Resistant to Skin Carcinogenesis." *Nat Med* 5, no. 7 (1999): 828–31.

30. G. A. Granger and T. W. Williams. "Lymphocyte Cytotoxicity In Vitro: Activation and Release of a Cytotoxic Factor." *Nature* 218, no. 148 (1968): 1253–54.

31. N. H. Ruddle et al. "An Antibody to Lymphotoxin and Tumor Necrosis Factor Prevents Transfer of Experimental Allergic Encephalomyelitis." *J Exp Med* 172, no. 4 (1990): 1193-1200.

32. C. F. Ware, P. C. Harris, and G. A. Granger. "Mechanisms of Lymphocyte-Mediated Cytotoxicity. II. Biochemical and Serologic Identification of a Precursor Lymphotoxin Form (Pre-LT) Produced by MLC-Sensitized Human T Lymphocytes In Vitro." *J Immunol* 126, no. 5 (1981): 1927–33.

33. J. L. Browning, et al. "Lymphotoxin Beta, a Novel Member of the TNF Family That Forms a Heteromeric Complex with Lymphotoxin on the Cell Surface." *Cell* 72, no. 6 (1993): 847–56.

34. J. L. Browning, et al. "Preparation and Characterization of Soluble Recombinant Heterotrimeric Complexes of Human Lymphotoxins Alpha and Beta." *J Biol Chem* 271, no. 15 (1996): 8618–26.

35. G. Mosialos, et al. "The Epstein-Barr Virus Transforming Protein LMP1 Engages Signaling Proteins for the Tumor Necrosis Factor Receptor Family." *Cell* 80, no. 3 (1995): 389–99.

36. H. Nakano, et al. "TRAF5, an Activator of NF-KappaB and Putative Signal Transducer for the Lymphotoxin-Beta Receptor." *J Biol Chem* 271, no. 25 (1996): 14661–64.

37. P. A. Koni, et al. "Distinct Roles in Lymphoid Organogenesis for Lymphotoxins Alpha and Beta Revealed in Lymphotoxin Beta-Deficient Mice." *Immunity* 6, no. 4 (1997): 491–500.

38. Q. Wu, et al. "Signal via Lymphotoxin-Beta R on Bone Marrow Stromal Cells Is Required for an Early Checkpoint of NK Cell Development." *J Immunol* 166, no. 3 (2001): 1684–89.

39. S. Miura, et al. "Molecular Cloning and Characterization of a Novel Glycoprotein, gp34, That Is Specifically Induced by the Human T-Cell Leukemia Virus Type I Transactivator p40tax." *Mol Cell Biol* 11, no. 3 (1991): 1313–25.

40. P. R. Baum, et al. "Molecular Characterization of Murine and Human OX40/OX40 Ligand Systems: Identification of a Human OX40 Ligand as the HTLV-1-Regulated Protein gp34." *Embo J* 13, no. 17 (1994): 3992–4001.

41. W. R. Godfrey, et al. "Identification of a Human OX-40 Ligand, a Costimulator of CD4+ T Cells with Homology to Tumor Necrosis Factor." *J Exp Med* 180, no. 2 (1994): 757–62.

42. D. J. Paterson, et al. "Antigens of Activated Rat T Lymphocytes Including a Molecule of 50,000 Mr Detected Only on CD4 Positive T Blasts." *Mol Immunol* 24, no. 12 (1987): 1281–90.

43. S. Mallett, S. Fossum, and A. N. Barclay. "Characterization of the MRC OX40 Antigen of Activated CD4 Positive T Lymphocytes—A Molecule Related to Nerve Growth Factor Receptor." *Embo J* 9, no. 4 (1990): 1063–68.

44. D. M. Calderhead, et al. "Cloning of Mouse Ox40: A T Cell Activation Marker That May Mediate T-B Cell Interactions." *J Immunol* 151, no. 10 (1993): 5261–71.

45. U. Latza, et al. "The human OX40 Homolog: cDNA Structure, Expression and Chromosomal Assignment of the ACT35 Antigen." *Eur J Immunol* 24, no. 3 (1994): 677–83.

46. E. Stuber, et al. "Cross-Linking of OX40 Ligand, a Member of the TNF/NGF Cytokine Family, Induces Proliferation and Differentiation in Murine Splenic B Cells." *Immunity* 2, no. 5 (1995): 507–21.

47. E. Stuber and W. Strober. "The T Cell-B Cell Interaction via OX40-OX40L Is Necessary for the T Cell-Dependent Humoral Immune Response." *J Exp Med* 183, no. 3 (1996): 979–89.

48. Y. Ohshima, et al. "Expression and Function of OX40 Ligand on Human Dendritic Cells." *J Immunol* 159, no. 8 (1997): 3838–48.

49. S. Kawamata, et al. "Activation of OX40 Signal Transduction Pathways Leads to Tumor Necrosis Factor Receptor-Associated Factor (TRAF) 2- and TRAF5-Mediated NF-KappaB Activation." *J Biol Chem* 273, no. 10 (1998): 5808–14.

50. A. Imura, et al. "The Human OX40/gp34 System Directly Mediates Adhesion of Activated T Cells to Vascular Endothelial Cells." *J Exp Med* 183, no. 5 (1996): 2185–95.

51. A. I. Chen, et al. "Ox40-Ligand Has a Critical Costimulatory Role in Dendritic Cell: T Cell Interactions." *Immunity* 11, no. 6 (1999): 689–98.

52. K. Murata, et al. "Impairment of Antigen-Presenting Cell Function in Mice Lacking Expression of OX40 Ligand." *J Exp Med* 191, no. 2 (2000): 365–74.

53. M. Kopf, et al. "OX40-Deficient Mice Are Defective in Th Cell Proliferation but Are Competent in Generating B Cell and CTL Responses after Virus Infection." *Immunity* 11, no. 6 (1999): 699–708.

54. R. J. Noelle, et al. "A 39-kDa Protein on Activated Helper T Cells Binds CD40 and Transduces the Signal for Cognate Activation of B Cells." *Proc Natl Acad Sci U S A* 89, no. 14 (1992): 6550–54.

55. R. J. Armitage, et al. "Molecular and Biological Characterization of a Murine Ligand for CD40." *Nature* 357, no. 6373 (1992): 80-82.

56. M. J. Yellin, et al. "A Human CD4- T Cell Leukemia Subclone with Contact-Dependent Helper Function." *J Immunol* 147, no. 10 (1991): 3389–95.

57. S. Lederman, et al. "Identification of a Novel Surface Protein on Activated CD4+ T Cells That Induces Contact-Dependent B Cell Differentiation (Help)." *J Exp Med* 175, no. 4 (1992): 1091–1101.

58. F. M. Uckun, et al. "Temporal Association of CD40 Antigen Expression with Discrete Stages of Human B-Cell Ontogeny and the Efficacy of Anti-CD40 Immunotoxins Against Clonogenic B-Lineage Acute Lymphoblastic Leukemia as well as B-Lineage Non-Hodgkin's Lymphoma Cells." *Blood* 76, no. 12 (1990): 2449–56.

59. A. H. Galy and H. Spits. "CD40 is Functionally Expressed on Human Thymic Epithelial Cells." *J Immunol* 149, no. 3 (1992): 775–82.

60. L. B. Clark, T. M. Foy, and R. J. Noelle. "CD40 and Its Ligand." *Adv Immunol* 63 (1996): 43–78.

61. M. Rothe, et al. "A Novel Family of Putative Signal Transducers Associated with the Cytoplasmic Domain of the 75 kDa Tumor Necrosis Factor Receptor." *Cell* 78, no. 4 (1994): 681–92.

62. B. R. Renshaw, et al. "Humoral Immune Responses in CD40 Ligand-Deficient Mice." *J Exp Med* 180, no. 5 (1994): 1889–1900.

63. J. Xu, et al. "Mice Deficient for the CD40 Ligand." *Immunity* 1, no. 5 (1994): 423–31.

64. T. Kawabe, et al. "The Immune Responses in CD40-Deficient Mice: Impaired Immunoglobulin Class Switching and Germinal Center Formation." *Immunity* 1, no. 3 (1994): 167–78.

65. T. Suda, et al. "Molecular Cloning and Expression of the Fas Ligand, a Novel Member of the Tumor Necrosis Factor Family." *Cell* 75, no. 6 (1993): 1169–78.

66. J. D. Mountz, et al. "Oncogene Expression in Autoimmune Mice." *J Mol Cell Immunol* 2, no. 3 (1985): 121–31.

67. R. D. Allen, et al. "Differences Defined by Bone Marrow Transplantation Suggest That lpr and gld Are Mutations of Genes Encoding an Interacting Pair of Molecules." *J Exp Med* 172, no. 5 (1990): 1367–75.

68. D. R. Green, et al. "Promotion and Inhibition of Activation-Induced Apoptosis in T-Cell Hybridomas by Oncogenes and Related Signals." *Immunol Rev* 142 (1994): 321–42.

69. T. Takahashi, et al. "Generalized Lymphoproliferative Disease in Mice, Caused by a Point Mutation in the Fas Ligand." *Cell* 76, no. 6 (1994): 969–76.

70. M. Hahne, et al. "Activated B Cells Express Functional Fas Ligand." *Eur J Immunol* 26, no. 3 (1996): 721–24.

71. A. D. Badley, et al. "Upregulation of Fas Ligand Expression by Human Immunodeficiency Virus in Human Macrophages Mediates Apoptosis of Uninfected T Lymphocytes." *J Virol* 70, no. 1 (1996): 199–206.

72. N. Itoh, et al. "The Polypeptide Encoded by the cDNA for Human Cell Surface Antigen Fas Can Mediate Apoptosis." *Cell* 66, no. 2 (1991): 233–43.

73. F. C. Kischkel, et al. "Cytotoxicity-Dependent APO-1 (Fas/CD95)-Associated Proteins Form a Death-Inducing Signaling Complex (DISC) with the Receptor." *Embo J* 14, no. 22 (1995): 5579–88.

74. A. Krueger, et al. "FLICE-Inhibitory Proteins: Regulators of Death Receptor-Mediated Apoptosis." *Mol Cell Biol* 21, no. 24 (2001): 8247–54.

75. P. H. Krammer. "CD95's Deadly Mission in the Immune System." *Nature* 407, no. 6805 (2000): 789–95.

76. N. Holler, et al. "Fas Triggers an Alternative, Caspase-8-Independent Cell Death Pathway Using the Kinase RIP as Effector Molecule." *Nat Immunol* 1, no. 6 (2000): 489–95.

77. R. M. Pitti, et al. "Genomic Amplification of a Decoy Receptor for Fas Ligand in Lung and Colon Cancer." *Nature* 396, no. 6712 (1998): 699–703.

78. S. Nagata and P. Golstein. "The Fas Death Factor." *Science* 267, no. 5203 (1995): 1449–56.

79. Y. Takahashi, H. Ohta, and T. Takemori. "Fas Is Required for Clonal Selection in Germinal Centers and the Subsequent Establishment of the Memory B Cell Repertoire." *Immunity* 14, no. 2 (2001): 181–92.

80. H. Stein, et al. "Evidence for the Detection of the Normal Counterpart of Hodgkin and Sternberg-Reed Cells." *Hematol Oncol* 1, no. 1 (1983): 21–29.

81. H. Stein, J. Gerdes, R. Schwarting, et al. "Three New Lymphoid Activation Antigens," in *Leukocyte Typing III*, eds A. J. McMichael et al. (Oxford: Oxford University, 1987), 574–75.

82. R. G. Goodwin, et al. "Molecular and Biological Characterization of a Ligand for CD27 Defines a New Family of Cytokines with Homology to Tumor Necrosis Factor." *Cell* 73, no. 3 (1993): 447–56.

83. K. Agematsu, et al. "Direct Cellular Communications between CD45R0 and CD45RA T Cell Subsets via CD27/CD70." *J Immunol* 154, no. 8 (1995): 3627–35.

84. A. Agathanggelou, et al. "Expression of Immune Regulatory Molecules in Epstein-Barr Virus-Associated Nasopharyngeal Carcinomas with Prominent Lymphoid Stroma. Evidence for a Functional Interaction between Epithelial Tumor Cells and Infiltrating Lymphoid Cells." *Am J Pathol* 147, no. 4 (1995): 1152–60.

85. D. Camerini, et al. "The T Cell Activation Antigen CD27 Is a Member of the Nerve Growth Factor/Tumor Necrosis Factor Receptor Gene Family." *J Immunol* 147, no. 9 (1991): 3165–69.

86. U. Schwab, et al. "Production of a Monoclonal Antibody Specific for Hodgkin and Sternberg-Reed Cells of Hodgkin's Disease and a Subset of Normal Lymphoid Cells." *Nature* 299, no. 5878 (1982): 65–67.

87. H. Stein, et al. "The Expression of the Hodgkin's Disease Associated Antigen Ki-1 in Reactive and Neoplastic Lymphoid Tissue: Evidence That Reed-Sternberg Cells and Histiocytic Malignancies Are Derived from Activated Lymphoid Cells." *Blood* 66, no. 4 (1985): 848–58.

88. O. Josimovic-Alasevic, et al. "Ki-1 (CD30) Antigen Is Released by Ki-1-Positive Tumor Cells In Vitro and In Vivo. I. Partial Characterization of Soluble Ki-1 Antigen and Detection of the Antigen in Cell Culture Supernatants and in Serum by an Enzyme-Linked Immunosorbent Assay." *Eur J Immunol* 19, no. 1 (1989): 157–62.

89. G. Pizzolo, et al. "Serum Levels of Soluble CD30 Molecule (Ki-1 Antigen) in Hodgkin's Disease: Relationship with Disease Activity and Clinical Stage." *Br J Haematol* 75, no. 2 (1990): 282–84.

90. H. Durkop, et al. "Molecular Cloning and Expression of a New Member of the Nerve Growth Factor Receptor Family That Is Characteristic for Hodgkin's Disease." *Cell* 68, no. 3 (1992): 421–27.

91. C. A. Smith, et al. "CD30 Antigen, a Marker for Hodgkin's Lymphoma, Is a Receptor Whose Ligand Defines an Emerging Family of Cytokines with Homology to TNF." *Cell* 73, no. 7 (1993): 1349–60.

92. J. F. Nawrocki, E. S. Kirsten, and R. I. Fisher. "Biochemical and Structural Properties of a Hodgkin's Disease-Related Membrane Protein." *J Immunol* 141, no. 2 (1988): 672–80.

93. M. A. Bowen, et al. "Structure and Expression of Murine CD30 and Its Role in Cytokine Production." *J Immunol* 156, no. 2 (1996): 442–49.

94. S. Aizawa, et al. "Cloning and Characterization of a cDNA for Rat CD30 Homolog and Chromosomal Assignment of the Genomic Gene." *Gene* 182, no. 1-2 (1996): 155–62.

95. C. Fonatsch, et al. "Assignment of the Human CD30 (Ki-1) Gene to 1p36." *Genomics* 14, no. 3 (1992): 825–26.

96. H. J. Gruss, et al. "Pleiotropic Effects of the CD30 Ligand on CD30-Expressing Cells and Lymphoma Cell Lines." *Blood* 83, no. 8 (1994): 2045–56.

97. A. Younes, et al. "CD30 Ligand Is Expressed on Resting Normal and Malignant Human B Lymphocytes." *Br J Haematol* 93, no. 3 (1996): 569–71.

98. A. Pinto, et al. "Human Eosinophils Express Functional CD30 Ligand and Stimulate Proliferation of a Hodgkin's Disease Cell Line." *Blood* 88, no. 9 (1996): 3299–3305.

99. P. Romagnani, et al. "High CD30 Ligand Expression by Epithelial Cells and Hassal's Corpuscles in the Medulla of Human Thymus." *Blood* 91, no. 9 (1998): 3323–32.

100. P. P. McDonald, et al. "CD30 Ligation Induces Nuclear Factor-Kappa B Activation in Human T Cell Lines." *Eur J Immunol* 25, no. 10 (1995): 2870–76.

101. S. Y. Lee, et al. "CD30/TNF Receptor-Associated Factor Interaction: NF-Kappa B Activation and Binding Specificity." *Proc Natl Acad Sci U S A* 93, no. 18 (1996): 9699–9703.

102. S. Aizawa, et al. "Tumor Necrosis Factor Receptor-Associated Factor (TRAF) 5 and TRAF2 Are Involved in CD30-Mediated NFkappaB Activation." *J Biol Chem* 272, no. 4 (1997): 2042–45.

103. G. Pizzolo, et al. "High Serum Level of the Soluble Form of CD30 Molecule in the Early Phase of HIV-1 Infection as an Independent Predictor of Progression to AIDS." *Aids* 8, no. 6 (1994): 74–45.

104. R. Gerli, et al. "High Levels of the Soluble Form of CD30 Molecule in Rheumatoid Arthritis (RA) Are Expression of CD30+ T Cell Involvement in the Inflamed Joints." *Clin Exp Immunol* 102, no. 3 (1995): 547–50.

105. R. Giacomelli, et al. "Circulating Levels of Soluble CD30 Are Increased in Patients with Systemic Sclerosis (SSc) and Correlate with Serological and Clinical Features of the Disease." *Clin Exp Immunol* 108, no. 1 (1997): 42–46.

106. G. Wang, et al. "High Plasma Levels of the Soluble Form of CD30 Activation Molecule Reflect Disease Activity in Patients with Wegener's Granulomatosis." *Am J Med* 102, no. 6 (1997): 517–23.

107. M. Okumura, et al. "Increased Serum Concentration of Soluble CD30 in Patients with Graves' Disease and Hashimoto's Thyroiditis." *J Clin Endocrinol Metab* 82, no. 6 (1997): 1757–60.

108. A. Gause, et al. "Clinical Significance of Soluble CD30 Antigen in the Sera of Patients with Untreated Hodgkin's Disease." *Blood* 77, no. 9 (1991): 1983–88.

109. P. L. Zinzani, et al. "Clinical Implications of Serum Levels of Soluble CD30 in 70 Adult Anaplastic Large-Cell Lymphoma Patients." *J Clin Oncol* 16, no. 4 (1998): 1532–37.

110. B. S. Kwon and S. M. Weissman. "cDNA Sequences of Two Inducible T-Cell Genes." *Proc Natl Acad Sci U S A* 86, no. 6 (1989): 1963–67.

111. C. A. Smith, T. Farrah, and R.G. Goodwin. "The TNF Receptor Superfamily of Cellular and Viral Proteins: Activation, Costimulation, and Death." *Cell* 76, no. 6 (1994): 959–62.

112. M. R. Alderson, et al. "Molecular and Biological Characterization of Human 4-1BB and Its Ligand." *Eur J Immunol* 24, no. 9 (1994): 2219–27.

113. R. H. Arch and C. B. Thompson. "4-1BB and Ox40 Are Members of a Tumor Necrosis Factor (TNF)-Nerve Growth Factor Receptor Subfamily That Bind TNF Receptor-Associated Factors and Activate Nuclear Factor KappaB." *Mol Cell Biol* 18, no. 1 (1998): 558–65.

114. S. R. Wiley, et al. "Identification and Characterization of a New Member of the TNF Family That Induces Apoptosis." *Immunity* 3, no. 6 (1995): 673–82.

115. R. M. Pitti, et al. "Induction of Apoptosis by Apo-2 Ligand, a New Member of the Tumor Necrosis Factor Cytokine Family." *J Biol Chem* 271, no. 22 (1996): 12687–90.

116. J. G. Emery, et al. "Osteoprotegerin Is a Receptor for the Cytotoxic Ligand TRAIL." *J Biol Chem* 273, no. 23 (1998): 14363–67.

117. F. Muhlenbeck, et al. "The Tumor Necrosis Factor-Related Apoptosis-Inducing Ligand Receptors TRAIL-R1 and TRAIL-R2 Have Distinct Cross-Linking Requirements for Initiation of Apoptosis and Are Non-redundant in JNK Activation." *J Biol Chem* 275, no. 41 (2000): 32208–13.

118. J. L. Bodmer, et al. "TRAIL Receptor-2 Signals Apoptosis through FADD and Caspase-8." *Nat Cell Biol* 2, no. 4 (2000): 241–43.

119. A. A. Kuang, et al. "FADD Is Required for DR4- and DR5-Mediated Apoptosis: Lack of Trail-Induced Apoptosis in FADD-Deficient Mouse Embryonic Fibroblasts." *J Biol Chem* 275, no. 33 (2000): 25065–68.

120. T. S. Griffith, et al. "Intracellular Regulation of TRAIL-Induced Apoptosis in Human Melanoma Cells." *J Immunol* 161, no. 6 (1998): 2833–40.

121. Y. Deng, Y. Lin, and X. Wu "TRAIL-Induced Apoptosis Requires Bax-Dependent Mitochondrial Release of Smac/DIABLO." *Genes Dev* 16, no. 1 (2002): 33–45.

122. T. F. Burns and W. S. El-Deiry. "Identification of Inhibitors of TRAIL-Induced Death (ITIDs) in the TRAIL-Sensitive Colon Carcinoma Cell Line SW480 Using a Genetic Approach." *J Biol Chem* 276, no. 41 (2001): 37879–86.

123. H. LeBlanc, et al. "Tumor-Cell Resistance to Death Receptor–Induced Apoptosis through Mutational Inactivation of the Proapoptotic Bcl-2 Homolog Bax." *Nat Med* 8, no. 3 (2002): 274–81.

124. N. Kayagaki, et al. "Expression and Function of TNF-Related Apoptosis-Inducing Ligand on Murine Activated NK Cells." *J Immunol* 163, no. 4 (1999): 1906–13.

125. K. Song, et al. "Tumor Necrosis Factor-Related Apoptosis-Inducing Ligand (TRAIL) Is an Inhibitor of Autoimmune Inflammation and Cell Cycle Progression." *J Exp Med* 191, no 7 (2000): 1095–1104.

126. E. Cretney, et al. "Increased Susceptibility to Tumor Initiation and Metastasis in TNF-Related Apoptosis-Inducing Ligand-Deficient Mice." *J Immunol* 168, no. 3 (2002): 1356–61.

127. N. A. Fanger, et al. "Human Dendritic Cells Mediate Cellular Apoptosis via Tumor Necrosis Factor-Related Apoptosis-Inducing Ligand (TRAIL)." *J Exp Med* 190, no. 8 (1999): 1155–64.

128. T. S. Griffith, et al. "Monocyte-Mediated Tumoricidal Activity via the Tumor Necrosis Factor-Related Cytokine, TRAIL." *J Exp Med* 189, no. 8 (1999): 1343–54.

129. M. J. Smyth et al. "Tumor Necrosis Factor-Related Apoptosis-Inducing Ligand (TRAIL) Contributes to Interferon Gamma-Dependent Natural Killer Cell Protection from Tumor Metastasis." *J Exp Med* 193, no. 6 (2001): 661–70.

130. K. Takeda, et al. "Involvement of Tumor Necrosis Factor-Related Apoptosis-Inducing Ligand in Surveillance of Tumor Metastasis by Liver Natural Killer Cells." *Nat Med* 7, no. 1 (2001): 94–100.

131. L. Zamai, et al. "Natural Killer (NK) Cell-Mediated Cytotoxicity: Differential Use of TRAIL and Fas Ligand by Immature and Mature Primary Human NK Cells." *J Exp Med* 188, no. 12 (1998): 2375–80.

132. N. Kayagaki, et al. "Type I Interferons (IFNs) Regulate Tumor Necrosis Factor-Related Apoptosis-Inducing Ligand (TRAIL) Expression on Human T Cells: A Novel Mechanism for the Antitumor Effects of Type I IFNs." *J Exp Med* 189, no. 9 (1999): 1451–60.

133. N. Kayagaki, et al. "Involvement of TNF-Related Apoptosis-Inducing Ligand in Human CD4+ T Cell-Mediated Cytotoxicity." *J Immunol* 162, no. 5 (1999): 2639–47.

134. K. Sato, et al. "Antiviral Response by Natural Killer Cells through TRAIL Gene Induction by IFN-Alpha/Beta." *Eur J Immunol* 31, no. 11 (2001): 3138–46.

135. A. Ashkenazi and V. M. Dixit. "Death Receptors: Signaling and Modulation." *Science* 281, no. 5381 (1998): 1305–8.

136. H. Walczak, et al. "Tumoricidal Activity of Tumor Necrosis Factor-Related Apoptosis-Inducing Ligand In Vivo." *Nat Med* 5, no. 2 (1999): 157–63.

137. A. Ashkenazi, et al. "Safety and Antitumor Activity of Recombinant Soluble Apo2 Ligand." *J Clin Invest* 104, no. 2 (1999): 155–62.

138. W. S. El-Deiry. "Insights into Cancer Therapeutic Design Based on p53 and TRAIL Receptor Signaling." *Cell Death Differ* 8, no. 11 (2001): 1066–75.

139. G. S. Wu, K. Kim, and W. S. el-Deiry. "KILLER/DR5, a Novel DNA-Damage Inducible Death Receptor Gene, Links the p53-Tumor Suppressor to Caspase Activation and Apoptotic Death." *Adv Exp Med Biol* 465 (2000): 143–51.

140. M. Nagane, H. J. Huang, and W. K. Cavenee. "The Potential of TRAIL for Cancer Chemotherapy." *Apoptosis* 6, no. 3 (2001): 191–97.
141. D. Lawrence, et al. "Differential Hepatocyte Toxicity of Recombinant Apo2L/TRAIL Versions." *Nat Med* 7, no. 4 (2001): 383–85.
142. B. Hilliard, et al. "Roles of TNF-Related Apoptosis-Inducing Ligand in Experimental Autoimmune Encephalomyelitis." *J Immunol* 166, no. 2 (2001): 1314–19.
143. K. Ichikawa, et al. "Tumoricidal Activity of a Novel Anti-human DR5 Monoclonal Antibody Without Hepatocyte Cytotoxicity." *Nat Med* 7, no. 8 (2001): 954–60.
144. B. R. Wong, et al. "TRANCE (Tumor Necrosis Factor [TNF]-Related Activation-Induced Cytokine), a New TNF Family Member Predominantly Expressed in T Cells, Is a Dendritic Cell-Specific Survival Factor." *J Exp Med* 186, no. 12 (1997): 2075–80.
145. D. M. Anderson, et al. "A Homologue of the TNF Receptor and Its Ligand Enhance T-Cell Growth and Dendritic-Cell Function." *Nature* 390, no. 6656 (1997): 175–79.
146. H. Yasuda, et al. "Identity of Osteoclastogenesis Inhibitory Factor (OCIF) and Osteoprotegerin (OPG): A Mechanism by Which OPG/OCIF Inhibits Osteoclastogenesis In Vitro." *Endocrinology* 139, no. 3 (1998): 1329–37.
147. D. L. Lacey, et al. "Osteoprotegerin Ligand Is a Cytokine That Regulates Osteoclast Differentiation and Activation." *Cell* 93, no. 2 (1998): 165–76.
148. B. R. Wong, et al. "TRANCE Is a Novel Ligand of the Tumor Necrosis Factor Receptor Family That Activates c-Jun N-Terminal Kinase in T Cells." *J Biol Chem* 272, no. 40 (1997): 25190–94.
149. H. Yasuda, et al. "Osteoclast Differentiation Factor Is a Ligand for Osteoprotegerin/ Osteoclastogenesis-Inhibitory Factor and Is Identical to TRANCE/RANKL." *Proc Natl Acad Sci U S A* 95, no. 7 (1998): 3597–3602.
150. L. Lum, et al. "Evidence for a Role of a Tumor Necrosis Factor-Alpha (TNF-Alpha)-Converting Enzyme-Like Protease in Shedding of TRANCE, a TNF Family Member Involved in Osteoclastogenesis and Dendritic Cell Survival." *J Biol Chem* 274, no. 19 (1999): 13613–18.
151. K. Fuller, et al. "TRANCE Is Necessary and Sufficient for Osteoblast-Mediated Activation of Bone Resorption in Osteoclasts." *J Exp Med* 188, no. 5 (1998): 997–1001.
152. Y. Y. Kong, et al. "Activated T Cells Regulate Bone Loss and Joint Destruction in Adjuvant Arthritis through Osteoprotegerin Ligand." *Nature* 402, no. 6759 (1999): 304–9.
153. J. E. Fata, et al. "The Osteoclast Differentiation Factor Osteoprotegerin-Ligand Is Essential for Mammary Gland Development." *Cell* 103, no. 1 (2000): 41–50.
154. W. S. Simonet, et al. "Osteoprotegerin: A Novel Secreted Protein Involved in the Regulation of Bone Density." *Cell* 89, no. 2 (1997): 309–19.
155. B. G. Darnay, et al. "Characterization of the Intracellular Domain of Receptor Activator of NF-KappaB (RANK). Interaction with Tumor Necrosis Factor Receptor-Associated Factors and Activation of NF-Kappab and c-Jun N-Terminal Kinase." *J Biol Chem* 273, no. 32 (1998): 20551–55.
156. B. R. Wong, et al. "TRANCE, a TNF Family Member, Activates Akt/PKB through a Signaling Complex Involving TRAF6 and c-Src." *Mol Cell* 4, no. 6 (1999): 1041–49.
157. B. G. Darnay and B. B. Aggarwal "Signal Transduction by TNF and TNF-Related Ligands and Their Receptors." *Ann of Rheum Dis* 58 (1999): S0–11.
158. V. Baud, et al. "Signaling by Proinflammatory Cytokines: Oligomerization of TRAF2 and TRAF6 Is Sufficient for JNK and IKK Activation and Target Gene Induction via an Amino-Terminal Effector Domain." *Genes Dev* 13, no. 10 (1999): 1297–1308.
159. G. Franzoso, et al. "Requirement for NF-KappaB in Osteoclast and B-Cell Development." *Genes Dev* 11, no. 24 (1997): 3482–96.
160. V. Iotsova, et al. "Osteopetrosis in Mice Lacking NF-KappaB1 and NF-KappaB2." *Nat Med* 3, no. 11 (1997): 1285–89.

161. E. Jimi, et al. "Osteoclast Differentiation Factor Acts as a Multifunctional Regulator in Murine Osteoclast Differentiation and Function." *J Immunol* 163, no. 1 (1999): 434–42.

162. M. Matsumoto, et al. "Involvement of p38 Mitogen-Activated Protein Kinase Signaling Pathway in Osteoclastogenesis Mediated by Receptor Activator of NF-Kappa B Ligand (RANKL)." *J Biol Chem* 275, no. 40 (2000): 31155–61.

163. W. C. Yeh, et al. "Early Lethality, Functional NF-KappaB Activation, and Increased Sensitivity to TNF-Induced Cell Death in TRAF2-Deficient Mice." *Immunity* 7, no. 5 (1997): 715–25.

164. W. C. Yang, et al. "Tec Kinases: A Family with Multiple Roles in Immunity." *Immunity* 12, no. 4 (2000): 373–82.

165. P. Soriano, et al. "Targeted Disruption of the c-src Proto-oncogene Leads to Osteopetrosis in Mice." *Cell* 64, no. 4 (1991): 693–702.

166. P. L. Schwartzberg, et al. "Rescue of Osteoclast Function by Transgenic Expression of Kinase-Deficient Src in src-/- Mutant Mice." *Genes Dev* 11, no. 21 (1997): 2835–44.

167. M. A. Lomaga, et al. "TRAF6 Deficiency Results in Osteopetrosis and Defective Interleukin-1, CD40, and LPS Signaling." *Genes Dev* 13, no. 8 (1999): 1015–24.

168. W. C. Dougall, et al. "RANK Is Essential for Osteoclast and Lymph Node Development." *Genes Dev* 13, no. 18 (1999): 2412–24.

169. Y. Y. Kong, et al. "OPGL Is a Key Regulator of Osteoclastogenesis, Lymphocyte Development and Lymph-Node Organogenesis." *Nature* 397, no. 6717 (1999): 315–23.

170. A. E. Hughes, et al. "Mutations in TNFRSF11A, Affecting the Signal Peptide of RANK, Cause Familial Expansile Osteolysis." *Nat Genet* 24, no. 1 (2000): 45–48.

171. C. Menaa, et al. "Enhanced RANK Ligand Expression and Responsivity of Bone Marrow Cells in Paget's Disease of Bone. *J Clin Invest* 105, no. 12 (2000): 1833–38.

172. L. C. Hofbauer, et al. "Estrogen Stimulates Gene Expression and Protein Production of Osteoprotegerin in Human Osteoblastic Cells." *Endocrinology* 140, no. 9 (1999): 4367–70.

173. M. Saika, et al. "17beta-estradiol Stimulates Expression of Osteoprotegerin by a Mouse Stromal Cell Line, ST-2, via Estrogen Receptor-Alpha." *Endocrinology* 142, no. (2001): 2205–12.

174. K. Yano, et al. "Immunological Characterization of Circulating Osteoprotegerin/ Osteoclastogenesis Inhibitory Factor: Increased Serum Concentrations in Postmenopausal Women with Osteoporosis." *J Bone Miner Res* 14, no. 4 (1999): 518–27.

175. H. A. Makhluf, et al. "Age-Related Decline in Osteoprotegerin Expression by Human Bone Marrow Cells Cultured in Three-Dimensional Collagen Sponges." *Biochem Biophys Res Commun* 268, no. 3 (2000): 669–72.

176. B. P. Lukert and L. G. Raisz. "Glucocorticoid-Induced Osteoporosis: Pathogenesis and Management." *Ann Intern Med* 112, no. 5 (1990): 352–64.

177. S. K. Lee and J. A. Lorenzo. "Parathyroid Hormone Stimulates TRANCE and Inhibits Osteoprotegerin Messenger Ribonucleic Acid Expression in Murine Bone Marrow Cultures: Correlation with Osteoclast-Like Cell Formation." *Endocrinology* 140, no. 8 (1999): 3552–61.

178. H. Ihara, et al. "Parathyroid Hormone-Induced Bone Resorption Does Not Occur in the Absence of Osteopontin." *J Biol Chem* 276, no. 16 (2001): 13065–71.

179. I. Itonaga, et al. "Rheumatoid Arthritis Synovial Macrophage-Osteoclast Differentiation Is Osteoprotegerin Ligand-Dependent." *J Pathol* 192, no. 1 (2000): 97–104.

180. P. Honore, et al. "Osteoprotegerin Blocks Bone Cancer-Induced Skeletal Destruction, Skeletal Pain and Pain-Related Neurochemical Reorganization of the Spinal Cord." *Nat Med* 6, no. 5 (2000): 521–28.

181. B. O. Oyajobi, et al. "Therapeutic Efficacy of a Soluble Receptor Activator of Nuclear Factor KappaB-IgG Fc Fusion Protein in Suppressing Bone Resorption and Hypercalcemia in a Model of Humoral Hypercalcemia of Malignancy." *Cancer Res* 61, no. 6 (2001): 2572–78.
182. P. J. Bekker, et al. "The Effect of a Single Dose of Osteoprotegerin in Postmenopausal Women." *J Bone Miner Res* 16, no. 2 (2001): 348–60.
183. A. C. Bharti, Y. Takada, and B. B. Aggarwal. "Curcumin (Diferuloylmethane) Inhibits Receptor Activator of NF-KappaB Ligand-Induced NF-KappaB Activation in Osteoclast Precursors and Suppresses Osteoclastogenesis." *J Immunol* 172, no. 10 (2004): 5940–47.
184. Y. Chicheportiche, et al. "TWEAK, a New Secreted Ligand in the Tumor Necrosis Factor Family That Weakly Induces Apoptosis." *J Biol Chem* 272, no. 51 (1997): 32401–10.
185. S. A. Marsters, et al. "Identification of a Ligand for the Death-Domain-Containing Receptor Apo3." *Curr Biol* 8, no. 9 (1998): 525–28.
186. C. N. Lynch, et al. "TWEAK Induces Angiogenesis and Proliferation of Endothelial Cells." *J Biol Chem* 274, no. 13 (1999): 8455–59.
187. M. Hahne, et al. "APRIL, a New Ligand of the Tumor Necrosis Factor Family, Stimulates Tumor Cell Growth." *J Exp Med* 188, no. 6 (1998): 1185–90.
188. H. B. Shu, W. H. Hu, and H. Johnson. "TALL-1 Is a Novel Member of the TNF Family That Is Down-Regulated by Mitogens." *J Leukoc Biol* 65, no. 5 (1999): 680–83.
189. Y. Zhai, et al. "VEGI, a Novel Cytokine of the Tumor Necrosis Factor Family, Is an Angiogenesis Inhibitor That Suppresses the Growth of Colon Carcinomas In Vivo." *FASEB J* 13, no. 1 (1999): 181–89.
190. T. S. Migone, et al. "TL1A Is a TNF-Like Ligand for DR3 and TR6/DcR3 and Functions as a T Cell Costimulator." *Immunity* 16, no. 3 (2002): 479–92.
191. A. Mukhopadhyay, et al. "Identification and Characterization of a Novel Cytokine, THANK, a TNF Homologue That Activates Apoptosis, Nuclear Factor-KappaB, and c-Jun NH2-Terminal Kinase." *J Biol Chem* 274, no. 23 (1999): 15978–81.
192. P. Schneider, et al. "BAFF, a Novel Ligand of the Tumor Necrosis Factor Family, Stimulates B Cell Growth." *J Exp Med* 189, no. 11 (1999): 1747–56.
193. S. A. Marsters, et al. "Interaction of the TNF Homologues BLyS and APRIL with the TNF Receptor Homologues BCMA and TACI." *Curr Biol* 10, no. 13 (2000): 785–88.
194. Y. Zhai, et al. "LIGHT, a Novel Ligand for Lymphotoxin Beta Receptor and TR2/HVEM Induces Apoptosis and Suppresses In Vivo Tumor Formation via Gene Transfer." *J Clin Invest* 102, no. 6 (1998): 1142–51.
195. D. N. Mauri, et al. "LIGHT, a New Member of the TNF Superfamily, and Lymphotoxin Alpha Are Ligands for Herpesvirus Entry Mediator." *Immunity* 8, no. 1 (1998): 21–30.
196. J. A. Harrop, et al. "Herpesvirus Entry Mediator Ligand (HVEM-L), a Novel Ligand for HVEM/TR2, Stimulates Proliferation of T Cells and Inhibits HT29 Cell Growth." *J Biol Chem* 273, no. 42 (1998): 27548–56.
197. A. L. Gurney, et al. "Identification of a New Member of the Tumor Necrosis Factor Family and Its Receptor, a Human Ortholog of Mouse GITR." *Curr Biol* 9, no. 4 (1999): 215–18.
198. B. Kwon, et al. "Identification of a Novel Activation-Inducible Protein of the Tumor Necrosis Factor Receptor Superfamily and Its Ligand." *J Biol Chem* 274, no. 10 (1999): 6056–61.
199. T. Kojima, et al. "TROY, a Newly Identified Member of the Tumor Necrosis Factor Receptor Superfamily, Exhibits a Homology with Edar and Is Expressed in Embryonic Skin and Hair Follicles." *J Biol Chem* 275, no. 27 (2000): 20742–47.
200. G. L. Sica, et al. "RELT, a New Member of the Tumor Necrosis Factor Receptor Superfamily, Is Selectively Expressed in Hematopoietic Tissues and Activates Transcription Factor NF-KappaB." *Blood* 97, no. 9 (2001): 2702–7.

201. G. Chen and D. V. Goeddel. "TNF-R1 Signaling: A Beautiful Pathway." *Science* 296, no. 5573 (2002): 1634–35.

202. J. M. Zapata and J. C. Reed. "TRAF1: Lord without a RING." *Sci STKE* 2002, no. 133 (2002): PE27.

203. M. Rothe, et al. "TRAF2-Mediated Activation of NF-Kappa B by TNF Receptor 2 and CD40." *Science* 269, no. 5229 (1995): 1424–27.

204. M. Prinz, et al. "Lymph Nodal Prion Replication and Neuroinvasion in Mice Devoid of Follicular Dendritic Cells." *Proc Natl Acad Sci U S A* 99, no. 2 (2002): 919–24.

205. A. M. Chinnaiyan and V. M. Dixit. "Portrait of an Executioner: The Molecular Mechanism of FAS/APO-1-Induced Apoptosis." *Semin Immunol* 9, no. 1 (1997): 69–76.

206. M. S. Sheikh, et al. "The Antiapoptotic Decoy Receptor TRID/TRAIL-R3 Is a p53-Regulated DNA Damage-Inducible Gene That Is Overexpressed in Primary Tumors of the Gastrointestinal Tract." *Oncogene* 18, no. 28 (1999): 4153–59.

207. R. D. Meng, et al. "The TRAIL Decoy Receptor TRUNDD (DcR2, TRAIL-R4) Is Induced by Adenovirus-p53 Overexpression and Can Delay TRAIL-, p53-, and KILLER/DR5-Dependent Colon Cancer Apoptosis." *Mol Ther* 1, no. 2 (2000): 130–44.

208. P. M. Chaudhary, et al. "Death Receptor 5, a New Member of the TNFR Family, and DR4 Induce FADD-Dependent Apoptosis and Activate the NF-KappaB Pathway." *Immunity* 7, no. 6 (1997): 821–30.

209. P. Schneider, et al. "TRAIL Receptors 1 (DR4) and 2 (DR5) Signal FADD-Dependent Apoptosis and Activate NF-KappaB." *Immunity* 7, no. 6 (1997): 831–36.

210. M. Krajewska, et al. "TRAF-4 Expression in Epithelial Progenitor Cells. Analysis in Normal Adult, Fetal, and Tumor Tissues." *Am J Pathol* 152, no. 6 (1998): 1549–61.

211. A. Matsushima, et al. "Essential Role of Nuclear Factor (NF)-KappaB-Inducing Kinase and Inhibitor of KappaB (IkappaB) Kinase Alpha in NF-KappaB Activation through Lymphotoxin Beta Receptor, but Not through Tumor Necrosis Factor Receptor I." *J Exp Med* 193, no. 5 (2001): 631–36.

212. Y. Hikichi, et al. "LIGHT, a Member of the TNF Superfamily, Induces Morphological Changes and Delays Proliferation in the Human Rhabdomyosarcoma Cell Line RD." *Biochem Biophys Res Commun* 289, no. 3 (2001): 670–77.

213. K. Misawa, et al. "Molecular Cloning and Characterization of a Mouse Homolog of Human TNFSF14, a Member of the TNF Superfamily." *Cytogenet Cell Genet* 89, no. 1–2 (2000): 89–91.

214. K. Y. Yu, et al. "A Newly Identified Member of Tumor Necrosis Factor Receptor Superfamily (TR6) Suppresses LIGHT-Mediated Apoptosis." *J Biol Chem* 274, no. 20 (1999): 13733–36.

215. H. Yamamoto, T. Kishimoto, and S. Minamoto. "NF-KappaB Activation in CD27 Signaling: Involvement of TNF Receptor-Associated Factors in Its Signaling and Identification of Functional Region of CD27." *J Immunol* 161, no. 9 (1998): 4753–59.

216. H. Akiba, et al. "CD27, a Member of the Tumor Necrosis Factor Receptor Superfamily, Activates NF-KappaB and Stress-Activated Protein Kinase/c-Jun N-Terminal Kinase via TRAF2, TRAF5, and NF-KappaB-Inducing Kinase." *J Biol Chem* 273, no. 21 (1998): 13353–58.

217. S. Ansieau, et al. "Tumor Necrosis Factor Receptor-Associated Factor (TRAF)-1, TRAF-2, and TRAF-3 Interact In Vivo with the CD30 Cytoplasmic Domain; TRAF-2 Mediates CD30-Induced Nuclear Factor Kappa B Activation." *Proc Natl Acad Sci U S A* 93, no. 24 (1996): 14053–58.

218. D. A. Francis, et al. "Induction of the Transcription Factors NF-Kappa B, AP-1 and NF-AT during B Cell Stimulation through the CD40 Receptor." *Int Immunol* 7, no. 2 (1995): 151–61.

219. I. K. Jang, et al. "Human 4-1BB (CD137) Signals Are Mediated by TRAF2 and Activate Nuclear Factor-Kappa B." *Biochem Biophys Res Commun* 242, no. 3 (1998): 613–20.

220. B. R. Wong, et al. "The TRAF Family of Signal Transducers Mediates NF-KappaB Activation by the TRANCE Receptor." *J Biol Chem* 273, no. 43 (1998): 28355-59.
221. B. G. Darnay, et al. "Activation of NF-KappaB by RANK Requires Tumor Necrosis Factor Receptor-Associated Factor (TRAF) 6 and NF-KappaB-Inducing Kinase. Identification of a Novel TRAF6 Interaction Motif." *J Biol Chem* 274, no. 12 (1999): 7724–31.
222. N. Nakagawa, et al. "RANK Is the Essential Signaling Receptor for Osteoclast Differentiation Factor in Osteoclastogenesis." *Biochem Biophys Res Commun* 253, no. 2 (1998): 395–400.
223. X. Z. Xia, et al. "TACI Is a TRAF-Interacting Receptor for TALL-1, a Tumor Necrosis Factor Family Member Involved in B Cell Regulation." *J Exp Med* 192, no. 1 (2000): 137–43.
224. M. Yan, et al. "Identification of a Receptor for BLyS Demonstrates a Crucial Role in Humoral Immunity." *Nat Immunol* 1, no. 1 (2000): 37–41.
225. G. Yu, et al. "APRIL and TALL-I and Receptors BCMA and TACI: System for Regulating Humoral Immunity." *Nat Immunol* 1, no. 3 (2000): 252–56.
226. A. Hatzoglou, et al. "TNF Receptor Family Member BCMA (B Cell Maturation) Associates with TNF Receptor-Associated Factor (TRAF) 1, TRAF2, and TRAF3 and Activates NF-Kappa B, elk-1, c-Jun N-Terminal Kinase, and p38 Mitogen-Activated Protein Kinase." *J Immunol* 165, no. 3 (2000): 1322–30.
227. Y. Laabi, A. Egle, and A. Strasser. "TNF Cytokine Family: More BAFF-ling Complexities." *Curr Biol* 11, no. 24 (2001): R1013–16.
228. J. S. Thompson, et al. "BAFF-R, a Newly Identified TNF Receptor That Specifically Interacts with BAFF." *Science* 293, no. 5537 (2001): 2108–11.
229. P. Schneider, et al. "TWEAK Can Induce Cell Death via Endogenous TNF and TNF Receptor 1." *Eur J Immunol* 29, no. 6 (1999): 1785–92.
230. S. R. Wiley, et al. "A Novel TNF Receptor Family Member Binds TWEAK and Is Implicated in Angiogenesis." *Immunity* 15, no. 5 (2001): 837–46.
231. A. Kumar, et al. "The Ectodermal Dysplasia Receptor Activates the Nuclear Factor-KappaB, JNK, and Cell Death Pathways and Binds to Ectodysplasin A." *J Biol Chem* 276, no. 4 (2001): 2668–77.
232. I. Thesleff and M. L. Mikkola. "Death Receptor Signaling Giving Life to Ectodermal Organs." *Sci STKE* 2002, no. 131 (2002): PE22.
233. M. Yan, et al. "Two-Amino Acid Molecular Switch in an Epithelial Morphogen That Regulates Binding to Two Distinct Receptors." *Science* 290, no. 5491 (2000): 523–27.

3 Blood–Brain Barrier Models for Investigating CNS Pathologies

Pierre-Olivier Couraud and Sandrine Bourdoulous

CONTENTS

3.1 INTRODUCTION

Microvascular endothelial cells of the central nervous system (CNS) substantially differ from other vascular endothelia in their capacity to strictly regulate the passage of molecules to and from the neural parenchyma. This selectivity resides in structural features unique to CNS endothelia, including expression of tight intercellular junctions that markedly limit paracellular permeability to hydrophilic molecules (Wolburg and Lippoldt 2002), a unique pattern of receptors and transporters that control the entry of nutriments into the CNS and multispecific drug export pumps that protect the CNS against a variety of hydrophobic compounds or xenobiotics (Schinkel 1999; Begley 2004). Together, these specific characters constitute the blood–brain barrier (BBB).

The BBB lies at a critical interface between the CNS and the rest of the organism that is central to the maintenance of cerebral homeostasis and normal neuronal function. However, several CNS pathologies are characterized by the infiltration of activated leukocytes, bacteria, or viruses (Couraud, P. O. et al. 2003), in particular multiple sclerosis (Raine 1994), meningitis, and HIV infection, respectively, whereas some neurological diseases, like stroke or Alzheimer's disease, are associated with brain endothelium dysfunction (Kalaria 1999). Unraveling the molecular mechanisms of BBB regulation or dysfunction would constitute a major breakthrough toward the understanding of the etiology of these diseases and the design of new potential therapeutic strategies.

It is the purpose of this chapter to present the major issues and recent achievements in the development of in vitro models of the BBB and to illustrate, through significant examples, the important input as well as expected results of these approaches for our understanding of BBB function and implication in various CNS pathologies.

3.2   BRAIN ENDOTHELIUM AND CNS PATHOLOGIES

3.2.1   INFECTIOUS DISEASES

Invasion of the CNS is a severe and frequently fatal event during the course of many infectious diseases. One of the key events in the pathogenesis of CNS infections is how bacterial pathogens or viruses interact with and cross the BBB to gain access to the central compartment (Couraud, P. O. et al. 2003).

In most cases, neuroinvasion occurs in the context of systemic disease and typically follows bacterial dissemination via the bloodstream. In this respect, intracellular and extracellular bacteria have not only developed different strategies to escape from complement-mediated serum killing and from the phagocytic activity of neutrophils before invading the brain parenchyma, but also use various mechanisms to enter the CNS. Whereas obligate and facultative intracellular bacterial pathogens may cross the BBB via the trans-endothelial migration of infected leukocytes (the so-called Trojan horse hypothesis), bacteria with predominantly extracellular life cycles, such as *Streptococcus pneumoniae*, *Neisseria meningitidis*, and *Escherichia coli K1*, may either migrate through inter-endothelial junctions or be internalized by brain endothelial cells and follow a transcytosis pathway (Nassif et al. 2002).

In a similar manner, retroviruses such as HIV-1 or HTLV-I are believed to invade the CNS through distinct alternative routes, including recirculation through the brain of monocytes or lymphocytes infected in peripheral organs and infection of, or macropinocytosis into, brain endothelial cells (Persidsky et al. 1999; Romero et al. 2000; Bussolino et al. 2001; Liu et al. 2002).

Most of these pathogens like *N. meningitidis* or HIV-1 are human-specific and animal models closely mimicking the human disease are not available, therefore limiting our knowledge of the precise molecular mechanism involved in their invasive processes and requiring the development of appropriate in vitro models of human brain endothelium.

3.2.2   Neuro-inflammatory and Demyelinating Diseases

Multiple sclerosis, at least its relapsing/remitting form, is associated with perivascular and parenchymal infiltration of the brain and spinal cord by mononuclear cells, predominantly activated T cells, and monocytes/macrophages. A well-described animal model of multiple sclerosis such as experimental allergic encephalomyelitis can be induced, in susceptible strains, by injection of a myelin protein or by adoptive transfer of activated myelin-specific T lymphocytes. These observations indicate that leukocyte recruitment to the CNS is a critical step in the pathogenesis of demyelinating neuro-inflammatory diseases (Raine 1994). Early in vitro experiments using cultured rodent brain endothelial cells largely contributed to our understanding that the migration of leukocytes across the BBB depends upon the expression by inflamed brain endothelium of cell adhesion molecules, such as ICAM-1 (CD54), VCAM-1 (CD106), and PECAM-1 (CD31) that are differentially involved in the trans-endothelial migration of distinct subsets of leukocytes (Yednock et al. 1992; Male et al. 1994). The particular role of VCAM-1 in the progression of experimental allergic encephalomyelitis and multiple sclerosis was recently highlighted, respectively by antibody-mediated inhibition of lymphocyte tethering to spinal cord microvascular endothelium (Vajkoczy et al. 2001) and the design of a clinical trial based on a monoclonal antibody against the VCAM-1 counter-receptor, the integrin VLA-4 expressed by activated lymphocytes (Polman and Uitdehaag 2003). Understanding the precise role of brain microvascular endothelial cells in leukocyte infiltration into the CNS might help control the otherwise intractable progression of this highly invalidating disease. Among important issues for addressing this disease without eliciting major side effects now appears to be the specific targeting of brain endothelium that obviously will require the identification of new BBB markers as potential clinical targets.

3.2.3   Neurological Diseases and Brain Tumors

Our ability to treat neurological diseases or brain tumors is greatly impaired by the highly restrictive permeability of the BBB for therapeutic drugs. Indeed, BBB endothelium expresses several drug export pumps of the ATP-binding cassette (ABC) transporter family, which recognize as ligands a number of xenobiotics such as chemotherapeutic anticancer agents, glucocorticoids, or anticonvulsive agents and drastically limit their bioavailability to the cerebral parenchyma (Schinkel 1999; Begley 2004). P-glycoprotein (P-gp, also called MDR1 or ABCB1), multidrug resistance–associated proteins (MRPs or ABCC1-8), and the breast cancer resistance protein (BCRP or ABCG2) (Eisenblatter et al. 2003; Zhang et al. 2003) are all known to be expressed by brain endothelium. Optimization of drug delivery to the CNS will thus include either transient inhibition of ABC transporters by specific inhibitors or understanding the regulatory mechanisms that control their expression by brain endothelium. In vitro models of human BBB expressing the appropriate ABC transporters would clearly contribute to unraveling these regulatory mechanisms. Moreover, substantial species differences in the selectivity of ABC transporters and their known regulators (Yamazaki et al. 2001; Bauer et al. 2004) further highlight our need of a model of the human BBB.

Whereas BBB dysfunction has long been recognized as a direct consequence of stroke and a precursor to poor outcome (Ayata and Ropper 2002), the hypothesis that vascular dysfunction in brains of individuals with Alzheimer's disease might contribute to the severity of the disease was more recently supported by observations that transport of the amyloid peptide Aβ across the BBB regulates its cerebral deposition (Zlokovic 2004). It has been recently demonstrated in murine models that peripheral Aβ binding agents including specific antibodies (Sigurdsson et al. 2001) promote transport of brain-derived Aβ into blood, thereby reducing amyloid load in the CNS. However, the unexpected difficulties encountered during the clinical trial designed on this basis highlight the need for the identification of the precise molecular mechanisms mediating this process in humans (Robinson et al. 2004). Together with the use of primate models of Alzheimer's disease, availability of a human model of BBB might help design new strategies to shift the Aβ transport equilibrium toward plasma.

3.3 IN VITRO MODELING OF BBB

Our capacity to model the BBB in vitro is closely linked to our understanding of the unique phenotype of brain endothelial cells in terms of expression and molecular organization of tight intercellular junctions and transport systems, which has been the subject of intensive investigations over the past few years (Wolburg and Lippoldt 2002; Begley 2004).

3.3.1 PRIMARY CULTURES OF BRAIN ENDOTHELIAL CELLS

The limitations imposed by whole animal experiments when studying BBB biology have led to the development of techniques to isolate and culture CNS-derived endothelial cells from various species (Joo 1993). Unfortunately, these cultured endothelial cells usually fail to develop functional tight junctions. However, a model based on cultured porcine brain endothelial cells in serum-free condition and in the presence of hydrocortisone was proposed and shown to permit BBB differentiation (Engelbertz et al. 2000). Alternatively, co-culture systems have been developed in which brain endothelial cells are grown on microporous filter membrane inserts in the presence of primary cultures of glial cells in order to mimic the anatomical and functional relationship between brain endothelium and surrounding astrocytes (Deli et al. 2004). Published reports on such co-culture systems, most of which are based on bovine brain endothelial cells (Rubin et al. 1991; Cecchelli et al. 1999), achieve a good correlation between in vitro permeability of standard molecules, over a wide range of hydrophilicity and the in vivo cerebral bioavailability of the same molecules. Such available models have proved to be of great value for both drug screening and basic studies on BBB biology. However, their use for immunological, microbiological, or virological studies remains difficult because of the existence of species barriers. Only very recently were similar models based on rat or mouse brain endothelial cells designed, which might broaden the scope of further investigations at the molecular level (M. Deli, personal communication; Perrière et al., submitted). Primary cultures of human brain endothelial cells were also proposed as tentative models for investigations on HIV-1–mediated encephalitis or *E. coli K1* invasion of the CNS (Persidsky et al. 1997; Kim 2000), but their rapid de-differentiation in culture greatly hampers their usefulness as validated in vitro models of the human BBB.

3.3.2   RAT BRAIN ENDOTHELIAL CELL LINES

Considering that establishment of BBB in vitro models based on primary cultures of brain endothelial cells is time consuming and requires substantial know-how from the investigator laboratory, limiting their availability and inter-laboratory standardization, we and others established a number of immortalized endothelial cell lines which maintain a stable phenotype in culture and constitute easily transferable models of brain endothelium. The rat brain endothelial RBE4, GP8, and GPNT cell lines have thus been extensively characterized and are now widely used as validated models of brain endothelium for biochemical, pharmacological, toxicological or immunological purposes (Roux et al. 1994; Abbott et al. 1995; Greenwood et al. 1996; Romero et al. 1997; Regina et al. 1999). In addition, co-culture systems of conditionally immortalized rat brain endothelial cells and astrocytes were also reported to recapitulate many BBB-specific characteristics, including expression of ABC transporters (Terasaki et al. 2003). However, these various cell lines appear to display a reduced complexity of tight junctions associated with a moderate permeability restriction, which limits their use in permeability regulation studies or drug screening.

3.3.3   TOWARD IN VITRO MODEL OF HUMAN BBB

Several conceptual as well as technical limitations have been preventing so far the immortalization of human brain endothelial cells. First, senescence naturally limits the proliferation of mammalian cells in culture, especially human cells, by shortening the telomere regions of chromosomes during cell division. Quite recently, however, exogenous expression of the catalytic subunit of human telomerase (hTERT) has been shown to prevent telomere shortening and has been recently proposed as a strategy for lifespan extension and in some cases immortalization of various cell types, including endothelial cells, without cell transformation (Yang et al. 1999; Gu et al. 2003). Second, a new generation of lentiviral vectors was shown to dramatically improve gene transduction in post-mitotic or slowly proliferating neural cells both in vitro and in vivo (Zennou et al. 2001). Benefiting from these significant technical breakthroughs, we then recently produced and extensively characterized an immortalized human brain endothelial cell line (hCMEC/D3) by sequentially infecting primary cultures of human brain endothelial cells with hTERT- and SV40 T antigen-encoding lentiviral vectors (Weksler et al. 2005). This cell line stably maintains the morphological characteristics of primary brain endothelial cells, expressing specific junctional proteins, cell surface adhesion molecules, and functional drug export ABC transporters. Moreover, hCMEC/D3 monolayers display highly restricted paracellular permeability to various compounds that can be correlated to their in vivo brain perfusion capacity, strongly suggesting that this cell line constitutes, even in the absence of co-cultured glial cells, the first fully characterized in vitro model of human BBB.

3.4 CONTRIBUTION OF IN VITRO BBB MODELS TO UNDERSTANDING OF CNS PATHOLOGIES

3.4.1 IN VITRO BBB MODELS AND MENINGITIS

The mechanisms mediating the crossing of the vascular endothelial barrier by *N. meningitidis*, resulting in fulminant meningitis, remain poorly described. We have shown that infection of cultured human endothelial cells by *N. meningitidis* elicits the localized polymerization of cortical actin, recruitment of the cytosolic protein ezrin that links membrane integral proteins such as CD44 or ICAM-1 to the actin cytoskeleton, and formation of membrane protrusions, reminiscent of epithelial microvilli that surround bacteria and provoke their internalization within endothelial intracellular vacuoles (Eugene et al. 2002). The formation of such membrane protrusions and *N. meningitidis* internalization within endothelial cells has been confirmed by postmortem examination of human brain samples, suggesting that this process is essential for the crossing of human endothelium via a transcytosis pathway (Nassif et al. 2002). We could further demonstrate that adhesion of *N. meningitidis* onto human endothelial cells induces the activation of the tyrosine kinase receptor ErbB2 and of the cytosolic tyrosine kinase Src, leading to the subsequent tyrosine phosphorylation of the actin-binding protein cortactin, which plays a key role in the internalization of *N. meningitidis* (Hoffmann et al. 2001). Ongoing transcriptomic and proteomic approaches aiming at the identification of the cellular receptor(s) of *N. meningitidis* will now benefit from the recent availability of the hCMEC/D3 human brain endothelial cell line.

E. coli K1, the most common extracellular bacterial pathogen causing neonatal meningitis, is able, like *N. meningitidis,* to adhere to brain endothelium and cross the BBB by a transcytosis pathway (Kim 2002). However, the molecular mechanisms involved in this process, which have been investigated using primary cultures of human brain endothelial cells, were shown to be quite different from those mediating *N. meningitidis* invasion, being elicited by interaction of various bacterial attributes with distinct cellular receptors, one of which, the 67-kDa laminin receptor, has been recently identified (Kim et al. 2004).

3.4.2 IN VITRO BBB MODELS AND MULTIPLE SCLEROSIS

Leukocyte adhesion to brain endothelium was known to be associated with intracellular signals, including cytoskeletal modification, protein phosphorylation, and calcium influx. In addition, it became increasingly clear, in particular from studies of leukocyte responses to integrin engagement, that in addition to enabling leukocytes to adhere to endothelium, adhesion molecules were also involved in intracellular signal transduction.

Taking advantage of the availability of the RBE4 and GP8 rat brain endothelial cell lines, we then intended to identify the intracellular signaling putatively coupled to endothelial adhesion molecules, with a special interest for ICAM-1 (Greenwood et al. 2002). We showed that ICAM-1 cross-linking (activation) in brain endothelial cells, or co-culture with T lymphocytes, triggers multiple signal transduction pathways: induction of intracellular calcium flux, activation of Rho family GTPases, and actin-cytoskeleton changes (Adamson et al. 1999; Etienne-Manneville et al. 2000). In addition, activation of ICAM-1 stimulates tyrosine kinases commonly activated downstream of integrin signaling, such as Src and the focal adhesion kinase and induces the tyrosine phosphorylation of proteins associated with the actin cytoskeleton (paxillin, p130 cas, cortactin) and the activation of MAPKinase cascades. Using functional assays, we established that ICAM-1 signaling is intimately and actively involved in facilitating lymphocyte diapedesis, since a pretreatment of the endothelial cells with a Rho GTPase inhibitor or an intracellular calcium chelator significantly reduces the transmigration of activated T-lymphocytes (Adamson et al. 1999; Etienne-Manneville et al. 2000).

These in vitro data were then further confirmed and extended by results indicating that treatment of mice with experimental allergic encephalomyelitis by inhibitors of Rho family GTPases largely prevented leukocyte infiltration into the CNS and dramatically limited the severity of the disease (Walters et al. 2002; Greenwood et al. 2003). Taken together, these observations document the fundamental interest of BBB in vitro models for unraveling the molecular mechanisms supporting leukocyte trans-endothelial migration and for the identification of new therapeutic targets in neuro-inflammatory diseases. Further studies will aim at understanding the capacity of brain endothelial cells to integrate multiple signalings from the various adhesion molecules (ICAM-1, VCAM-1, PECAM-1) known to be involved in the infiltration of activated lymphocytes or monocytes into the cerebral parenchyma (Greenwood et al. 2002).

3.4.3   IN VITRO BBB MODELS FOR PROTEIN DISCOVERY

Interestingly, in vitro models based on primary cultures of brain endothelial cells or brain endothelial cell lines were recently used for validation of newly discovered proteins that directly affect BBB function: two examples will be mentioned here.

Whereas the functional interaction between brain endothelial cells and astrocytes has been recognized for many years as a critical factor for the establishment and maintenance of BBB in situ, the identity of the astrocytic factors involved so far remains a matter of debate. Several candidates were proposed, including glial-derived neurotrophic factor, but only very recently was *src*-suppressed C-kinase substrate (SSeCKS) discovered. Using primary cultures of human brain endothelial cells, Lee et al. (2003) demonstrated that astrocyte-derived SSeCKS, acting through the production of the vascular stabilizing factor angiopoietin-1 (Thurston et al. 2000), can reduce endothelial paracellular permeability and increase the expression of several tight junction-associated proteins, strongly suggesting that astrocyte-derived SSeCKS actively contributes in situ to BBB integrity. This discovery might have important consequences regarding the development of therapeutic strategies for brain diseases related to BBB disruption.

As mentioned above, the breast cancer resistance protein (BCRP or ABCG2) (Eisenblatter et al. 2003; Zhang et al. 2003) was recently identified as an ABC transporter expressed at the BBB. Indeed, among several ABC transporter sequences expressed in human brain cDNA libraries, only ABCG2 was found to be highly expressed in cultured human brain endothelial cells. In addition, overexpression of human ABCG2 in immortalized rat brain endothelial cells resulted in enhanced polarized abluminal to luminal transport of various substrates. ABCG2 is upregulated in both glioblastoma microvasculature and parenchymal tissue and these findings suggest that this ABC transporter is actively involved in modulating drug delivery to the brain and in conferring drug resistance to glioblastomas (Zhang et al. 2003).

Such examples underline the important contribution of in vitro BBB models, together with complementary in vivo approaches, to protein discovery and understanding of BBB physiology.

3.5 CONCLUSIONS AND PERSPECTIVES

In summary, the BBB constitutes a dynamic interface responsible for brain homeostasis, which strictly controls intracerebral infiltration of immune cells or pathogens and is thus directly involved in various CNS diseases, including neuro-inflammatory, infectious, neurological, or tumoral diseases. For more than 15 years establishment of in vitro BBB models has been the subject of intensive investigations and has benefited more recently from the substantial progress accomplished in our understanding of the molecular structure of the BBB, including organization of the tight junction-associated multi-protein complexes and identification of multispecific drug export pumps.

A number of such models are now available, based on primary cultures of brain endothelial cells or immortalized brain endothelial cell lines, often in co-culture with glial cells. Besides fully validated bovine and porcine models, which are already used for drug screening as well as basic studies on BBB biology, rat and mouse models are presently under development, which, through in vitro culture of brain endothelial cells from genetically modified animals, will soon extend the scope of potential applications of BBB modeling to molecular studies of brain endothelium dysfunctions in various pathological states. We believe that the availability of various in vitro BBB models, from diseased and control animals, will make possible the differential global analysis of the BBB-related gene and protein expression profiles in normal or pathological situations in the future.

Forthcoming human models will ultimately provide a unique opportunity to perform high throughput screening of CNS drug candidates with an improved predictive value on efficacy in humans. These models may also constitute unique tools to design new strategies for drug delivery to the CNS via identification and targeting of membrane proteins specifically expressed by brain endothelium, or via specific inhibition of drug export pumps expressed at the human BBB. Finally, in parallel with other approaches such as laser capture microscopy-assisted microdissection of human brain vessels from post-mortem tissue, human BBB models may allow in depth investigation of pathogenetic mechanisms in those human neurologic diseases, including infection by human-specific pathogens, characterized by vascular involvement.

REFERENCES

Abbott, N. J., P. O. Couraud, et al. 1995. Studies of an immortalized brain endothelial cell line: Characterization, permeability and transport. In *New concepts of a blood-brain barrier,* eds. J. Greenwood, D. J. Begley, M. Segal, and S. Lightman. New-York: Plenum Press.

Adamson, P., S. Etienne, et al. 1999. Lymphocyte migration through brain endothelial cell monolayers involves signaling through endothelial ICAM-1 via a rho-dependent pathway. *J Immunol* 162(5):2964–73.

Ayata, C., and A. H. Ropper. 2002. Ischaemic brain oedema. *J Clin Neurosci* 9(2):113–24.

Bauer, B., A. M. S. Hartz, et al. 2004. Pregnane X receptor up-regulation of P-glycoprotein expression and transport function at the blood-brain barrier. *Mol Pharmacol* 66(3):413–19.

Begley, D. J. 2004. ABC transporters and the blood-brain barrier. *Curr Pharm Des* 10(12):1295–1312.

Bussolino, F., S. Mitola, et al. 2001. Interactions between endothelial cells and HIV-1. *Int J Biochem Cell Biol* 33(4):371–90.

Cecchelli, R., B. Dehouck, et al. 1999. In vitro model for evaluating drug transport across the blood-brain barrier. *Adv Drug Deliv Rev* 36(2–3):165–78.

Couraud, P. O., X. Nassif, et al. 2003. Mechanisms of infiltration of immune cells, bacteria and viruses through brain endothelium. *Adv Mol Cell Biol* 31:255–68.

Deli, M. A., C. S. Abraham, et al. 2004. Permeability studies on in vitro blood-brain barrier models: physiology, pathology and pharmacology. *Cell Mol Neurobiol* 24(5):59–127.

Eisenblatter, T., S. Huwel, et al. 2003. Characterisation of the brain multidrug resistance protein (BMDP/ABCG2/BCRP) expressed at the blood-brain barrier. *Brain Res* 971(2):221–31.

Engelbertz, C., D. Korte, et al. 2000. The development of in vitro models for the blood-brain and blood-CSF barriers. In *The blood-brain barrier and drug delivery to the CNS.* eds. D. Begley, M. W. Bradbury, and J. Kreuter, 33–63. New York: Marcel Dekker.

Etienne-Manneville, S., J. B. Manneville, et al. 2000. ICAM-1-coupled cytoskeletal rearrangements and transendothelial lymphocyte migration involve intracellular calcium signaling in brain endothelial cell lines. *J Immunol* 165(6):3375–83.

Eugene, E., I. Hoffmann, et al. 2002. Microvilli-like structures are associated with the internalization of virulent capsulated *Neisseria meningitidis* into vascular endothelial cells. *J Cell Sci* 115(Pt 6):1231–41.

Greenwood, J., S. Etienne-Manneville, et al. 2002. Lymphocyte migration into the central nervous system: implication of ICAM-1 signalling at the blood-brain barrier. *Vasc Pharm* 38:315–22.

Greenwood, J., G. Pryce, et al. 1996. SV40 large T immortalised cell lines of the rat blood-brain and blood-retinal barriers retain their phenotypic and immunological characteristics. *J Neuroimmunol* 71(1–2):51–63.

Greenwood, J., C. E. Walters, et al. 2003. Lovastatin inhibits brain endothelial cell Rho-mediated lymphocyte migration and attenuates experimental autoimmune encephalomyelitis. *FASEB J* 17:905–7.

Gu, X., J. Zhang, et al. 2003. Brain and retinal vascular endothelial cells with extended life span established by ectopic expression of telomerase. *Invest Ophthalmol Vis Sci* 44(7):3219–25.

Hoffmann, I., E. Eugene, et al. 2001. Activation of ErbB2 receptor tyrosine kinase supports invasion of endothelial cells by *Neisseria meningitidis*. *J Cell Biol* 155(1):133–43.

Joo, F. 1993. The blood-brain barrier in vitro: the second decade [see comments]. *Neurochem Int* 23(6):499–521.

Kalaria, R. N. 1999. The blood-brain barrier and cerebrovascular pathology in Alzheimer's disease. *Ann N Y Acad Sci* 893:113–25.

Kim, K. J., J. W. Chung, et al. 2004. 67-kDa laminin receptor promotes internalization of cytotoxic necrotizing factor-1-expressing *Escherichia coli* K1 into human brain microvascular endothelial cells. *J Biol Chem* 280(2):1360–68. Epub 2004 Oct 29.

Kim, K. S. 2000. *E. coli* invasion of brain microvascular endothelial cells as a pathogenetic basis of meningitis. New York: Kluwer Academic/Plenum Publishers.

Kim, K. S. 2002. Strategy of *Escherichia coli* for crossing the blood-brain barrier. *Infect Immun* 186:S220–24.

Lee, S. W., W. J. Kim, et al. 2003. SSeCKS regulates angiogenesis and tight junction formation in blood-brain barrier. *Nat Med* 9(7):900–6.

Liu, N. Q., A. S. Lossinsky, et al. 2002. Human imunodeficiency virus type 1 enters brain microvascular endothelia by macropinocytosis dependent on lipid rafts and the mitogen-activated protein kinase signaling pathway. *J Virol* 76(13):6689–6700.

Male, D., J. Rahman, et al. 1994. Lymphocyte migration into the CNS modelled in vitro: roles of LFA-1, ICAM-1 and VLA-4. *Immunology* 81(3):366–72.

Nassif, X., S. Bourdoulous, et al. 2002. How do extracellular pathogens cross the blood-brain barrier? *Trends Microbiol* 10(5):227–32.

Perriere, N., Demeuse, P., Garcia, E., Regina, A., Debray, M., Andreux, J. P., Couvreur, P., Scherrmann, J. M., Temsamani, J., Couraud, P. O. et al. (2005). Puromycin-based purification of rat brain capillary endothelial cell cultures. Effects on the expression of blood-brain barrier-specific properties. *J Neurochem* 93, 279–89,

Persidsky, Y., A. Ghorpade, et al. 1999. Microglial and astrocyte chemokines regulate monocyte migration through the blood-brain barrier in human immunodeficiency virus-1 encephalitis. *Am J Pathol* 155(5):1599–1611.

Persidsky, Y., M. Stins, et al. 1997. A model for monocyte migration through the blood-brain barrier during HIV-1 encephalitis. *J Immunol* 158(7):3499–3510.

Polman, C. H., and B. M. Uitdehaag. 2003. New and emerging treatment options for multiple sclerosis. *Lancet Neurol* 2(9):563–66.

Raine, C. S. 1994. The Dale E. McFarlin Memorial Lecture: the immunology of the multiple sclerosis lesion. *Ann Neurol* 36(Suppl):S61–72.

Regina, A., I. A. Romero, et al. 1999. Dexamethasone regulation of P-glycoprotein activity in an immortalized rat brain endothelial cell line, GPNT. *J Neurochem* 73(5):1954–63.

Robinson, S. R., G. M. Bishop, et al. 2004. Lessons from the AN1792 Alzheimer vaccine: lest we forget. *Neurobiol Aging* 25(5):609–15.

Romero, I. A., M. C. Prevost, et al. 2000. Interactions between brain endothelial cells and human T-cell leukemia virus type 1-infected lymphocytes: mechanisms of viral entry into the central nervous system. *J Virol* 74(13):6021–30.

Romero, I. A., R. J. Rist, et al. 1997. Metabolic and permeability changes caused by thiamine deficiency in immortalized rat brain microvessel endothelial cells. *Brain Res* 756(1–2):133–40.

Roux, F., O. Durieu-Trautmann, et al. 1994. Regulation of gamma-glutamyl transpeptidase and alkaline phosphatase activities in immortalized rat brain microvessel endothelial cells. *J Cell Physiol* 159(1):101–13.

Rubin, L. L., D. E. Hall, et al. 1991. A cell culture model of the blood-brain barrier. *J Cell Biol* 115(6):1725–35.

Schinkel, A. H. 1999. P-glycoprotein, a gatekeeper in the blood-brain barrier. *Adv Drug Delivery Rev* 36:179–94.

Sigurdsson, E., H. Scholtzova, et al. 2001. Immunization with a nontoxic/nonfibrillar amyloid-beta homologous peptide reduces Alzheimer's disease-associated pathology in transgenic mice. *Am J Pathol* 159:439–47.

Terasaki, T., S. Ohtsuki, et al. 2003. New approaches to in vitro models of blood-brain barrier drug transport. *Drug Discovery Today* 8:944–54.

Thurston, G., J. S. Rudge, et al. 2000. Angiopoietin-1 protects the adult vasculature against plasma leakage. *Nature Med* 6(4):460–63.

Vajkoczy, P., M. Laschinger, et al. 2001. Alpha4-integrin-VCAM-1 binding mediates G protein-independent capture of encephalitogenic T cell blasts to CNS white matter microvessels. *J Clin Invest* 108(4):557–65.

Walters, C. E., G. Pryce, et al. 2002. Inhibition of Rho GTPases with protein prenyltransferase inhibitors prevents leukocyte recruitment to the central nervous system and attenuates clinical signs of disease in an animal model of multiple sclerosis. *J Immunol* 168(8):4087–94.

Weksler, B. B., E. Subileau, et al. 2005. Blood-brain barrier specific properties of a human adult brain endothelial cell line. *FASEB J* 19:1872–74.

Wolburg, H., and A. Lippoldt. 2002. Tight junctions of the blood-brain barrier: development, composition and regulation. *Vasc Pharm* 38:323–37.

Yamazaki, M., W. E. Neway, et al. 2001. In vitro substrate identification studies for p-glycoprotein-mediated transport: species difference and predictability of in vivo results. *J Pharmacol Exp Ther* 296(3):723–35.

Yang, J., E. Chang, et al. 1999. Human endothelial cell life extension by telomerase expression. *J Biol Chem* 274(37):26141–48.

Yednock, T. A., C. Cannon, et al. 1992. Prevention of experimental autoimmune encephalomyelitis by antibodies against alpha 4 beta 1 integrin. *Nature* 356(6364):63–66.

Zennou, V., C. Serguera, et al. 2001. The HIV-1 DNA flap stimulates HIV vector-mediated cell transduction in the brain. *Nat Biotechnol* 19(5):446–50.

Zhang, W., J. Mojsilovic-Petrovic, et al. 2003. The expression and functional characterization of ABCG2 in brain endothelial cells and vessels. *FASEB J* 17(14):2085–87. Epub 2003 Sep 4.

Zlokovic, B. V. 2004. Clearing amyloid through the blood-brain barrier. *J Neurochem* 89:807–11.

4 Identification of Activation of Latent TGFβ as Principal In Vivo Function of Integrin αvβ6

Dean Sheppard

CONTENTS

4.1   INTRODUCTION

Integrins are heterodimeric transmembrane receptors that were first identified based on their ability to bind and mediate cell adhesion to distinct components of the extracellular matrix (1,2). Classified as "adhesion receptors" integrins were initially thought of largely as a glorified form of cellular glue. However, it soon became clear that these receptors existed in multiple affinity states that can be regulated by cellular signals—that is, integrins are "reversible cellular glue" controlled by what have been called "inside-out signals." Furthermore, it became clear that integrins, like most other cellular receptors, are themselves capable of initiating cellular signals that profoundly affect cell behavior, termed "outside-in signaling" (3). In this chapter, I describe the circuitous steps that led to the discovery of an entirely different role of an integrin: activation of the cytokine transforming growth factor β from latent extracelluar stores, and presentation of the active cytokine to receptors on adjacent cells.

4.2 IDENTIFICATION OF INTEGRIN β6 SUBUNIT

In the mid 1980s, I was appointed to direct a new center, the Lung Biology Center, which had as its goal the application of cell and molecular approaches to understanding common lung diseases. At the time, I had no training or experience in either cell or molecular biology, having spent all of my fellowship training and time on the UCSF faculty studying in vivo mechanisms of airway narrowing in patients and in animal models of asthma. My first task as director of this new center was to recruit new faculty who were outstanding cell or molecular biologists. A year after the center opened and moved into its new facilities, I decided to spend a sabbatical year in the lab of one of the center's new faculty members, Robert Pytela, whose major focus at the time was the cloning and characterization of genes encoding integrin subunits. As a post-doctoral fellow in Errki Ruoslahti's laboratory, Robert had played a central role in identifying, cloning, and functionally characterizing the first known integrin receptors for fibronectin and vitronectin (4,5), and had then gone on to clone the αM subunit of the integrin αMβ2. Frustrated with the slow pace of traditional cloning methods, Robert was about to embark on a more efficient strategy to clone additional integrins just at the time I began my sabbatical.

My rationale for joining Robert's lab was rather naïve. From the time I had spent studying in vivo models of asthma, it was clear to me that our models were inadequate and focused only on acute and reversible events. In contrast, the human disease, which I was all too familiar with from my outpatient practice, was chronic and persistent. I was therefore interested in identifying biological pathways that might underlie disease persistence. Since it had been known for many years that patients with asthma had dramatic changes in the amount and composition of extracellular matrix in the airway wall, I reasoned (hoped?) that signals transmitted from the abnormal matrix to airway epithelial cells might contribute to disease persistence. Since integrins had recently been identified as the principal family of receptors for components of the extracellular matrix, it seemed reasonable to spend a sabbatical year learning something about these receptors. I was especially interested in integrins on pulmonary epithelial cells, since these are the most numerous cells in the lung and are in direct contact with the external environment. Not coincidentally, my brief foray into cell biology prior to joining Robert's lab utilized polarized cultures of airway epithelial cells as an experimental model system.

Understandably, most of what was known about integrins at that time was based on studies in nonadherent cells (platelets and leukocytes) and in fibroblasts (1,2,6). During my sabbatical year in Robert's lab, David Erle, Robert, and I attempted to identify new members of the integrin α and β subunit families expressed in lung cells using what at the time seemed like a radical and risky approach—homology PCR. This approach, which of course soon became quite routine and with the completion of many mammalian genomes is now virtually obsolete, turned out to be very well suited to finding new integrins (7,8). The primers Robert designed were located in highly conserved regions that the now-solved crystal structure of an integrin demonstrate are critical for interaction of integrins with their ligands (9,10). David's principal focus was on using this approach to identify leukocyte integrins, and I focused on epithelial cells. Among the sequences we identified by PCR were the

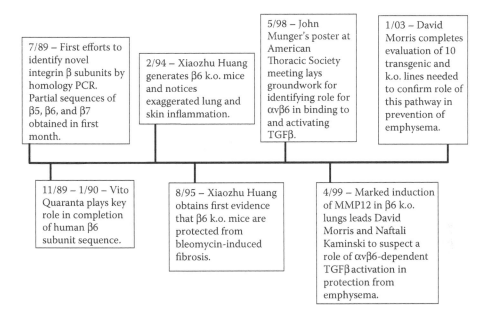

FIGURE 4.1   Timeline of discovery of significance of integrin-mediated activation of TGFβ.

β7 subunit (11,12), a component of lymphocyte integrins critical for homing to Peyer's patches and adhesion to intestinal epithelium, the α8 subunit, and the β5 (7), β6 (8), and α9 subunits (13), which have kept my laboratory busy since that time. Remarkably, within the first few weeks of starting this project, we used the first pair of degenerate β subunit primers Robert designed to obtain partial sequences of β5, β6, and β7. We had no idea at the time that our first truly original insight into the biological meaning of these three weeks of work would have to wait another nine years (Fig. 4.1).

In 1989 we were able to obtain only partial integrin subunit sequences by homology PCR. Because my lab had been studying primary cultures of guinea pig airway epithelial cells, these were the source of my initial cDNA. However, after extending one ~350 base novel β subunit sequence to ~1700 bases, I was unable to complete the 3′ and 5′ sequences. At that point we benefited from the openness and sharing that have been notable features of the integrin field since its inception. Vito Quaranta, who knew Robert Pytela from their post-doctoral fellowships in La Jolla, had also been working to clone epithelial integrin subunits and for that purpose had created a cDNA library from a pancreatic carcinoma cell line (FG-2) that he and David Cheresh had shown expressed a novel integrin β subunit they called βx (14). In a remarkable example of scientific generosity, Vito offered to use our 1700 base novel β subunit sequence (and another novel 360 base sequence) to screen his library. We agreed that he would send us any clones that hybridized with the longer fragment (which turned out to be β6) and that he would follow up on any clones that hybridized with the shorter fragment (which turned out to be β5). Within a few weeks, Vito sent us 20 overlapping cDNA clones that contained the full-length sequence of human β6! He also utilized the predicted cytoplasmic domain sequence to synthesize a cytoplasmic domain peptide, immunize rabbits, and generate a very useful polyclonal antiserum that was invaluable for identifying and characterizing the β6 subunit protein and for demonstrating that it forms an integrin heterodimer with the previously identified αv subunit. In the end, "βx" really described two new integrin β subunits, β5 and β6, which are both highly expressed partners of αv in FG-2 cells, merge together as a single broad band on SDS-PAGE, and together explain the functional effects ascribed to αvβx in the original description.

4.3   IDENTIFYING αvβ6 LIGANDS AND CELLULAR FUNCTIONS

Since we began our studies of the αvβ6 integrin from sequence identification, we were initially faced with the task of studying an orphan receptor, with no known ligands and no clear biological function. Since the largest group of known integrin ligands were components of the extracellular matrix, we started with the assumption that ECM components would also be the ligand(s) for αvβ6. In retrospect, while the resultant experiments were informative, many of our early findings were probably somewhat misleading as a consequence of this initial bias. Since Robert Pytela and others had successfully used affinity chromatography to identify ligands for other integrins, we began by passing surface labeled lysates from αvβ6-expressing cells (FG-2 cells from Vito Quaranto) over columns composed of ECM proteins (vitronectin, fibronectin, and type I collagen) cross-linked to Sepharose. Our initial pilot studies were performed with ECM–Sepharose columns generously provided by UCSF colleague Randy Kramer, who had made extra beads as part of a search (successful) for ligands for another new integrin, α7β1, and was happy to share them with us. From these studies, Mike Busk learned that αvβ6 could bind directly to fibronectin in an RGD- and divalent cation-dependent fashion and we thought we had identified αvβ6 as an alternative fibronectin receptor (15).

Over the next few years we expressed the full-length and a variety of mutant forms of the β6 subunit in a variety of mammalian cells and examined a wide range of cell behaviors in response to what we considered to be the principal αvβ6 ligand (fibronectin) (16–21). The major goal of these experiments was to try to understand why nature would go to the trouble of evolving more than one integrin receptor for fibronectin. Many of our results were, in retrospect, predictable based on the behavior of other members of the integrin family—results that were largely determined by the limitations of our experimental imagination. We did find one novel function of αvβ6: enhancement of cell growth in three-dimensional culture, an effect that depends on an 11 amino acid carboxyl terminal extension that is unique to the β6 subunit (16,20). However, we have yet to determine how this sequence leads to enhanced cell growth and why this effect is only manifested in three-dimensional growth environments.

4.4   IN VIVO FUNCTIONS OF αvβ6

The above in vitro experiments would never have led us to the most important function of αvβ6 or the identification of its preferred ligand. These insights came from our attempts to understand the initially puzzling phenotype of mice expressing a null mutation of the β6 subunit (and thus lacking αvβ6) that Xiaozhu Huang generated in the lab (22). The first thing we noticed about these mice was that homozygous knockout animals all had transient areas of baldness over the tops of their heads. These lesions were morphologically characterized by infiltration of the dermis with large numbers of macrophages. Examination of the lungs of these animals revealed increased numbers of alveolar macrophages as well, and scattered aggregates of lymphocytes and occasional eosinophils and neutrophils in the pulmonary interstitium. In both the skin and the lungs, this exaggerated inflammation depended on environmental triggers—minor local trauma in the skin and exposure to off-gassing from cage bedding material in the lungs. Thus, although we confirmed that αvβ6 itself was restricted in its expression to epithelial cells, the major in vivo phenotype was impaired resolution of inflammation, suggesting that this integrin plays a role in epithelial-mediated negative regulation of tissue inflammation. We also reported an association between knockout of the β6 subunit gene and a functional abnormality, airway hyperresponsiveness (AHR), commonly associated with asthma, the disease I had initially set out to study. This association seemed plausible, since airway hyper-responsiveness had been shown to be a downstream consequence of airway inflammation in several models, and the β6 knockout mice clearly had chronic airway inflammation. However, this observation was made in a mixed genetic background of 129 × C57bl/6 mice. When we subsequently bred these mice back to each pure genetic background, the association with AHR disappeared. However, wild-type 129 mice had substantially greater airway responsiveness than wild-type C57bl/6 mice. Since the β6 knockout allele that we selected homozygous knockout animals for was transmitted from an initial 129 chromosome, it is likely that the apparent AHR in β6 knockout mice was due to another 129-strain gene (or genes) in linkage disequilibrium with the β6 subunit gene on mouse chromosome 2.

The exaggerated inflammation in the lungs and skin of β6 knockout mice has been a consistent feature in every genetic background we have examined, and could be rescued by transgenic overexpression of the human β6 gene in lung epithelial cells of β6 knockout mice (23). We thus sought to identify how αvβ6 could negatively regulate inflammation. One possibility was that ligation of the integrin by ligands (e.g., fibronectin) that accumulate at sites of injury might signal epithelial cells to secrete one or more anti-inflammatory cytokines or decrease production of one or more pro-inflammatory cytokines. We therefore put considerable effort into attempting to identify αvβ6-mediated effects on expression or secretion of pro- and anti-inflammatory cytokines without meaningful success.

As a pulmonary physician, I was especially interested in understanding the basis of the pulmonary inflammation in these mice. Failing to identify the mechanism underlying the exaggerated lung inflammation, I thought at least we could take advantage of this phenotype to enhance the utility of existing murine models of chronic lung disease. As a first effort, we decided to examine the effects of intra-tracheal treatment with bleomycin, a commonly used model of pulmonary fibrosis. At that time, it was widely accepted that pulmonary fibrosis was a direct consequence of pulmonary inflammation, and bleomycin was well established as a potent stimulus to early inflammation and later fibrosis. We reasoned that the β6 knockout mice would likely develop exaggerated and prolonged inflammation, and would consequently develop more severe and perhaps persistent pulmonary fibrosis. As is often the case, the failure of this hypothesis turned out to be our biggest break in understanding the in vivo function of αvβ6. As we had predicted, β6 knockout mice did indeed develop enhanced and prolonged pulmonary inflammation. However, the big surprise was that these mice were dramatically protected from the subsequent development of pulmonary fibrosis (24).

4.5 MAKING CONNECTION TO TGFβ

The combination of exaggerated inflammation, but a virtual absence of consequent fibrosis, pointed very strongly toward the participation of one or more members of the transforming growth factor β (TGFβ) family of proteins downstream of αvβ6. TGFβ had been well established as a central mediator of tissue fibrosis in multiple organs, including the lung (25–27). Indeed, blocking antibodies against TGFβ (26), administration of molecules (e.g., decorin) capable of sequestering TGFβ (28), soluble forms of the high affinity TGFβII receptor (29), and inhibitors of TGFβ signaling (e.g., SMAD7 [30]) had all been shown to potently inhibit tissue fibrosis in various in vivo models. Furthermore, the striking tissue inflammation seen in TGFβ1 knockout mice strongly implicated TGFβ family members as important negative regulators of tissue inflammation (31).

We thus undertook a series of studies designed to identify effects of αvβ6 expression or ligation on expression of TGFβ family members and their receptors. As with our earlier effort to identify changes in expression of other pro- or anti-inflammatory cytokines downstream of αvβ6, these efforts were entirely unsuccessful. Again, our big break came through a very fruitful collaboration. Nearly two years after we began seeking ways to connect αvβ6 and TGFβ, I saw a poster at the American Thoracic Society meeting by another pulmonary physician, John Munger, showing that two members of the integrin family, αvβ1 and αvβ5, directly bind to an N-terminal fragment of the TGFβ gene product called the latency-associated peptide (LAP) (32). John had read our original description of exaggerated lung and skin inflammation in the β6 knockout mice, and even without knowing that these animals were protected from pulmonary fibrosis he considered the possibility, that had been completely lost on us, that αvβ6 might bind to LAP and activate latent complexes of TGFβ1. John knew that TGFβ1 and TGFβ3 LAP contained the linear tri-peptide RGD that serves as a recognition sequence for several members of the integrin family (2) (including αvβ6 [15,21]). However, although John was gratified that he had found two integrins that did indeed bind to LAP, his results were disappointing, since he was not able to identify αvβ6 binding to his LAP column and he was unable to show that either integrin that did bind led to TGFβ activation.

Fortunately (for my own involvement with this project), John had made one unpredictable mistake in the design of his initial experiments. Recognizing that αvβ6 is restricted in its expression to epithelial cells, John performed his initial experiments by surface labeling a lung epithelial cell line, A549, and passing the labeled lysate over an LAP column. However, A549 cells, despite their pulmonary epithelial origin, do not express αvβ6. John sent my lab LAP–Sepharose and stable TGFβ–luciferase reporter cells, and we sent him several β6-transfected cell lines and blocking antibodies. Together, our labs were able to rapidly determine that TGFβ1 LAP is a much better ligand for αvβ6 than any of the extracellular matrix proteins we had been studying, and that the interaction of αvβ6 with TGFβ1 LAP does indeed activate latent complexes (24).

One especially attractive aspect of this mechanism for activating TGFβ is that it appears to be very tightly spatially restricted. When αvβ6-expressing cell lines and TGFβ reporter cells were plated on opposite sides of microporous filters, little or no TGFβ activity was detectable (24). Furthermore, when we co-transfected NIH3T3 cells with both the β6 subunit and a TGFβ-luciferase reporter and then plated these cells at a series of densities, TGFβ activity was completely lost at densities that did not allow direct cell-to-cell contact. These data strongly suggest that αvβ6 can only present active TGFβ to receptors on adjacent cells in direct contact with the integrin-expressing cells, and that there is little or no release of free, active TGFβ by this process.

4.6   IDENTIFYING IN VIVO FUNCTIONS FOR αvβ6-DEPENDENT TGFβ ACTIVATION

Our initial efforts to identify the ligands and biologic functions of αvβ6 were hampered by the backward process we used to identify this integrin (beginning with sequence rather than function) and by the artificial experimental systems available at the start of these studies. In contrast, once we had mice lacking αvβ6, efforts to identify biologically significant functions were much more straightforward. After we knew that a principal in vivo function of αvβ6 was activation of latent TGFβ, it became possible to use the β6 knockout mice to screen for previously unexpected in vivo functions of both the integrin and of TGFβ. The first interesting and somewhat surprising example of this strategy grew out of our initial observations about protection from bleomycin-induced pulmonary fibrosis. Many pulmonary scientists are skeptical about the relevance of the bleomycin model to human pulmonary fibrosis because in a number of ways it more closely resembles the late consequences of another pulmonary disorder, acute lung injury. Indeed, occasional patients treated with bleomycin develop acute lung injury. In mice, if the animals are analyzed at 5 days after bleomycin treatment, rather than the 30–60 days needed to develop full-blown pulmonary fibrosis (at least in 129-strain mice), the pathology and functional impairments closely resemble the early phases of acute lung injury. One prominent feature at this time point is a marked increase in extravasation of protein-rich fluid into the alveolar spaces (pulmonary edema). Since acute lung injury (even more so than for pulmonary fibrosis) was thought to be a direct consequence of the toxic products released from recruited leukocytes, we were confident that the β6 knockout mice would develop exaggerated pulmonary edema. However, rather than demonstrating the expected enhancement of pulmonary edema in response to bleomycin, β6 knockout mice were completely protected (33).

One potentially trivial explanation for the protection of β6 knockout mice from bleomycin-induced pulmonary edema was compensatory induction of nonspecific protective pathways in response to chronic low-level pulmonary inflammation. To address this possibility, we examined the effects of treatment with a TGFβ blocking drug (a TGFβRII-immunoglobulin Fc chimera) on bleomycin-induced pulmonary edema in wild-type mice. Since wild-type mice did not have pre-existing lung inflammation, protection by blocking TGFβ would both rule out this trivial explanation for our finding and directly implicate TGFβ as an effector of pulmonary edema. The result was quite clear cut: blocking TGFβ was as effective as loss of the β6 subunit in protecting mice from bleomycin-induced pulmonary edema. Furthermore, TGFβ blockade was equally effective in blocking the pulmonary edema induced by bacterial endotoxin (33).

The identification of αvβ6-dependent TGFβ activation as a key step in the induction of pulmonary edema begged the question of how TGFβ produced this effect. One possible mechanism had previously been described by Peter Vincent's laboratory. In an in vitro cell culture system, they had shown that TGFβ can directly increase the permeability across endothelial monolayers (34). Since the alveolar epithelium is also an important barrier to alveolar flooding, Jean-Francois Pittet examined the effects of TGFβ on the permeability across confluent polarized cultures of rat alveolar epithelial cells and found an increase in epithelial permeability as well (33). Finally, since pulmonary edema depends on a relative increase in fluid movement into the alveolar spaces compared to the rate of re-absorption across the epithelium, Jean-Francois also evaluated the effects of TGFβ on epithelial sodium re-absorption, a critical rate-limiting step in fluid movement out of the alveoli. TGFβ also reduced Na uptake, apparently by decreasing the expression of the heterotrimeric Na channel, ENaC, on the apical surfaces of these cells (35). Thus, active TGFβ, not previously known to be a contributor to the development of pulmonary edema, actually has three parallel effects on the cells that normally protect the alveoli from flooding, and these likely work in concert to promote pulmonary edema.

4.7   IDENTIFICATION OF ROLE FOR αvβ6-DEPENDENT TGFβ ACTIVATION IN PREVENTION OF PULMONARY EMPHYSEMA

To more broadly search for potential roles for αvβ6-dependent TGFβ–activation in vivo, Naftali Kaminski performed expression array studies on the lungs of wild-type and β6 knockout mice (36). For this purpose, we utilized Affymetrix microarrays with probes recognizing ~6000 murine genes. A relatively small number of genes were expressed at dramatically higher levels in β6 knockout lungs. The most induced gene, matrix metalloprotease 12 (MMP12) was especially interesting to us because its expression is restricted to macrophages. That allowed us to distinguish real induction of expression from an artefactual consequence of cellular recruitment. We knew from cell count data that β6 knockout animals had approximately twice as many pulmonary macrophages as wild-type mice, but MMP12 was induced more than 20-fold. Indeed, by Taqman analysis, alveolar macrophages obtained from β6 knockout mice expressed approximately 200-fold higher levels of MMP12 mRNA than did macrophages from wild-type animals (37).

Steve Shapiro's lab had previously shown that MMP12 is induced by cigarette smoke in alveolar macrophages and that mice lacking MMP12 were protected from the emphysema caused by chronic exposure to cigarette smoke (38). This finding raised the possibility that β6 knockout mice, whose macrophages chronically produced large amounts of MMP12, might develop spontaneous emphysema. At that point, we had been studying these mice for more than six years, and several pulmonary scientists and pathologists had looked at sections from the lungs of β6 knockout mice without noticing any evidence of emphysema. However, because of the high costs of housing mice at UCSF, we had always studied these animals within the first few months after birth. Since emphysema is a slowly progressive disease (e.g., cigarette smoke–induced emphysema in mice requires six months of chronic exposure to smoke), David Morris, then a post-doctoral fellow in the lab, reasoned that we might need to let the β6 knockout mice get older to detect the increases in alveolar size characteristic of emphysema. This turned out to be correct. At two months of age alveolar size was the same in wild-type and β6 knockout mice, but by six months β6 knockout alveoli were significantly larger, and by 14 months of age the emphysema in these animals was readily apparent (37). In a tour-de-force of animal breeding and patience, David showed that the emphysema in β6 knockout mice was completely dependent on MMP12 (MMP12 knockout mice were protected) and could be rescued by transgenic expression (in alveolar epithelial cells) only of forms of the β6 subunit capable of activating TGFβ (38). Moreover, he showed that inducible transgenic expression of active TGFβ in lung epithelial cells could also reverse MMP12 induction. Thus, the results of an initial experiment that could only be described as a fishing expedition allowed us to identify an unexpected link between an epithelial integrin, TGFβ, and its downstream signaling pathways and protection from pulmonary emphysema. Emphysema occurs in only approximately one in eight heavy cigarette smokers, and the causes of this differential sensitivity are largely unknown. Each of the molecules in this pathway can now be added to the list of candidates to help explain this heterogeneity in the development of emphysema.

4.8 ROLES OF αvβ6-DEPENDENT TGFβ ACTIVATION IN OTHER ORGANS

One benefit of modern biology is that it is relatively easy to make contact with and share reagents and ideas with scientists around the country and the world. We have now shared β6 knockout mice with more than 20 other laboratories in an effort to most efficiently determine which in vivo roles of TGFβ depend on αvβ6-dependent activation and which do not. Thus far these collaborative efforts have led to the identification of two additional roles for this pathway. One of these came from a collaboration with Pamela Knight and Hugh Miller, who had shown that expression of a specific mast cell granule chymase (mast cell protease 1, MCP-1) that is induced in intestinal mucosal mast cells in response to parasitic infection, was highly responsive to TGFβ. Pam came to my lab for a few weeks to infect β6 knockout mice with the nematode Nippostrongyloidis and showed that not only was the induction of MCP-1 dramatically inhibited, so was the marked increase in mucosal mast cell number that normally accompanies nematode infection (39). Thus, αvβ6-dependent TGFβ activation plays a key role in mucosal mast cell homeostasis.

TABLE 4.1
Demonstrated In Vivo Functions of αvβ6-Mediated TGF-β Activation

Function	Experimental Observations	Reference
Resolution of epithelial inflammation	Skin and lung inflammation in β6 knockout mice	Huang et al. 1996 (22)
Pulmonary fibrosis	Protection from bleomycin-induced pulmonary in β6 knockout mice	Munger et al. 1999 (24)
Renal fibrosis	Protection from tubulointerstitial fibrosis induced by unilateral ureteral obstruction in β6 knockout mice	Ma et al. 2003 (40)
Acute lung injury	Protection from pulmonary edema in response to intracheal bleomycin or bacterial endotoxin in β6 knockout mice and/or by TGFβ blockade	Pittet et al. 2001 (33)
Intestinal mast cell homeostasis	Absence of mast cell hyperplasia and mast cell protease-1 induction in response to Nippostrongyloidis infection in β6 knockout mice.	Knight et al. 2002 (39)
Protection from pulmonary emphysema	β6 knockout mice develop spontaneous emphysema that is prevented by loss of MMP12, transgenic expression of human b6 in alveolar epithelial cells or inducible pulmonary expression of active TGFβ.	Morris et al. 2003 (37)

A second new role for αvβ6-dependent TGFβ activation was identified by Li-Jun Ma and Agnes Fogo, who used these animals in a model of tubulointerstitial renal fibrosis induced by unilateral ureteral obstruction (UUO). As we observed in pulmonary fibrosis, β6 knockout mice were significantly protected from UUO-induced renal fibrosis (40). It thus seems likely that this pathway could play a role in tissue fibrosis in other epithelial organs in which αvβ6 is expressed (Table 4.1).

4.9   SUMMARY

I suspect that the identification of the in vivo significance of αvβ6-dependent TGFβ activation followed a relatively common pattern of scientific discovery. The initial line of investigation that eventually led to this discovery was based on a hypothesis that 14 years of inquiry has not provided any support for: that ligation of epithelial integrins by components of the extracellular matrix is responsible for the persistence of chronic lung diseases such as asthma. In fact, this line of investigation has thus far provided little direct insight into the mechanisms underlying persistence of asthma, the disease that motivated me to pursue it. However, probably more as a consequence of serendipity than initial insight, the starting hypothesis did have a buried grain of truth. It does now appear that at least one epithelial integrin (αvβ6) makes important contributions to at least three chronic lung diseases and that interaction with ligand is critical in each case. If we had made more progress on the initial hypothesis, or had doggedly pursued it despite lack of progress, we certainly would have missed the more important identification of a novel mechanism by which an integrin can alter the activity of its ligand, and thus of its surrounding milieu, in addition to simply detecting and responding to that milieu. As is often the case, identification of that process was done by proceeding through a series of misinterpretations and failed experiments and was ultimately successful due to equal measures of good luck and extensive help from other labs.

REFERENCES

1. R. O. Hynes. "Integrins: A Family of Cell Surface Receptors." *Cell* 48, no. 4 (1987): 549–54.
2. E. Ruoslahti, and M. D. Pierschbacher. "New Perspectives in Cell Adhesion: RGD and Integrins." *Science* 238, no. 4826 (1987): 491–97.
3. E. A. Clark, and J. S. Brugge. "Integrins and Signal Transduction Pathways: The Road Taken." *Science* 268, no. 5208 (1995): 233–39.
4. R. Pytela, M. D. Pierschbacher, and E. Ruoslahti. "A 125/115-kDa Cell Surface Receptor Specific for Vitronectin Interacts with the Arginine-Glycine-Aspartic Acid Adhesion Sequence Derived from Fibronectin." *Proc. Natl. Acad. Sci. USA* 82 (1985): 5766–70.
5. R. Pytela, M. D. Pierschbacher, and E. Ruoslahti. "Identification and Isolation of a 140 kd Cell Surface Glycoprotein with Properties Expected of a Fibronectin Receptor." *Cell* 40 (1985): 191–98.
6. M. E. Hemler, C. Huang, and L. Schwarz. "The VLA Protein Family. Characterization of Five Distinct Cell Surface Heterodimers Each with a Common 130,000 Molecular Weight Beta Subunit." *J Biol Chem* 262, no. 7 (1987): 3300–9.
7. D. J. Erle, D. Sheppard, J. Breuss, C. Rüegg, and R. Pytela. "Novel Integrin α and β Subunit cDNAs Identified Using the Polymerase Chain Reaction." *Am J Resp Cell Mol Biol* 5 (1991): 170–77.
8. D. Sheppard, C. Rozzo, L. Starr, V. Quaranta, D. J. Erle, and R. Pytela. "Complete Amino Acid Sequence of a Novel Integrin β Subunit (β6) Identified in Epithelial Cells Using the Polymerase Chain Reaction." *J Biol Chem* 265, no. 20 (1990): 11502–7.
9. J. P. Xiong, T. Stehle, R. Zhang, A. Joachimiak, M. Frech, S. L. Goodman, and M. A. Arnaout. "Crystal Structure of the Extracellular Segment of Integrin Alpha Vbeta3 in Complex with an Arg-Gly-Asp Ligand." *Science* 296, no. 5565 (2002): 151–55.
10. J. P. Xiong, T. Stehle, B. Diefenbach, R. Zhang, R. Dunker, D. L. Scott, A. Joachimiak, S. L. Goodman, and M. A. Arnaout. "Crystal Structure of the Extracellular Segment of Integrin Alpha Vbeta3." *Science* 294, no. 5541 (2001): 339–45.
11. C. Ruegg, A. A. Postigo, E. E. Sikorski, E. C. Butcher, R. Pytela, and D. J. Erle. "Role of Integrin Alpha 4 Beta 7/Alpha 4 Beta P in Lymphocyte Adherence to Fibronectin and VCAM-1 and in Homotypic Cell Clustering." *J Cell Biol* 117, no. 1 (1992): 179–89.
12. D. J. Erle, C. Ruegg, D. Sheppard, and R. Pytela. "Complete Amino Acid Sequence of an Integrin Beta Subunit (Beta 7) Identified in Leukocytes." *J Biol Chem* 266, no. 17 (1991): 11009–16.
13. E. L. Palmer, E. L., C. Ruegg, R. Ferrando, R. Pytela, and D. Sheppard. "Sequence and Tissue Distribution of the Integrin Alpha 9 Subunit, a Novel Partner of Beta 1 That Is Widely Distributed in Epithelia and Muscle." *J Cell Biol* 123, no. 5 (1993): 1289–97.
14. D. A. Cheresh, J. W. Smith, H. M. Cooper, and V. Quaranta. "A Novel Vitronectin Receptor Integrin (Alpha v Beta x) Is Responsible for Distinct Adhesive Properties of Carcinoma Cells." *Cell* 57, no. 1 (1989): 59–69.
15. M. Busk, R. Pytela, and D. Sheppard. "Characterization of the Integrin αvβ6 as a Fibronectin-Binding Protein." *J Biol Chem* 267 (1992): 5790–96.

16. M. Agrez, A. Chen, R. Cone, R. Pytela, and D. Sheppard. "The Alpha v Beta 6 Integrin Promotes Proliferation of Colon Carcinoma Cells Through a Unique Region of the Beta 6 Cytoplasmic Domain." *J Cell Biol* 127 (1994): 547–56.

17. R. I. Cone, A. Weinacker, A. Chen, and D. Sheppard. "Effects of Beta Subunit Cytoplasmic Domain Deletions on the Recruitment of the Integrin Alpha v Beta 6 to Focal Contacts." *Cell Adhes Commun* 2 (1994): 101–13.

18. A. Weinacker, A. Chen, M. Agrez, R. I. Cone, S. Nishimura, E. Wayner, R. Pytela, and D. Sheppard. "Role of the Integrin Alpha v Beta 6 in Cell Attachment to Fibronectin. Heterologous Expression of Intact and Secreted Forms of the Receptor." *J Biol Chem* 269, no. 9 (1994): 6940–48.

19. J. Chen, T. Maeda, K. Sekiguchi, and D. Sheppard. "Distinct Structural Requirements for Interaction of the Integrins Alpha 5 Beta 1, Alpha v Beta 5, and Alpha v Beta 6 with the Central Cell Binding Domain in Fibronectin." *Cell Adhes Commun* 4, no. 4–5 (1996): 237–50.

20. R. B. Dixit, A. Chen, J. Chen, and D. Sheppard. "Identification of a Sequence within the Integrin Beta6 Subunit Cytoplasmic Domain That Is Required to Support the Specific Effect of Alpha v Beta 6 on Proliferation in Three-Dimensional Culture." *J Biol Chem* 271 (1996): 25976–80.

21. Y. Yokosaki, Monis, H., Chen, J. and D. Sheppard. "Differential Effects of the Integrins $\alpha 9\beta 1$, $\alpha v\beta 3$, and $\alpha v\beta 6$ on Cell Proliferative Responses to Tenascin. Roles of the β Subunit Extracellular and Cytoplasmic Domains." *J Biol Chem* 271 (1996): 24144–50.

22. X. Z. Huang, J. F. Wu, D. Cass, D. J. Erle, D. Corry, S. G. Young, R. V. J. Farese, and D. Sheppard. "Inactivation of the Integrin Beta 6 Subunit Gene Reveals a Role of Epithelial Integrins in Regulating Inflammation in the Lung and Skin." *J Cell Biol* 133 (1996): 921–28.

23. X. Z. Huang, J. F. Wu, W. Zhu, R. Pytela, and D. Sheppard. "Expression of the Human Integrin Beta6 Subunit in Alveolar Type II Cells and Bronchiolar Epithelial Cells Reverses Lung Inflammation in Beta6 Knockout Mice." *Am J Respir Cell Mol Biol* 19 (1998): 636–42.

24. J. S. Munger, X. Z. Huang, H. Kawakatsu, M. J. D. Griffiths, S. L. Dalton, J. F. Wu, J. F. Pittet, et al. "The Integrin $\alpha v\beta 6$ Binds and Activates Latent TGFβ1: A Mechanism for Regulating Pulmonary Inflammation and Fibrosis." *Cell* 96 (1999): 319–28.

25. A. Roberts, M. B. Sporn, R. K. Assoian, J. M. Smith, N. S. Roche, L. M. Wakefield, U. I. Heine, et al. "Transforming Growth Factor Type Beta: Rapid Induction of Fibrosis and Angiogenesis In Vivo and Stimulation of Collagen Formation In Vitro." *Proceedings of the National Academy of Sciences of the United States of America* 83 (1986): 4167–71.

26. S. Giri, D. M. Hyde, and M. A. Hollinger. "Effect of Antibody to Transforming Growth Factor-Beta on Bleomycin Induced Accumulation of Lung Collagen in Mice." *Thorax* 48 (1993): 959–66.

27. A. B. Roberts, and M. B. Sporn. "Physiological Actions and Clinical Applications of Transforming Growth Factor-Beta (TGF-beta)." *Growth Factors* 8, no. 1 (1993): 1–9.

28. W. A. Border, N. A. Noble, T. Yamamoto, J. R. Harper, Y. Yamaguchi, M. D. Pierschbacher, and E. Ruoslahti. "Natural Inhibitor of Transforming Growth Factor-Beta Protects Against Scarring in Experimental Kidney Disease." *Nature* 360, no. 6402 (1992): 361–64.

29. Q. Wang, Y. Wang, D. M. Hyde, P. J. Gotwals, V. E. Koteliansky, S. T. Ryan, and S. N. Giri. "Reduction of Bleomycin Induced Lung Fibrosis by Transforming Growth Factor Beta Soluble Receptor in Hamsters." *Thorax* 54, no. 9 (1999): 805–12.

30. A. Nakao, M. Fujii, R. Matsumura, K. Kumano, Y. Saito, K. Miyazono, and I. Iwamoto. "Transient Gene Transfer and Expression of Smad7 Prevents Bleomycin-Induced Lung Fibrosis in Mice." *J Clin Invest* 104, no. 1 (1999): 5–11.

31. M. M. Shull, I. Ormsby, A. B. Kier, S. Pawlowski, R. J. Diebold, M. Yin, R. Allen, et al. "Targeted Disruption of the Mouse Transforming Growth Factor-Beta 1 Gene Results in Multifocal Inflammatory Disease." *Nature* 359 (1992): 693–99.
32. J. S. Munger, J. G. Harpel, F. G. Giancotti, and D. B. Rifkin. "Interactions Between Growth Factors and Integrins: Latent Forms of Transforming Growth Factor-β Are Ligands for the Integrin αvβ1." *Mol Biol Cell* 9 (1998): 2627–38.
33. J. F. Pittet, M. J. Griffiths, T. Geiser, N. Kaminski, S. L. Dalton, X. Huang, L. A. Brown, et al. "TGF-Beta Is a Critical Mediator of Acute Lung Injury." *J Clin Invest* 107, no. 12 (2001): 1537–44.
34. V. I. Hurst, P. L. Goldberg, F. L. Minnear, R. L. Heimark, and P. A. Vincent. "Rearrangement of Adherens Junctions by Transforming Growth Factor-Beta1: Role of Contraction." *Am J Physiol* 276, no. 4 Pt 1 (1999): L582–95.
35. J. Frank, J. Roux, H. Kawakatsu, G. Su, A. Dagenais, Y. Berthiaume, M. Howard, et al. "TGF-β1 Decreases αENaC Expression and Alveolar Epithelial Vectorial Sodium and Fluid Transport via an ERK 1/2-Dependent Mechanism." *J Biol Chem* 278, no. 45 (2003): 43939–50.
36. N. Kaminski, J. D. Allard, J. F. Pittet, F. Zuo, M. J. Griffiths, D. Morris, X. Huang, D. Sheppard, and R. A. Heller. "Global Analysis of Gene Expression in Pulmonary Fibrosis Reveals Distinct Programs Regulating Lung Inflammation and Fibrosis." *Proc Natl Acad Sci U S A* 97, no. 4 (2000): 1778–83.
37. D. G. Morris, X. Huang, N. Kaminski, Y. Wang, S. D. Shapiro, G. Dolganov, A. Glick, and D. Sheppard. "Loss of Integrin Alpha v Beta 6-Mediated TGF-Beta Activation Causes MMP12-Dependent Emphysema." *Nature* 422, no. 6928 (2003): 169–73.
38. R. D. Hautamaki, D. K. Kobayashi, R. M. Senior, and S. D. Shapiro. "Requirement for Macrophage Elastase for Cigarette Smoke-Induced Emphysema in Mice." *Science* 277 (1997): 2002–4.
39. P. A. Knight, S. H. Wright, J. K. Brown, X. Huang, D. Sheppard, and H. R. Miller. "Enteric Expression of the Integrin Alpha (v) Beta (6) Is Essential for Nematode-Induced Mucosal Mast Cell Hyperplasia and Expression of the Granule Chymase, Mouse Mast Cell Protease-1." *Am J Pathol* 161, no. 3 (2002): 771–79.
40. L.-J. Ma, H. Yang, H. Gaspert, G. Carlesso, J. Davidson, D. Sheppard, and A. Fogo. "Transforming Growth Factor-β Dependent and Independent Pathways of Induction of Tubulointerstitial Fibrosis in β6-/- Mice." *Am J Pathol* 163 (2003): 1261–73.

5 Purification of Basic Fibroblast Growth Factor by Heparin Affinity

Michael Klagsbrun and Yuen Shing

CONTENTS

5.1 INTRODUCTION

In an attempt to determine how cell proliferation was regulated, a number of laboratories began to isolate peptide growth factors some 25–30 years ago. Several groups, including our own, were focused on vascular growth factors, proteins that could stimulate the proliferation of endothelial cells in vitro and angiogenesis in vivo. It took us more than five years to purify a mitogen for 3T3 and endothelial cells, first from bovine cartilage, then rat chondrosarcoma, and finally human hepatoma cells. This mitogen turned out to be basic fibroblast growth factor (basic FGF), first described by Denis Gospodarowicz but not yet fully purified. The breakthrough in successful purification came about when we found that our mitogen had a very strong affinity for heparin. The introduction of heparin affinity chromatography resulted in a rapid two-step purification scheme with a high yield. It is now recognized that many growth factors are heparin binding and that these interactions are of biological significance.

5.2 CARTILAGE-DERIVED GROWTH FACTOR

Growth factor research began in our laboratory as early as 1975. Platelet-derived growth factor (PDGF) (1) and fibroblast growth factor (FGF) had just been identified

(2). It seemed that the growth factor field was rapidly expanding into an exciting area of research. At that time there was an intensive attempt in our department to purify angiogenesis inhibitors from cartilage, an avascular tissue that was readily available from local slaughterhouses. We reasoned that young calf cartilage might be a good source of growth factors since this was a tissue that was growing rapidly in newborn calves. To undertake this analysis, it was necessary to extract the cartilage and to develop a biological assay to measure growth factor activity.

The calf scapulae were digested overnight with protease-free collagenase to liberate the chondrocytes. The chondrocytes in turn were lysed to prepare the starting material. We used a growth factor assay in which BALB/c 3T3 cells were grown in 96 well plates to confluence and allowed to remain without further media change to become quiescent. Incorporation of tritiated thymidine into DNA was measured in a scintillation counter. The quiescent cells usually incorporated about 1000–2000 CPM. However, the addition of aliquots of cartilage lysate to the 3T3 cells resulted in incorporation of up to 100,000–150,000 CPM, indicating the presence of potent growth factor activity. Isolated bovine chondrocytes were also used as targets for the cartilage-derived growth factor (CDGF). These findings began our growth factor research program (3).

An attempt was made to purify CDGF. Initial analysis by gel filtration chromatography indicated that CDGF had a molecular mass of 11–13 kDa and was a cationic protein. Chondrocyte cell fractionation indicated that CDGF was in part nuclear. A complex biochemical purification scheme including ion exchange chromatography, preparative isoelectric focusing, and gel filtration chromatography yielded a pure protein upon SDS gel chromatography and Coomassie blue staining with a molecular mass of 16,400 kDa (4). However, the CDGF was effective only at 3–5 µg/mL, raising suspicion about this protein's purity since growth factors are typically active in the low ng/mL level. Further purification was necessary.

5.3 CHONDROSARCOMA-DERIVED GROWTH FACTOR

It became clear that obtaining large amounts of cartilage was rate limiting in terms of purifying sufficient amounts of CDGF for sequencing. To circumvent this problem, we obtained in 1981 the Swarm rat cartilage tumor, chondrosarcoma. This tumor could be transplanted from rat to rat yielding a large supply of tumor tissue. Tumor extracts were purified by the methods of ion exchange and gel filtration as developed for the purification of CDGF. A chondrosarcoma-derived growth factor (ChDGF) was identified that appeared to have the same biochemical properties as CDGF and was most probably the same protein.

5.4 PURIFICATION BY HEPARIN AFFINITY

It became apparent that our purification schemes for CDGF and ChDGF were not efficient. The yield from these procedures was very poor, less than 1%, and there was never enough protein remaining for sequencing, which at that time required microgram amounts. At about this time several investigators in our department were

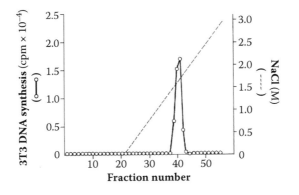

FIGURE 5.1 Heparin affinity chromatography. A chondrosarcoma tumor extract, partially purified by BioRex 70 cation exchange chromatography, was applied to a heparin–Sepharose column. Fractions were collected and assayed for the ability to stimulate tritiated thymidine incorporation into DNA. ChDGF (basic FGF) elutes at 1.8 M NaCl.

using heparin affinity columns for purification of cartilage-derived angiogenesis inhibitors. We reasoned that since CDGF and ChDGF were cationic, they might adhere to the negatively charged heparin. We took a column that had been used in the cartilage-derived angiogenesis inhibitor studies and eluted it with 0.5 M NaCl and then 2 M NaCl. To our surprise, potent CDGF activity was eluted at 2 M, but not at 0.5 M NaCl, suggesting that the strong binding of CDGF was much more than just a cation exchange effect. Indeed, we found that PDGF, a highly basic growth factor with a pI of about 10, eluted at 0.5 M NaCl as do most cationic proteins. We next set up NaCl elution gradients. Both CDGF and ChDGF were eluted with about 1.8 M NaCl (Fig. 5.1). CDGF and ChDGF did not show similar strong binding to chondroitin sulfate indicating heparin specificity. These results suggested strongly that CDGF and ChDGF had very strong affinities for heparin. Importantly, heparin affinity was a conduit to rapid purification and high yields. In fact, we were able to purify ChDGF to homogeneity in only two steps, a Bio-Rex 70 cation exchange column to remove bulk protein and one heparin column with an elution gradient of 0.1–2.0 M NaCl (5). SDS-PAGE analysis after heparin chromatography showed only one silver-stained ChDGF band with a molecular mass of about 18 kDa (Fig. 5.2). Subsequently, CDGF was also purified by heparin affinity as an 18 kDa protein and this purification was achieved by heparin affinity alone (6).

To test whether ChDGF/CDGF had any potential angiogenesis activity, we collaborated with the Folkman lab, which had recently isolated endothelial cell (EC) clones (7). The fractions eluted from the heparin affinity columns used to purify ChDGF in the 3T3 cell assay were added to EC. The EC elution profiles were found to overlap with the 3T3 profiles. ChDGF stimulated the number of EC about twofold. Subsequently, we showed that a pure single band ChDGF stimulated angiogenesis in the cornea assay (8). When we looked back on these results, we realized that we had been one of the first, if not the first, to purify an angiogenesis factor to homogeneity.

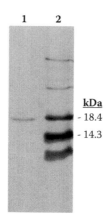

FIGURE 5.2 SDS PAGE gel electrophoresis and silver stain. Heparin-affinity purified ChDGF is a single protein with a molecular mass of 18 kDa.

5.5 HEPARIN-BINDING GROWTH FACTOR IS EQUIVALENT TO BASIC FGF

The term fibroblast growth factor (FGF) was first introduced in 1974 by Denis Gospodarowicz to describe a 13.3-kDa cationic protein (pI of 9.6) found in pituitary and brain that stimulated the division of 3T3 fibroblasts. FGF was subsequently demonstrated to be an EC mitogen. Later, it was shown that there are two related genes, acidic FGF and basic FGF. Interestingly, acidic FGF, despite being anionic, had a strong affinity for heparin and eluted at 1.1 M NaCl. Despite a number of early reports, basic FGF was not purified to homogeneity until 1985 (9). For FGF reviews see (10–13).

In 1984, we found that human hepatoma cells were an excellent source of our heparin-binding growth factor activity and that, furthermore, these cells could be grown in huge amounts in fermenters at the Monsanto Company. Accordingly, we began large scale purification of what we called human hepatoma-derived growth factor (HDGF, equivalent to CDGF, ChDGF). Antibodies to basic FGF cross-reacted with HDGF, and furthermore, HDGF was found to contain peptide sequences homologous to sequences of basic FGF (14). As it turned out, ChDGF, CDGF, HDGF, and over 30 other growth factors, when newly analyzed for heparin affinity, were actually found to be either basic or acidic FGF (10). We had problems in rapidly sequencing basic FGF because our HDGF preparation was N-terminally blocked. Our purification was carried out under neutral conditions and it turned out that we had isolated a 154-amino acid protein. However, several other labs used acid extraction methods that inadvertently cleaved off eight N-terminal amino acids, enabling them to determine the N-terminal sequence by Edman degradation (9).

5.6 HEPARIN AFFINITY BIOLOGY

We also noticed that heparin affinity was a biologically relevant phenomenon (15). Many growth factors were subsequently found to bind heparin and were classified as heparin-binding growth factors (HBGFs). These included vascular endothelial growth factor (VEGF), heparin-binding EGF (HB-EGF), hepatocyte growth factor (HGF), and many others. Many cell types express heparin sulfate proteoglycans (HSPG) on the cell surface and bind HBGFs. An example of biological significance is the HSPG requirement for basic FGF to bind to its FGF receptor (16). In 1987, Israel Vlodavsky came to our laboratory on sabbatical. He had been a pioneer in the preparation of extracellular matrix (ECM) and had shown that ECM could support the proliferation of fibroblasts and EC. We found that the EC ECM was a rich source of basic FGF, that the basic FGF was bound to HSPG, and that this insoluble form of basic FGF could support EC proliferation (17). Heparanase or excess heparin could release the basic FGF from ECM. These results suggested that ECM basic FGF was a storage form that could be released when needed, perhaps as a response to injury. In addition, stored basic FGF could also be released from tumors by tumor-derived proteinases and heparanases, thereby enhancing angiogenesis.

5.7   BASIC FGF TODAY

Whereas it took us months to purify microgram amounts, recombinant technology has made it possible to generate unlimited amounts of basic FGF. These days basic FGF is used in a variety of ways. For example, it is a standard component of EC culture media. As an angiogenesis factor, basic FGF has efficacy in treating wounds, ischemia and stroke, and other pathological conditions. Angiogenesis inhibitors are tested for the ability to inhibit basic FGF and VEGF-mediated angiogenesis in cornea, chorioallantoic membranes and mouse models. Antibodies to basic FGF are used diagnostically to monitor cancer patients, with urine levels being high in those with cancer and lower in patients in remission (18). Basic FGF also has nonvascular functions, such as neuronal protection.

In summary, the efforts of many laboratories led to the purification of basic and acidic FGF, the cloning of their genes, and identification of FGF receptors. In addition, it has been discovered that acidic and basic FGF belong to a large family comprising over 20 members (reviewed in 19). Availability of these FGF family members in pure form has greatly facilitated our ability to determine the biological properties and to understand the significance of these growth factors.

REFERENCES

1. R. Ross, J. Glomset, B. Kariya, and L. Harker. "A Platelet-Dependent Serum Factor That Stimulates the Proliferation of Arterial Smooth Muscle Cells In Vitro." *Proc Natl Acad Sci USA* 71 (1974): 1207–10.
2. D. Gospodarowicz. "Localisation of a Fibroblast Growth Factor and Its Effect Alone and with Hydrocortisone on 3T3 Cell Growth." *Nature* 249 (1974): 123–27.
3. M. Klagsbrun, R. Levenson, R. Langer, S. Smith, and C. Lillehei. "The Stimulation of DNA Synthesis and Cell Division in Chondrocytes and 3T3 Cells by a Growth Factor Isolated from Cartilage." *Exp Cell Res* 105 (1977): 99–108.
4. M. Klagsbrun and S. Smith. "Purification of a Cartilage-Derived Growth Factor." *J Biol Chem* 255 (1980): 10859–66.

5. Y. Shing, J. Folkman, R. Sullivan, C. Butterfield, J. Murray, and M. Klagsbrun. "Heparin Affinity: Purification of a Tumor-Derived Capillary Endothelial Cell Growth Factor." *Science* 223 (1984): 1296–99.

6. R. Sullivan and M. Klagsbrun. "Purification of Cartilage-Derived Growth Factor by Heparin Affinity Chromatography." *J Biol Chem* 260 (1985): 2399–2403.

7. J. Folkman, C. C. Haudenschild, and B. R. Zetter. "Long-Term Culture of Capillary Endothelial Cells." *Proc Natl Acad Sci USA* 76 (1979): 5217–21.

8. Y. Shing, J. Folkman, C. Haudenschild, D. Lund, R. Crum, and M. Klagsbrun. "Angiogenesis Is Stimulated by a Tumor-Derived Endothelial Cell Growth Factor." *J Cell Biochem* 29 (1985): 275–87.

9. F. Esch, A. Baird, N. Ling, N. Ueno, F. Hill, L. Denoroy, R. Klepper, D. Gospodarowicz, P. Bohlen, and R. Guillemin. "Primary Structure of Bovine Pituitary Basic Fibroblast Growth Factor (FGF) and Comparison with the Amino-Terminal Sequence of Bovine Brain Acidic FGF." *Proc Natl Acad Sci USA* 82, Oct (1985): 6507–11.

10. J. Folkman and M. Klagsbrun. "Angiogenesis Factors." *Science* 235 (1987): 442–47.

11. D. Gospodarowicz, N. Ferrara, L. Schweigerer, and G. Neufeld. "Structural Characterization and Biological Functions of Fibroblast Growth Factor." *Endocr Rev* 8 (1987): 95–114.

12. A. Baird and P. A. Walicke. "Fibroblast Growth Factors." *Br Med Bull* 45 (1989): 438–52.

13. M. Klagsbrun. "The Fibroblast Growth Factor Family: Structural and Biological Properties." *Prog Growth Factor Res* 1 (1989): 207–35.

14. M. Klagsbrun, J. Sasse, R. Sullivan, S. Smith, and J. A. Smith. "Human Tumor Cells Synthesize an Endothelial Cell Growth Factor That Is Structurally Related to Basic Fibroblast Growth Factor." *Proc Natl Acad Sci USA* 83 (1986): 2448–52.

15. M. Klagsbrun. "The Affinity of Fibroblast Growth Factors (FGFs) for Heparin: FGF-Heparan Sulfate Interactions in Cells and Extracellular Matrix." *Curr Opin Cell Biol* 2 (1990): 857–63.

16. A. Yayon, M. Klagsbrun, J. D. Esko, P. Leder, and D. M. Ornitz. "Cell Surface, Heparin-Like Molecules Are Required for Binding of Basic Fibroblast Growth Factor to Its High Affinity Receptor." *Cell* 64 (1991): 841–48.

17. I. Vlodavsky, J. Folkman, R. Sullivan, R. Fridman, R. Ishai-Michaeli, J. Sasse, and M. Klagsbrun. "Endothelial Cell-Derived Basic Fibroblast Growth Factor: Synthesis and Deposition into Subendothelial Extracellular Matrix." *Proc Natl Acad Sci USA* 84 (1987): 2292–96.

18. M. Nguyen, H. Watanabe, A. E. Budson, J. P. Richie, D. F. Hayes, and J. Folkman. "Elevated Levels of an Angiogenic Peptide, Basic Fibroblast Growth Factor, in the Urine of Patients with a Wide Spectrum of Cancers." *J Natl Cancer Inst* 86 (1994): 356–61.

19. D. M. Ornitz and N. Itoh. "Fibroblast Growth Factors." *Genome Biol* 2 (2001): 3005.1–12.

6 How It Came to Pass That Interferon Was Used as an Angiogenesis Inhibitor

Bruce R. Zetter

CONTENTS

I never intended to study endothelial cells. Having received a PhD in microbiology, I wanted to learn mammalian cell biology and spent two wonderful years in the lab of John Buchanan at M.I.T. studying the role of proteases in cellular growth control under the tutelage of Lan Bo Chen, who was also working for Buchanan. At the time (1975), the topic of the nature and function of growth factors was very compelling. In fact, it compelled me to leave for the Salk Institute, right next to the beach in San Diego, and start a second fellowship in the lab of Denis Gospodarowicz. Denis had only recently published the first article on the purification of what he called fibroblast growth factor (FGF) (1) and was testing every cell he could get his hands on to find out which ones responded best to FGF. As I joined the lab, the answer emerged that endothelial cells were among the most responsive cells to growth stimulation by FGF.

The ability to stimulate endothelial cell growth led the lab to grow large quantities of endothelial cells in culture, mostly derived from cow aortas, and we all started to investigate basic questions of endothelial biology that had been previously difficult to study. These included the questions of why platelets did not adhere to intact endothelial cells, why endothelial cells did not overgrow each other when confluent, and whether endothelial cells from different types of blood vessels had different properties. Unintentionally, the ability of FGF to stimulate endothelial cell growth propelled the Gospodarowicz lab into the forefront of those investigating endothelial cell biology.

Eventually, I needed a real job. Impressed by his pioneering articles on tumor angiogenesis, I applied for a faculty position in the department chaired by Judah Folkman at Boston Children's Hospital. I always suspected that he hired me more for my ability to purify FGF then for anything else, since Folkman was at the time deeply committed to the task of culturing capillary endothelial cells and he perhaps thought that I could help (or at the very least bring in some FGF). Capillary endothelial cells, we reasoned, would be the best cells to use to study the growth of new capillaries that occurred as new blood vessels invaded growing tumors. The task was daunting since these cells had never been cultured successfully. My minimal contribution was to suggest that we apply collagen to the dishes on which we would grow the cells. That worked but we got cultures that were overgrown with pericytes, the mesenchymal cells that surround many small vessels. Folkman, a trained surgeon, decided to put a micromanipulator into a tissue culture hood and physically smash every offending pericyte using a small glass capillary tube that had been fired to have a rounded tip. He would then wash out the culture (leaving the endothelial cells attached to the dish) and add new culture medium. He called the entire process the "weed and feed" method of endothelial cell culture. The only problem was that the capillary endothelial cells still did not grow well. We wanted to add FGF but didn't have enough, so we used culture medium that had been conditioned by incubation with tumor cells (tumor cell conditioned medium). This did the trick and the capillary endothelial cells began to grow beautifully, unencumbered by any wayward pericytes (2).

That left me with the question of what aspect of capillary endothelial cell biology I should study. Taking the tack of using the best available tools, I remembered a new cell migration assay that I had seen on a visit to Lan Bo Chen when he was a fellow at Cold Spring Harbor. Lan Bo was working closely with Guenther Albrecht-Buehler who was interested in how cells migrated right after they had divided. Albrecht-Buehler had devised a new migration assay based on the principle that moving cells would ingest small particles as they moved and thus leave a "phagokinetic track" showing where and how far they had migrated over a period of time (3). I learned this assay from Frank Solomon at M.I.T. (who had learned it earlier from Albrecht-Buehler) and quantified it using an early image analyzer to measure the size of the tracks left by the cells as they moved. Using the assay and the capillary endothelial cells we had cultured, I was able to show that angiogenic factors released by tumor cells were able to stimulate capillary endothelial cell migration (4). The work neatly showed that endothelial cell migration was likely to be stimulated during the process of tumor angiogenesis.

I never intended to study interferon. Danielle Brouty-Boye came to do a sabbatical in our department. Danielle had been working with Ion Gresser in Villejuif, France, and in their lab were pioneers in the purification of interferon. At the time, you could not call a supplier and order interferon, and the few academic labs that could purify it had too little to share. But Danielle not only had interferon, she had brought some with her to study its effects on endothelial cells. In other words, Danielle had interferon and I had capillary endothelial cells moving around in a dish. We also shared a tissue culture hood so we were often working side by side. It did not take us long to figure out that it might be interesting if she dropped a little bit of interferon into my endothelial cell cultures. To look at the effects, we used the phagokinetic migration assays. The results were not only dramatic but also reproducible. Every time capillary endothelial cells smelled a whiff of interferon, they stopped dead in their tracks. We reasoned that interferon might therefore be an angiogenesis inhibitor and promptly published our results (5). At the time, there were no angiogenesis inhibitors used in humans, and we hoped that when the day came that there was more interferon available, it could be used to stop angiogenesis in tumors and other diseases.

This might have been the end of the story. Fortunately, at least one person read the paper. His name was Carl White and he was a pulmonologist at Denver Children's Hospital. White was looking for a way to treat a 12-year-old patient who had a hemangioma, a non-metastatic tumor comprised principally of blood vessels and endothelial cells, growing in his lungs. Such tumors in the lung were nearly always fatal. At a lecture in Denver in 1987, seven years after the publication of our interferon paper, White saw a slide describing the effects of interferon on endothelial cell motility. He then looked up our paper and had an insight that interferon might be a new approach to the treatment of hemangiomas since these tumors consisted almost solely of endothelial cells. After consulting with Folkman, securing IRB approval, and securing a supply of interferon from the supplier, White embarked on what would become a long-term treatment of his patient with injections of interferon. Although there was little evidence of success at first, the tumor slowly regressed and the child returned to good health (6).

The next hemangioma patient was treated with interferon at Boston Children's Hospital by Folkman and his colleagues and, after a second success, they embarked on a larger clinical trial which showed considerable success of interferon treatment in children and adults with hemangiomas (7). What was of interest to me was that the treatment regimen consisted of relatively small doses of interferon given over a long period of time—a regimen that differed considerably from the more common use of cytotoxic chemotherapies at high doses and for short periods. The low-dose, long-term dosing schedule has since been shown to be successful for a variety of anti-angiogenesis agents. Its adoption seems to stem directly from White's early treatment of his patient with interferon.

Interferon has been used in thousands of patients with vascular hemangiomas and has been an effective treatment in the majority of these patients (8). Its use stemmed from the combination of a basic science observation and the receptive minds of inspired clinicians. More recently, however, toxicity issues emerged that have reduced the use of the drug for this indication. It is my hope that the same kind of unexpected observation that led to this use of interferon will take place again, in another laboratory, and give rise to the next generation of treatment for hemangiomas.

REFERENCES

1. D. Gospodarowicz and J. S. Moran. "Stimulation of Division of Sparse and Confluent 3T3 Cell Populations by a Fibroblast Growth Factor, Dexamethasone, and Insulin." *Proc Natl Acad Sci U S A* 71, no. 11 (1974): 4584–88.
2. J. Folkman, C. C. Haudenschild, and B. R. Zetter. "Long-Term Culture of Capillary Endothelial Cells." *Proc Natl Acad Sci U S A* 76, no. 10 (1979): 5217–21.
3. G. Albrecht-Buehler. "The Angular Distribution of Directional Changes of Guided 3T3 Cells." *J Cell Biol 80, no. 1* (1979): 53–60.
4. B. R. Zetter. "Migration of Capillary Endothelial Cells Is Stimulated by Tumour-Derived Factors." *Nature* 285, no. 5759 (1980): 41–43.
5. D. Brouty-Boye and B. R. Zetter. "Inhibition of Cell Motility by Interferon." *Science* 208, no. 4443 (1980): 516–18.

6. C. W. White, H. M. Sondheimer, E. C. Crouch, H. Wilson, and L. L. Fan. "Treatment of Pulmonary Hemangiomatosis with Recombinant Interferon Alfa-2a." *N Engl J Med* 320, no. 18 (1989): 1197–1200.
7. R. A. Ezekowitz, J. B. Mulliken, and J. Folkman. "Interferon Alfa-2a Therapy for Life-Threatening Hemangiomas of Infancy." *N Engl J Med* 326, no. 22 (1992): 1456–63.
8. D. J. Lindner. "Interferons as Antiangiogenic Agents." *Curr Oncol Rep* 4, no. 6 (2002): 510–14.

7 Use of Phage Display to Discover Inhibitors of Cell Migration

Mikael Björklund, Michael Stefanidakis,
Tanja-Maria Ranta, Aino Kangasniemi,
Terhi Ruohtula, and Erkki Koivunen

CONTENTS

7.1 BACKGROUND

There are only 20 amino acids and a few modifications of them found in nature. Yet these few building blocks can generate a huge diversity of chains of different composition and length whereupon life is based. It can be estimated that only a minority of this diversity has evolved into proteins encoded by the genomes. There might be very specific rules that govern peptide-mediated interactions during protein folding and protein–protein interactions, but these may be so complex that we do not know them yet.

Phage display is a method where a library of random peptides is built on the surfaces of phage particles, and peptides capable of performing a desired interaction can be isolated (1). We got interested in this method early on as it provided a tool to study peptide diversity and to select those peptides that are biologically active or relevant. One of the first applications was to study the peptide-binding specificity of an integrin, a cell surface receptor mediating cell adhesion (2). The method worked well as indicated by the isolation of RGD-containing peptides, which are typical integrin ligand sequences. These findings raised several questions but also provided ideas for the future work. The major question was how a single peptide motif becomes enriched during a biopanning, as a big protein like integrin should bind a variety of peptide motifs. The answer lies in the peptide libraries used and the natural peptide-mediated interactions that clearly seem to be limited. A library of hexapeptides reveals only some peptide interactions that are characteristic for short sequences such as ionic interactions. We found that by using more complex and longer peptides constrained by disulfides it was possible to isolate specific ligands to members of the integrin family (3). Furthermore, novel sequences were isolated that were not present in natural proteins, suggesting that the peptide diversity of the new phage libraries exceeded that of the genomes. The question was also raised, are some peptides too toxic or so active that evolution has abandoned them for the proteins? Can some peptides isolated by phage display even be potential drug leads?

7.2 DISCOVERY OF SPECIFIC GELATINASE INHIBITOR PEPTIDE CTT

These questions were in my mind when I (E. Koivunen) left Erkki Ruoslahti's laboratory in San Diego and returned to my home country where I set up my own laboratory at the University of Helsinki in 1995. At that time phage display had already been used for various applications but not for discovery of proteinase inhibitors. We knew that proteinases could degrade phage proteins, weakening the phage infectivity so that it was not an easy task to enrich "peptide inhibitors." Based on our previous experience on tumor cell migration and proteinases, we were mostly interested in matrix metalloproteinases (MMP) -2 and -9, also called gelatinases, the expression of which often correlates with cancer metastasis. Furthermore, there were no specific gelatinase inhibitors available at that time—another motivation for our research. By that time, we had constructed a series of peptide libraries with sizes ranging from 5-mer to 20-mer and used a pool of them in biopannings. As is usually the case, the first experiments failed; pro-MMPs or their trypsin-activated forms gave no phage enrichment. Fortunately, there was still one more MMP form worth trying: MMP activated with aminophenylmercuric acetate. Phage display has the advantage that the biopanning takes only a few days, and on a microtiter well format one can easily try several protocols. Suddenly we got a 1000-fold phage enrichment by using a highly purified MMP-9 subjected to the well-known chemical activation that induces an autocatalytic MMP-9 fragmentation. Some of the fragments or exposed MMP domains bound the phage. We still needed to manually sequence the phage DNA in those days; it was a state of the art until the fast and reliable sequencing robots took the task.

The real work begins after 100 or so phage clones are sequenced. The MMP-9 biopanning gave more than 20 different sequences and it was not immediately apparent which of them, if any, would inhibit the enzymatic activity. We thus prepared synthetic peptides based on the phage sequences, and tested them one by one on gelatinase activity assays. This tedious work showed that peptides containing the HWGF motif inhibit the MMP-2 and MMP-9 gelatinases but not other members of the MMP family. The HWGF-containing peptides also inhibited the migration of several cell lines in in vitro assays, the most active peptide being CTTHWGFTLC (CTT). HWGF was not the major motif enriched by the biopanning and CTT may have gone unnoticed unless we had carried out a fifth round of panning to search for tight binders. Altogether, it took four years before the paper describing this work was published (4). We showed that CTT was also active in vivo and slowed tumor progression in mouse models, apparently due to specific blocking of gelatinases. CTT was one of the first phage display peptides examined for in vivo tumor targeting in a mouse model and the peptide showed strong tumor homing in comparison to the normal tissues.

7.3   MULTIPLE USES FOR CTT PEPTIDE

The potency of the CTT peptide in vitro and in vivo suggested that this peptide would be a highly useful tool for tumor targeting and to study gelatinase functions. We were very interested in seeing if the CTT peptide could also be used to target other cargo than the bacteriophage. Liposomes represent a good starting point for the development of tumor-targeted drugs as they are generally non-immunogenic and can be loaded with various molecules. We coated liposomes with the CTT peptide and used them successfully to target gelatinase-expressing cancer cells in vitro (5). In these proof-of-principle experiments a fluorescent marker was used to measure the targeting efficiency, but in principle any drug or toxin could now be incorporated into liposomes and delivered into tumors using the CTT peptide. To our delight, other researchers have also adopted the use of CTT as a targeting peptide. The CTT peptide was used to re-target adenoviruses for a therapeutic gene delivery in a rabbit restenosis model (6). As a result of the incorporation of the CTT peptide, the natural tissue tropism of adenoviruses was modified, and a beneficial effect on the arterial wall was observed. Such remarkable results clearly indicate a clinical potential for CTT as a targeting peptide.

There are also multiple non-targeting applications for the CTT peptide based on the gelatinase specificity. To name a few, the CTT peptide has been used to localize gelatinases in tissue samples (7) and to evaluate the contribution of gelatinases in various biological processes (8–10). Thus, the CTT peptide appears to be a handy "Swiss knife" whenever the gelatinases are studied. As a spinoff from our academic studies, we established a company called CTT Cancer Targeting Technologies Ltd. to continue the development of CTT peptide-based cancer diagnostics and drugs.

7.4   INHIBITION OF LEUKOCYTE ADHESION AND MIGRATION BY β_2 INTEGRIN LIGAND PEPTIDE LLG-C4

At the same time we initiated the MMP studies, we also began to search for peptide ligands to the leukocyte β_2 integrins. The β_2 integrins are important for many aspects of leukocyte biology, including their migration, but no small molecule inhibitors were yet available in 1995. In this project, there were many false starts and wrong turns by frustrated students and it took until 2001 and 2003 for our first papers to be published (11,12). Also here, the key was to continue panning with different protocols until a "good-looking" peptide was discovered. Researchers can often judge the value of the peptide by just looking at its sequence. The peptide CPCFLLGCC (LLG-C4) was such an immediate hit as it had four cysteines and thus was likely a very compact structure in the peptide world. Leukocytes, but not the cells lacking the β_2 integrins, indeed strongly adhered to the LLG-C4 peptide when it was coated on microtiter wells (11). During the experiments we observed that some batches of synthetic peptides were considerably more active than the others. The explanation behind this phenomenon was that the four cysteines had two choices to form disulfide bonds. Indeed, when the disulfide bond formation was directed during peptide synthesis, the conformer with the disulfide-bonding C(1-8;3-9) was found to be 20 times more active than the other conformer with C(1-9;3-8) disulfides. Furthermore, structural analysis of the peptide containing the active disulfide configuration revealed that the structure of this peptide was much more compact than that of the less active peptide. The LLG-C4 peptide was able to inhibit leukocyte adhesion and migration; antibody inhibition experiments showed that this effect was mainly due to inhibition of the $\alpha_M\beta_2$ and $\alpha_X\beta_2$ integrins (11).

The LLG-C4 peptide was discovered by biopanning on a purified $\alpha_M\beta_2$ integrin. The major active ligand-binding site locates to the so-called I domain of the α subunit. The I domains of different integrins have been available for years, but biopannings on these domains performed by us and others had failed. However, over the years the quality and peptide diversity of our libraries had improved, and when active α_M I domains were available, it was a reason to try again. This time it worked. Nonspecific phages were removed by prior incubation on wells coated with GST alone. The phage library was then transferred to wells coated with an α_M I domain GST fusion protein. We did not re-find LLG-C4, but another peptide motif that had been discovered before.

Most of the peptides isolated in biopanning with the I domain contained a common DDGW or EDGW motif despite the known ligand-binding promiscuity of the α_M I domain. The peptide ADGACILWMDDGWCGAAG (DDGW) was the most potent of these peptides. Surprisingly, the I domain-binding peptide motif was highly similar to a peptide mimic of the RGD-peptide binding site (13). However, we now had this motif in the reverse situation, as the integrin ligand. This finding reawakened our early thoughts about the diversity of the naturally occurring protein–protein interactions, which had remained rather dormant as new peptide motifs had been constantly found. Such a recurrence of an old motif in a new situation may indicate that there are only a limited number of protein–protein interactions, and due to the rapid increase in the use of phage display we may soon have a better understanding of the number of allowed interactions. What immediately becomes evident is that nature will not waste any good interactions or ingredients but uses them elegantly in appropriate situations.

7.5   CONVERGENCE OF RESEARCH LINES—DDGW PEPTIDE INHIBITS proMMP-9 BINDING TO $\alpha_M\beta_2$ INTEGRIN

Another surprise came when we searched for possible proteins that could use a DDGW-like motif in the α_M I domain binding. We identified that the gelatinases contain a similar motif in their catalytic domain. MMPs and integrins have been linked together, as shown by the finding that the C terminal hemopexin-like domain of MMP-2 can directly interact with $\alpha_v\beta_3$ integrins and that inhibition of this interaction inhibits angiogenesis (14). It has also been found that MMP-1 interacts with the $\alpha_2\beta_1$ integrin, presumably by the same mechanism (15). However, we found that proMMP-2 and proMMP-9 bind to the α_M I domain using the catalytic domain, and this interaction is inhibited with the DDGW peptide and also the CTT peptide. In contrast, the LLG-C4 peptide could not efficiently compete with this interaction. The ability of the CTT peptide to inhibit the gelatinase-I domain interaction further indicated that the interaction required the catalytic domain of gelatinases and not the C terminal domain as with the MMP-2/$\alpha_v\beta_3$ interaction. This data was further strengthened by the dissection of proMMP-9 into small peptides. The DDGW-like sequence in the catalytic domain was the only site binding to the α_M I domain. The DDGW and the MMP-derived DDGW-like peptide also inhibited leukocyte migration in vitro and in vivo, suggesting that gelatinases, especially MMP-9 and β_2 integrins, cooperate in leukocyte migration (12,16).

7.6   CATALYTIC AND NONCATALYTIC INHIBITION OF GELATINASE-MEDIATED CELL MIGRATION AND INVASION

Highly encouraged by the results obtained, we performed biopanning with MMP-9 again, this time with the proenzyme form. No CTT peptide-like sequences occurred; instead two other motifs were enriched. Using recombinant MMP-9 domains, we found that the new peptides, PPC and CRV, bound to the collagen/gelatin-binding domain (CBD) and the C terminal hemopexin/vitronectin-like domain, respectively (17). The CBD is a unique substrate-binding module for the gelatinases. Not surprisingly, we found that the PPC peptide inhibited gelatinase activity. Thus, PPC is an absolutely specific gelatinase inhibitor and cannot inhibit other MMPs. Interestingly, we noticed that two extracellular matrix proteins, fibronectin and vitronectin, have PPC-like sequences and we could confirm that MMP-9 indeed recognizes these sequences.

CRV was a highly interesting peptide. First of all, it showed clear selectivity toward MMP-9, whereas CTT and PPC bind both gelatinases. Furthermore, CRV did not significantly inhibit gelatinase activity in vitro, but it was as potent an inhibitor of cell migration and invasion as CTT. CRV also reduced human tumor xenograft growth in a mouse model (17). Interestingly, CRV peptide had significant sequence similarity to the integrin β subunit, especially the β_5 integrin. The CRV-like sequence is located in the stalk of the integrin, where the activation-dependent epitopes reside, suggesting that this interaction requires integrins in the extended, active

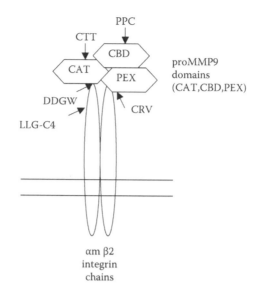

FIGURE 7.1   A schematic model of proMMP-9 interaction with αMβ2 integrin. The binding sites of the phage display peptides are depicted.

conformation. We found that MMP-9 interacts with the $\alpha_v\beta_5$ integrin. This inter-action was inhibited by the MMP-9 C terminal domain and a recombinant integrin domain from the stalk of the β_5 chain. This finding was conceptually important as it has been shown that the MMP-2 binding to $\alpha_v\beta_3$ integrin does not compete with the RGD-containing ligands such as vitronectin.

Figure 7.1 summarizes our findings so far on the interactions of integrins and proMMP-9 and depicts the binding site for each of the CTT, LLG-C4, RGD, DDGW, PPC, and CRV peptides. The cell surface activation of proMMP-9 requires a pro-teinase cascade, the components of which are at least urokinase (uPA), its receptor uPAR, and plasminogen. We propose the term invadosome for this complex as it is required for the migration of invasive tumor cells. We observe the binding of proMMP-9 to the integrin and suggest that the role of the MMP-integrin complex is to locate proMMP-9 activation in proximity of a matrix ligand/substrate, the deg-radation of which is required for cell migration. This is a "peptidoscopic" view of the MMP-integrin complex obtained by phage display and as the research advances, we may see a more thorough view of all the components that take part in the cell migration machinery. What is most encouraging is that phage display helps to reveal biologically relevant interactions.

7.7   REDUCTION OF WORKLOAD BY RECOMBINANT EXPRESSION OF PHAGE DISPLAY PEPTIDES

As exemplified by the discovery of the CTT peptide, functional analysis of the phage display peptides still remains a considerable task. Furthermore, chemical synthesis of a series of peptides can in the long run become very expensive and time consuming. Binding assays of peptides as a fusion with glutathione-S-transferase (GST) have proven valuable, but it is often difficult to obtain a sufficiently high concentration of the fusion peptides to demonstrate enzyme inhibition or inhibition of cell migration. To circumvent these hurdles, we developed a recombinant method to produce phage display-derived peptides in a soluble form by expressing them as a fusion with inteins (18). Inteins were chosen as they can autocatalytically cleave the fused protein in the presence of thiols or appropriate pH and temperature (19,20). Indeed, the pH- and temperature-sensitive cleavage was found to be ideal for the production of disulfide-containing peptides (18). This method allowed us to rapidly produce multiple peptides and assay them for activity and solubility. The use of this approach has greatly expedited our research as we are now able to select the best water-soluble, high-affinity peptides among all the sequences isolated from a phage display selection. This method was used to select the best binders to the α_M integrin I domain and the MMP-9 domains (12,17). Furthermore, we have started to extend the peptide diversity of phage display libraries by incorporation of unnatural amino acids (21). We hope that our approach will make the path from a peptide to a drug a less challenging one.

7.8 CONCLUSIONS

During these years we have learned many of the secrets behind a successful phage selection and we hope that this knowledge will transform into a deeper understanding of the mechanisms of cell migration. Finally, in the future when enough phage display selections have been performed, we may also understand the rules governing protein–protein interactions.

REFERENCES

1. G. P. Smith. "Filamentous Fusion Phage: Novel Expression Vectors That Display Cloned Antigens on the Virion Surface." *Science* 228, no. 4705 (1985): 1315–17.
2. E. Koivunen, D. A. Gay, and E. Ruoslahti. "Selection of Peptides Binding to the Alpha 5 Beta 1 Integrin from Phage Display Library." *J Biol Chem* 268, no. 27 (1993): 20205–10.
3. E. Koivunen, B. Wang, and E. Ruoslahti. "Isolation of a Highly Specific Ligand for the Alpha 5 Beta 1 Integrin from a Phage Display Library." *J Cell Biol* 124, no. 3 (1994): 373–80.
4. E. Koivunen, W. Arap, H. Valtanen, A. Rainisalo, O. P. Medina, P. Heikkila, C. Kantor, et al. "Tumor Targeting with a Selective Gelatinase Inhibitor." *Nat Biotechnol 17, no. 8* (1999): 768–74.
5. O. P. Medina, T. Soderlund, L. J. Laakkonen, E. K. Tuominen, E. Koivunen, and P. K. Kinnunen. "Binding of Novel Peptide Inhibitors of Type IV Collagenases to Phospholipid Membranes and Use in Liposome Targeting to Tumor Cells In Vitro." *Cancer Res* 61, no. 10 (2001): 3978–85.
6. M. P. Turunen, H. L. Puhakka, J. K. Koponen, M. O. Hiltunen, J. Rutanen, O. Leppanen, A. M. Turunen, et al. "Peptide-Retargeted Adenovirus Encoding a Tissue Inhibitor of Metalloproteinase-1 Decreases Restenosis after Intravascular Gene Transfer." *Mol Ther* 6, no. 3 (2002): 306–12.

7. E. Pirila, P. Maisi, T. Salo, E. Koivunen, and T. Sorsa. "In Vivo Localization of Gelatinases (MMP-2 and -9) by In Situ Zymography with a Selective Gelatinase Inhibitor." *Biochem Biophys Res Commun* 287, no. 3 (2001): 766–74.

8. S. Cheng and D. H. Lovett. "Gelatinase A (MMP-2) Is Necessary and Sufficient for Renal Tubular Cell Epithelial-Mesenchymal Transformation." *Am J Pathol* 162, no. 6 (2003): 1937–49.

9. C. W. Franzke, K. Tasanen, H. Schacke, Z. Zhou, K. Tryggvason, C. Mauch, P. Zigrino, et al. "Transmembrane Collagen XVII, an Epithelial Adhesion Protein, Is Shed from the Cell Surface by ADAMs." *EMBO J* 21, no. 19 (2002): 5026–35.

10. C. Fernandez-Patron, K. G. Stewart, Y. Zhang, E. Koivunen, M. W. Radomski, and S. T. Davidge. "Vascular Matrix Metalloproteinase-2-Dependent Cleavage of Calcitonin Gene-Related Peptide Promotes Vasoconstriction." *Circ Res* 87, no. 8 (2000): 670–76.

11. E. Koivunen, T. M. Ranta, A. Annila, S. Taube, A. Uppala, M. Jokinen, G. van Willigen, E. Ihanus, and C. G. Gahmberg. "Inhibition of Beta(2) Integrin-Mediated Leukocyte Cell Adhesion by Leucine-Leucine-Glycine Motif-Containing Peptides." *J Cell Biol* 153, no. 5 (2001): 905–16.

12. M. Stefanidakis, M. Bjorklund, E. Ihanus, C. G. Gahmberg, and E. Koivunen. "Identification of a Negatively Charged Peptide Motif within the Catalytic Domain of Progelatinases That Mediates Binding to Leukocyte Beta 2 Integrins." *J Biol Chem* 278 (2003): 34674–84.

13. R. Pasqualini, E. Koivunen, and E. Ruoslahti. "A Peptide Isolated from Phage Display Libraries Is a Structural and Functional Mimic of an RGD-Binding Site on Integrins." *J Cell Biol* 130, no. 5 (1995): 1189–96.

14. P. C. Brooks, S. Silletti, T. L. von Schalscha, M. Friedlander, and D. A. Cheresh. "Disruption of Angiogenesis by PEX, a Noncatalytic Metalloproteinase Fragment with Integrin Binding Activity." *Cell* 92, no. 3 (1998): 391–400.

15. T. P. Stricker, J. A. Dumin, S. K. Dickeson, L. Chung, H. Nagase, W. C. Parks, and S. A. Santoro. "Structural Analysis of the Alpha(2) Integrin I Domain/Procollagenase-1 (Matrix Metalloproteinase-1) Interaction." *J Biol Chem* 276, no. 31 (2001): 29375–81.

16. M. Stefanidakis, T. Ruohtula, N. Borregaard, C. G. Gahmberg, and E. Koivunen. "Intracellular and Cell Surface Localization of a Complex Between AlphaMbeta2 Integrin and Promatrix Metalloproteinase-9 Progelatinase in Neutrophils." *J Immunol* 172 (2004): 7060–68.

17. M. Bjorklund, P. Heikkila, and E. Koivunen. "Peptide Inhibition of Catalytic and Noncatalytic Activities of Matrix Metalloproteinase-9 Blocks Tumor Cell Migration and Invasion." *J Biol Chem* 279, no. 28 (2004): 29589–97.

18. M. Björklund, H. Valtanen, H. Savilahti, and E. Koivunen. "Use of Intein-Directed Peptide Biosynthesis to Improve Serum Stability and Bioactivity of a Gelatinase Inhibitory Peptide." *Comb Chem High Throughput Screen* 6 (2003): 29–35.

19. F. B. Perler and E. Adam. "Protein Splicing and Its Applications." *Curr Opin Biotechnol* 11, no. 4 (2000): 377–83.

20. T. C. Evans Jr., J. Benner, and M. Q. Xu. "The Cyclization and Polymerization of Bacterially Expressed Proteins Using Modified Self-Splicing Inteins." *J Biol Chem* 274, no. 26 (1999): 18359–63.

21. M. Bjorklund and E. Koivunen. "Steps Towards Phage Display Libraries with an Extended Amino Acid Repertoire." *Letters in Drug Design & Discovery* 1, no. 2 (2004): 163–67.

8 Combinatorial Mapping of Vascular Zip Codes by In Vivo Phage Display

Renata Pasqualini and Wadih Arap

CONTENTS

8.1 INTRODUCTION

Differential protein expression in vascular endothelium associated with normal or diseased tissues offers the potential for developing novel diagnostic, imaging, and therapeutic strategies. In essence, our research program uses combinatorial library selection (peptide and antibody-based) to discover, validate, and exploit the vascular biochemical diversity of endothelial cell surfaces toward a new vascular-targeted pharmacology. Such targeting technologies may lead to the development of ligand-directed therapeutics and imaging agents for application in the treatment of cancer, obesity, and other medical conditions that show distinct vascular attributes. Translational applications through clinical trials, which are scheduled to start soon at our own institution and elsewhere, will ultimately determine the value of this strategy.

Organ-specific trafficking of cells and proteins through the vascular and lymphatic systems supports the notion that the endothelia express selective molecular addresses that are tissue-specific. Prostate and breast cancer, for instance, metastasize to the axial skeleton with very high frequency, whereas leukocytes rapidly accumulate within sites of inflammation. Despite intense investigation in the past 30 years, not much is known or well understood about the molecules involved in such interactions.

Over a decade ago, we conceived that the molecular diversity of the vasculature could be efficiently fingerprinted at the molecular level, if a combinatorial library of ligands with different binding capabilities could be injected and allowed to distribute, so that probes selectively homing to a given organ or tissue could be recovered and identified. This idea led to the development of in vivo phage display.

We developed in vivo phage display by selecting phage capable of homing to different vascular beds after administration of a phage display random peptide library. Our screenings are based on the use of peptides that are expressed on the surface of bacteriophage. Our extensive previous work has firmly established that peptide libraries can be used to probe organ- and tumor-specific vascular homing. In order to construct such libraries, random oligonucleotides are individually fused to cDNAs encoding a phage surface protein, producing a large collection of phage particles (10^9 transducing units) displaying unique peptides. In the in vivo procedure, phage capable of homing to certain organs or tumors following an intravenous injection are selected from the libraries; the ability of individual peptides to target a tissue can also be analyzed by this method.

A novel vascular address system, akin to zip codes, makes it possible to target organ-specific blood vessels and newly formed blood vessels (angiogenesis) in tumors. We have isolated peptides that can home to normal blood vessels or sites of angiogenesis through the circulation via this vascular address system. Every normal or diseased organ appears to display a unique signature on its blood vessels that our probes can use as a target. We have also developed complementary methods of assessing the distribution of probes targeted to the vasculature, their tissue specificity, and their target cells.

Furthermore, our system provides an innovative way of identifying endothelial cell surface markers expressed in vivo. Thus, in addition to providing novel tools for selective vascular targeting of therapies, we perfected methodologies to further our understanding of tumor endothelium specificity and enhanced the ability to define the role endothelial cell markers play in angiogenesis. Vascular targeting peptides bind to different receptors that are selectively expressed on the vasculature of a given tissue. Some of these vascular markers are vascular proteases that not only serve as receptors for circulating ligands but also modulate angiogenesis. Vascular receptors are also involved in tumor cell homing during the metastatic process, as we originally hypothesized. Targeted delivery of cytotoxic chemotherapy, proapoptotic peptides, and cytokines to receptors in the angiogenic vasculature showed marked therapeutic efficacy in tumor-bearing mouse models.

One can rely on isolating ligands and receptors in mice to find the putative human homologues. However, even when such homologues are found, the question remains as to whether they actually target human vasculature. In humans, cell surface receptors may have a very different distribution and function than they have in mice. The potential of in vivo phage display to identify vascular targeting peptides has hardly been fully explored. Thus, a logical extension of our work was to isolate motifs that home in vivo to human blood vessels.

The ability to target therapies to the appropriate location in the body has been a long-standing goal of researchers and physicians. If it were possible to deliver drugs selectively to intended targets, many therapeutic compounds that are effective, but have untoward effects, could find a use or could be used more effectively. For example, higher concentrations of the agent would be reached in the target tissues and fewer side effects would result. Targeting could also make it possible to use drugs for which reaching therapeutically effective concentrations in the diseased tissue is a problem. Unfortunately, there are only a few situations in which targeted drug delivery is actually possible. Antibodies prepared against various antigens, and infiltrating lymphocytes have been used for tissue targeting. However, they are not particularly effective, and other ways of drug targeting are often invasive.

Over the past decade, we have developed a method that provides a new way of discovering molecular "addresses" for targeted therapies. The method is based on in vivo screening of large libraries of compounds for molecules that have selective tissue affinities. We use peptide libraries displayed on phage, but in principle other types of libraries could also be employed for this purpose. The method, at least when phage libraries are used, primarily but not exclusively, targets tissue-specific differences in endothelial cells. There are many indications where vascular beds in different organs and tissues differ from one another, and such differences serve cellular trafficking in the body. Thus, lymphocytes home in particular compartments of the lymphoid system guided by specific address molecules in the endothelia of those compartments, while other blood leukocytes recognize sites of inflammation through endothelial markers induced by inflammatory signals. Unlike earlier work, the two methodologies used here directly select for molecules (peptidomimetics or monoclonal antibodies) capable of homing into tumors and other target tissues in vivo. They may, therefore, uncover new markers for the blood or lymphatic vessels of various tissues, providing new means for selective targeting of therapies.

The vasculature serving sites of injury has many special features; most importantly, it continuously forms new blood vessels by angiogenesis. Angiogenic blood vessels express molecules that are not present in mature vasculature. Certain cell adhesion molecules such as alpha v integrins and various receptors for angiogenic growth factors are among such known markers. Moreover, tumor vasculature is anatomically distinct in that it contains special fenestrations, and this may also be reflected in molecular differences. Finally, a central issue to be addressed is the relative contribution of "conventional" versus "hematogenic" (de novo) angiogenesis, with the latter arising from progenitor endothelial cells.

Surprisingly, despite major advances from the Human Genome Project, the molecular diversity of receptors in the human vasculature is not entirely clear. Although ligands and receptors isolated in animal models have been useful to identify putative human homologs, it is unlikely that targeted delivery will always be achieved in humans through such an approach. Data from the Mouse and Human Genome Projects indicate that the higher complexity of the human species relative to other mammalian species derives from expression patterns of proteins at different tissue sites, levels, or times rather from a greater number of genes. Indeed, striking examples of species-specific differences in gene expression within the human vascular network have recently surfaced. Such differences in protein expression patterns and ligand–receptor accessibility caution that vascular proteomics results obtained in animal models must be carefully evaluated before extrapolation to human studies. Thus, selection of phage display random peptide libraries in humans may reduce costly late-stage clinical trial failures by shifting decisions to earlier stages of the drug development process. Given this rationale, we reasoned it would be possible to map the vasculature in patients by direct selection of combinatorial peptide libraries. We have already shown that phage capable of selectively targeting organs or tumors after intravenous injection into mice can be screened, selected, and isolated. Peptides synthesized based on the sequences displayed by phage that "home" in the designated vascular beds bind to the same structures in these tissues as the phage. Finally, the peptides of interest serve at least two non-mutually exclusive purposes.

First, the peptide sequences themselves are leads to be developed into peptidomimetic drugs. Second, the identification of the corresponding receptor(s) will enable further targeted modulation by other drugs. One might envision that optimization of a vascular-targeted pharmacology would allow the best potential outcomes. Peptidomimetic-based agents with these targeting properties could selectively deliver drugs, genes, proteins, or cells into the vasculature of a target site or organ. Valuable data on distinguishing characteristic of different vascular beds, including those undergoing angiogenesis, are bound to ensue. Indeed, a panel of ligand peptidomimetics and their corresponding receptors is currently being translated into clinical applications.

On the other hand, far less is known about the lymphatic endothelium. Indeed, only a handful of "specific" surface receptors have been identified—let alone validated—in lymphatic endothelial cells and the degree of molecular diversity is still largely unknown. Thus, we propose to use such libraries to identify molecules that are specifically or selectively expressed in the lymphatic vasculatures of individual healthy and diseased tissues. The lymphatic endothelium cell layer, which can be primarily targeted through this methodology, is likely to differ in properties, but relatively few molecular differences have been identified at this time. Our method, in vivo combinatorial screening of phage libraries, makes it possible to identify such lymphatic ligand–receptor molecular systems in high throughput screening.

If successful, a completed receptor-based map of the human vascular and lymphatic endothelium will have broad implications for the development of targeted drugs. Thus far, we have developed methodology to allow cell-free, in vitro, and in vivo selection of libraries of random peptides to identify ligands that home to specific human vascular beds. These strategies revealed a vascular address system that allows targeting of tissue-specific and angiogenesis-related receptors expressed in blood vessels. Targeted delivery of drugs, imaging vectors, and proteins to specific receptors has already been accomplished in animal models. Translational clinical trials are scheduled to start soon.

8.2 HUMAN VASCULAR MAPPING PROJECT

Despite major advances from the Human Genome Project, the molecular diversity of receptors in the human vasculature remains largely unknown. Although ligands and receptors isolated in animal models have been useful to identify putative human homologues, it is unlikely that targeted delivery will always be achieved in humans through such an approach. Data from the Mouse and Human Genome Projects indicate that higher complexity of the human species relative to other mammalian species derives from expression patterns of proteins at different tissue sites, levels, or times rather than from a greater number of genes. Indeed, striking examples of species-specific differences in gene expression within the human vascular network have surfaced (reviewed in Hajitou et al. 2006a; Sergeeva et al. 2006). We had previously shown that phage capable of selectively homing to isogenic and xenograft tumors after intravenous administration in mice could be selected and isolated (Arap et al. 1998; Rajotte et al. 1998; Koivunen et al. 1999; Ellerby et al. 1999; Arap et al. 2000; Pasqualini et al. 2000). However, experiments in tumor-bearing mice must be very carefully evaluated and validated before extrapolation to human applications. Indeed, differences in protein expression patterns and ligand–receptor accessibility caution that vascular proteomics results obtained in animal models are often not recapitulated in humans. Thus, direct selection of phage display peptide libraries in patients may reduce costly late-stage clinical trial failures by shifting decisions to earlier stages in the drug development process.

Given this rationale—upon the relocation of our group to the University of Texas M. D. Anderson Cancer Center (MDACC) in 2000—we have reasoned it might be possible to map the vasculature directly in patients by using combinatorial phage display application in vivo. Thus, we proposed to use combinatorial libraries displaying peptides or antibody fragments to identify molecules that are specifically or selectively expressed in the vasculature of individual tissues or organ systems under normal or pathological conditions. Endothelial cells, which are primarily targeted by our methodologies, are known to differ in their morphological properties in different tissues, but relatively few corresponding molecular differences (termed human vascular zip codes) had been identified. Our methodology, direct selection of combinatorial libraries in cancer patients, may enable accelerated identification, validation, and prioritization of such molecules. In 2002, we reported the first combinatorial peptide library selection in a cancer patient (Arap et al. 2002). Because these experiments were performed in terminal wean or brain-dead human subjects, we have also created and reported our institutional ethical guidelines (Pentz et al. 2003) for human experimentation; our internal guidelines were later extended and harmonized with those from transplantation medicine to serve as a proposal for national guidelines for end-of-life-research (Pentz et al. 2005).

Future research can be academically divided into three areas of investigation: (1) Translation of the original targeting findings into clinical applications through trials in cancer patients, (2) improvement of targeting technology and ethics guidelines, and (3) the discovery of new relevant human targets for intervention.

1. Since the validation of the interleukin-11 receptor (Arap et al. 2002) as the first human vascular tumor target (zip code) in a large cohort of prostate cancer patients (Zurita et al., 2004), a targeted proapoptotic peptidomimetic drug based on these findings (termed BMTP-11) is in pre-IND phase at MDACC. Indeed, a sophisticated "phase zero" clinical trial with pharmaco-

genomics and targeted molecular imaging endpoints will start within 12–18 months post Food and Drug Administration (FDA) consultation.

2. In a related line of targeting research, we have been focusing on technological improvements to allow synchronous combinatorial selection to multiple human tissues/organs in a single selection. We have recently validated combinatorial targeting refinement in mouse models (Kolonin et al. 2006a). Enhancements planned for this large project ("The Human Vascular Mapping Project 2.0") include

    • Very high-throughput phage DNA analysis with new host bacteria-free sequencing (for instance, by using 454-based technology, among others).

    • In order to perform combinatorial selections in patients earlier than at the end-of-life, we will approach the Recombinant Advisory Committee (RAC) of the National Institutes of Health (NIH) for logistics and permits required.

    • We have a created an internal database for peptides selected from libraries. We have been in the process of launching a comprehensive relational database. This virtual tool will be Oracle-compatible and accessible online as an academic resource.

    • Integration with genomics, proteomics, mass spectrometry, real-time PCR, gene, protein, and antibody arrays.

3. With enabling technological and ethics frameworks in place, we performed three rounds of serial combinatorial selections in cancer patients with bone marrow metastasis. In as yet unpublished work, we have selected, isolated, and validated two unrecognized zip codes to allow vascular target delivery in patients; one ligand-receptor system in tumor blood vessels is required for human cancer cell site-specific metastasis to human bone marrow, and another one in white fat blood vessels is involved in human obesity.

8.3 TARGETED MOLECULAR IMAGING

Our group has quite recently reported the design, generation, and construction of AAV/phage (termed AAVP) particles (Hajitou et al. 2006b; Hajitou et al. 2007; Soghomonyan et al., 2007) for targeted molecular-genetic imaging. These hybrid vectors containing prokaryotic and eukaryotic cis-genomic elements have the potential to integrate ligand-directed targeting and molecular-genetic imaging. In a related line of research, we have used labeled targeted peptide motifs themselves as imaging tools (Yao et al. 2005; Marchiò et al., Cancer Cell, 2004; Arap et al., Cancer Cell, 2004; Zurita et al. 2004; Cardó-Vila et al. 2003; Chen et al. 2004; Mintz et al. 2003). In pilot experiments, AAVP-based molecular-genetic imaging appears to be superior in side-by-side comparison because it provides prediction of therapeutic response in addition to only monitoring (unpublished results). Thus, we plan to focus primarily on the development of AAVP-based molecular-genetic imaging (Hajitou et al. 2006b; Hajitou et al. 2007; Soghomonyan et al. 2007). Finally, we have also designed and developed self-assembled biocompatible networks of phage-gold as nanotechnology-based molecular sensors and reporters (Souza et al. 2006a; Souza et al. 2006b); this new methodology can be incorporated and it is quite synergistic with AAVP.

8.3.1 FUTURE PLANNED RESEARCH

There are several areas of research planned: (1) To use prototypes of this new class of targeted hybrid vectors for therapy and for molecular-genetic imaging, (2) to develop AAVP-based library applications, (3) to create other chimeric prokaryotic–eukaryotic vectors, and ultimately (4) to generate an "imaging transcriptome" for human tumors.

1. We will use prototypes of this new class of targeted hybrid vectors for therapy and for molecular-genetic imaging. Specifically,
   - Discovery of new specific ligand motifs that target tumor endothelium is planned.
   - Study the attributes of another form of targeted AAV, directly selected from AAV-display libraries (Müller et al. 2003).
   - In as yet unpublished work, AAVP-based tumor antivascular gene therapy with targeted TNF delivery is ongoing.
   - In as yet unpublished work, we have showed that is possible to predict drug response in sarcoma with targeted AAVP molecular imaging.
   - Integrate the principles of self-assembled biocompatible networks to AAVP to create a reverse transducing matrix.
2. The use of AAVP-based combinatorial peptide libraries for direct patient selections will be required for patient settings such as preoperative and metastatic tumors. As such, the steps toward this goal are as follows:
   - Design, generation, and production of targeted AAVP prototypes and AAVP libraries in Good Manufacturing Practices (GMP) facilities for patient applications.
   - RAC approval for long-term transduction in cancer patients.
   - Proof-of-concept with a reporter/suicide gene (i.e., HSV-*tk*).
3. We will create other hybrid vectors with the combined biological attributes of bacteriophage and animal viruses. The generation of a double-stranded DNA construct with elements of adenovirus and lambda phage is ongoing.
4. The incorporation of transcriptional targeting (through tissue-specific or radiation-induced promoters) to ligand-directed AAVP-targeting may enable one to determine a gene status without tissue biopsy.

8.4 FINGERPRINTING CIRCULATING ANTIBODIES FROM CANCER PATIENTS

Phage random peptide libraries were initially developed and used in the 1980s to map binding sites of immunoglobulins. This technology has been used to explore the diversity of the humoral immune response. Often, the selected peptides correspond to either primary sequences found in disease-associated antigens or conformational mimic peptides of such antigens. We have epitope-mapped ("fingerprinted") the circulating pool of human antibodies elicited against tumors, which led to the isolation of disease targets in prostate cancer patients (Mintz et al. 2003) and in ovarian cancer patients (Vidal et al. 2004). We later demonstrated that the immunogenic tumor target glucose-regulated protein-78 (GRP78) is expressed in the cell surface and it enables internalization of circulating ligands (Arap et al. 2004).

Essentially, one can identify the antigens against which the disease-related autoantibodies are reacting. Since human cancer is a heterogeneous disease and its clinical behavior can be unpredictable, grouping tumors based on the individual humoral response of patients may prove relevant. This discovery platform could benefit patients in the short term if profiles of reactivity against a marker (or a panel of markers) can provide clues as to what kind of personalized therapy should be attempted. This work also ties in with efforts to map the human vasculature (see discussion in the previous section), because cancer patients can mount a humoral immune response against vascular "zip codes"—receptors that are differentially expressed in the tumor blood vessels. In addition, targeting tumor antigens may lead to the development of targeted vaccines against tumors.

We have also shown that the targeting of antigens to lymphatic tissue in vivo modulates immunity (Trepel et al. 2001). To test this hypothesis, the humoral response elicited by phage vaccination was measured, and lymph node-targeted phage were more antigenic than control untargeted phage; the effect was specific and inhibited by co-administration of the lymph-node targeting synthetic peptides displayed.

8.4.1   FUTURE PLANNED RESEARCH

Among the areas of cancer research planned are horizontal follow-up of the humoral immune response in patients over time.

- In as yet unpublished work, we have recently shown that serially measured autoantibodies from prostate cancer patients can serve as early markers and prognostic factors for the disease (Mintz et al., submitted).
- We have also been planning to combine such antibodies with protein arrays for fast identification of a panel of tumor-related antigens.

8.5   TARGETING MOUSE MODELS OF HUMAN DISEASE IN VIVO

Aside from discovery and validation of tumor vascular zip codes in experimental tumor-bearing mouse models such as aminopeptidase A (Marchiò et al. 2004) and aminopeptidase N (CD13; Rangel et al. 2007) among many others, we have also been interested in nonmalignant diseases with an angiogenic component including (1) obesity, (2) retinopathies, and (3) malformations.

1. Obesity is an increasingly prevalent human condition in developed societies. Despite major progress in the understanding of the molecular mechanisms leading to obesity, no safe and effective treatment has yet been found. We reported a novel anti-obesity therapy based on targeted induction of apoptosis in the vasculature of adipose tissue. We isolated a

motif that homes to white fat vasculature. We show that the peptide targets prohibitin, which we established as a vascular marker of adipose tissue. Targeting a pro-apoptotic peptide to prohibitin in the adipose vasculature caused ablation of white fat. Resorption of established white adipose tissue and normalization of metabolism resulted in obesity reversal without detectable adverse effects. Because prohibitin is expressed in human white fat, this work may lead to development of targeted drugs for obese patients (Kolonin et al. 2004).

2. Proliferative retinopathies: We have used a mouse model of oxygen-induced retinopathy to discover mechanisms of disease in retinitis pigmentosa and retinopathy of prematurity in mice and human patients (Lahdenranta et al. 2001). We next hypothesized that ligands from combinatorial libraries that target tumor vasculature may serve to target angiogenic vasculature in retinal diseases. Treatment of mice with a peptidomimetic drug reduced oxygen-induced retinal angiogenesis by selectively inducing activated endothelial cell apoptosis (Lahdenranta et al. 2007). These targeted drugs may be used against retinal diseases.

3. Congenital malformations: An estimated 3% of children in developed countries are born with nongenetic birth defects. However, the nature and mechanisms of teratogenesis are poorly understood. The placenta, which controls delivery of nutrients to the fetus, is a major target for embryotoxic drugs. To gain insight into molecular mechanisms of transplacental transport, we screened a phage-display random library for peptides that bind to placental receptors in pregnant mice. We have shown that a ligand peptide targets the yolk sac, is transported into the embryo, and blocks materno-fetal immunoglobulin transport. Pregnant mice receiving peptide had hemorrhagic necrosis in the placenta and disrupted embryo development (Kolonin 2002).

8.5.1 Future Planned Research

- Obesity: We are in pre-IND phase for our targeted peptidomimetic drug against obesity. In as yet unpublished work, we have found a natural circulating ligand to the receptor prohibitin; we have also shown that such a ligand-receptor system is functional in humans. Thus, translation into clinical applications appears highly possible. As part of a follow-up investigation, we are about to start a trial in obese baboons.

- Proliferative retinopathies: We have another line of research in which a neuropilin-binding small peptidomimetic cyclic motif is also effective against retinopathy of prematurity.

- Congenital malformations: The placental vasculature is a non-endothelial surface. High-throughput screening for embryotoxic ligands that target placental receptors could be developed to systematically identify and perhaps avoid exposure to teratogenic drugs in patients.

8.6   RATIONAL DEVELOPMENT OF
SMALL PEPTIDOMIMETIC DRUGS

In previous work, we have devised and optimized a new approach for the screening of cell surface-binding peptides from phage libraries. This method, Biopanning and Rapid Analysis of Selective Interactive Ligands (termed BRASIL) is based on a single-step differential centrifugation in which a cell suspension incubated with phage in an aqueous upper phase is centrifuged through a non-miscible organic lower phase. BRASIL is more sensitive and more specific than standard methods that rely on repeated washing steps. As a proof of principle, we screened human endothelial cells stimulated with vascular endothelial growth factor. We have built a peptide-based ligand-receptor map of binding sites within the VEGF family. Finally, we have shown that a selected peptide (CPQPRPLC) is a novel functional mimic motif that binds specifically to vascular endothelial growth factor receptor-1 and to neuropilin-1 (Giordano et al. 2001). As a follow-up, we have used NMR-based spectroscopy to understand the structural basis of the interaction between our mimic motif and the VEGF receptors. We have then shown that receptor binding is mediated by the motif Arg-Pro-Leu, thus forming a candidate small drug lead (Giordano et al. 2005). Finally, we have shown that BRASIL is useful to isolate targeting ligands and to probe cell surface receptor diversity with phage libraries in the NCI-60 cell panel (Kolonin et al. 2006b).

8.6.1   FUTURE PLANNED RESEARCH

- We are currently developing the neuropilin-binding peptide motif Arg-Pro-Leu as a cyclic retro-inverted small mimetic, $_D$(Cys-Leu-Pro-Arg-Cys). This targeted drug has shown anti-angiogenic effects in several experimental models of angiogenesis (Giordano et al., submitted).
- In a related line of research, we have fingerprinted therapeutic monoclonal antibodies (such as cetuximab) to discover epitopes that also yielded a cyclic retro-inverted small mimetic with EGFR decoy attributes (Cardó-Vila et al., submitted).
- In as yet unpublished work, we have shown that BRASIL-based cell surface profiling of the NCI-60 cancer cell lines with a combinatorial peptide library yielded a large panel of candidate target ligand motifs for validation and receptor identification.

8.7   HYBRIDOMA-FREE GENERATION OF
MONOCLONAL ANTIBODIES

Production of monoclonal antibodies requires immortalization of splenocytes by somatic fusion to a myeloma cell line partner ("hybridoma"). Although hybridomas can be immortal, they may depend on a feeder cell layer and may be genetically unstable. Since the inception of hybridoma technology, efforts to improve efficiency and stability of monoclonal antibody-producing cell lines have not brought about substantial progress. Moreover, suitable human multiple myeloma-derived cell lines for the production of human antibodies have been very difficult to develop.

In previous work, we reported a strategy that simplifies the generation of antibodies and eliminates the need for hybridomas. We demonstrated that splenocytes derived from transgenic mice harboring a mutant temperature-sensitive simian virus-40 large tumor antigen under the control of a mouse major histocompatibility promoter (termed *H-2K^b*-tsA58 or immortomouse) are conditionally immortal at permissive temperatures and produce monoclonal antibodies (Pasqualini and Arap 2004).

8.7.1 FUTURE PLANNED RESEARCH

We are genetically engineering a conditional ImmortoXenomouse strain that will produce hybridoma-free human antibodies. This reagent may well enable personalized monoclonal antibody-based therapy. If so, this approach will become a method of choice for generation and production of monoclonal antibodies with potential advantages in high-throughput target discovery and/or antibody-based immunotherapy.

8.8 CONCLUSIONS

We focus on discovery and evaluation of functional protein–protein interactions in the context of human disease for the development of ligand-directed targeted agents. Over the past seven years, we have identified targets against human cancer and obesity. If successful, a receptor-based human vascular map will have broad implications for the development of vascular-targeted agents. Thus far, we have been applying these platforms on an academic scale. The HHMI funding mechanism would allow our program to vastly expand our research endeavors.

REFERENCES

Arap, W., W. Haedicke, M. Bernasconi, R. Kain, D. Rajotte, S. Krajewski, H. M. Ellerby, D. E. Bredesen, R. Pasqualini, and E. Ruoslahti. 2000. Targeting the prostate for destruction through a vascular address. *Proc Natl Acad Sci U S A* 99:1527–31.
Arap, W., M. G. Kolonin, M. Trepel, J. Lahdenranta, M. Cardó-Vila, R. J. Giordano, P. J. Mintz, et al. 2002. Steps toward mapping the human vasculature by phage display. *Nat Med* 8:121–27.
Arap, M. A., J. Lahdenranta, P. J. Mintz, A. Hajitou, A. S. Sarkis, W. Arap, and R. Pasqualini. 2004. Cell surface expression of the stress response chaperone GRP78 enables tumor targeting by circulating ligands. *Cancer Cell* 6:275–84.
Arap, W., R. Pasqualini, and E. Ruoslahti. 1998. Cancer treatment by targeted drug delivery to tumor vasculature in a mouse model. *Science* 279:377–80.
Cardó-Vila, M., W. Arap, and R. Pasqualini. 2003. αvβ5 Integrin-dependent programmed cell death triggered by a peptide mimic of Annexin V. *Mol Cell* 11:1151–62.
Chen, L., A. J. Zurita, P. U. Ardelt, R. J. Giordano, W. Arap, and R. Pasqualini. 2004. Design and validation of a bifunctional ligand display system for receptor targeting. *Chem Biol* 11:1081–91.
Ellerby, H. M., W. Arap, L. M. Ellerby, R. Kain, R. Andrusiak, G. D. Rio, S. Krajewski, et al. 1999. Anti-cancer activity of targeted pro-apoptotic peptides. *Nat Med* 5:1032–38.

Giordano, R. J., C. D. Anobom, M. Cardó-Vila, J. Kalil, A. P. Valente, R. Pasqualini, F. C. L. Almeida, and W. Arap. 2005. Structural basis for the interaction of a vascular endothelial growth factor mimic peptide motif and its corresponding receptors. *Chem Biol* 12:1075–83.

Giordano, R. J., M. Cardó-Vila, J. Lahdenranta, R. Pasqualini, and W. Arap. 2001. Biopanning and rapid analysis of selective interactive ligands. *Nat Med* 7:1249–53.

Hajitou, A., R. Pasqualini, and W. Arap. 2006a. Vascular targeting: Recent advances and therapeutic perspectives. *Trends Cardiovasc Med* 16:80–88.

Hajitou, A., R. Rangel, M. Trepel, S. Soghomonyan, J. G. Gelovani, M. M. Alauddin, R. Pasqualini, and W. Arap. 2007. Design and construction of targeted AAVP vectors for mammalian cell transduction. *Nat Protoc* 3:523–31.

Hajitou, A., M. Trepel, C. E. Lilley, S. Soghomonyan, M. M. Alauddin, F. C. Marini III, B. H. Restel, et al. 2006b. A hybrid vector for ligand-directed tumor targeting and molecular imaging. *Cell* 125:385–98.

Koivunen, E., W. Arap, H. Valtanen, A. Rainisalo, O. P. Medina, P. Heikkila, C. Kantor, et al. 1999. Tumor targeting with a selective gelatinase inhibitor. *Nat Biotechnol* 17:768–74.

Kolonin, M. G., L. Bover, J. Sun, A. J. Zurita, K. A. Do, J. Lahdenranta, M. Cardó-Vila, et al. 2006b. Ligand-directed surface profiling of human cancer cells with combinatorial peptide libraries. *Cancer Res* 66:34–40.

Kolonin, M. G., R. Pasqualini, and W. Arap. 2002. Teratogenicity induced by targeting a placental immunoglobulins transporter. *Proc Natl Acad Sci U S A* 99:13055–60.

Kolonin, M. G., P. K. Saha, L. Chan, R. Pasqualini, and W. Arap. 2004. Reversal of obesity by targeted ablation of adipose tissue. *Nat Med* 10:625–32.

Kolonin, M. G., J. Sun, K. A. Do, C. I. Vidal, Y. Ji, K. A. Baggerly, R. Pasqualini, and W. Arap. 2006a. Synchronous selection of homing peptides for multiple tissues by in vivo phage display. *FASEB J* 20:979–81.

Lahdenranta, J., R. Pasqualini, R. O. Schlingemann, M. Hagedorn, W. B. Stallcup, C. D. Bucana, R. L. Sidman, W. Arap. 2001. An anti-angiogenic state in mice and humans with retinal photoreceptor cell degeneration. *Proc Natl Acad Sci U S A* 98:10368–73.

Lahdenranta, J., R. L. Sidman, R. Pasqualini, and W. Arap. 2007. Treatment of hypoxia-induced retinopathy with targeted proapoptotic peptidomimetic in a mouse model of disease. *FASEB J* 21:3272–78.

Marchió, S., J. Lahdenranta, R. O. Schlingemann, D. Valdembri, P. Wesseling, M. A. Arap, A. Hajitou, et al. 2004. Aminopeptidase A is a functional target in angiogenic blood vessels. *Cancer Cell* 5:151–62.

Mintz, P., J. Kim, K. A. Do, X. Wang, R. G. Zinner, M. Cristofanilli, M. A. Arap, et al. 2003. Fingerprinting the circulating repertoire of antibodies from cancer patients. *Nat Biotechnol* 21:57–63.

Müller, O. J., F. Kaul, M. D. Weitzman, R. Pasqualini, W. Arap, J. A. Kleinschmidt, and M. Trepel. 2003. Random peptide libraries displayed on adeno-associated virus to select for targeted gene therapy vectors. *Nat Biotechnol* 21:1040–46.

Pasqualini, R., and W. Arap. 2004. Hybridoma-free generation of monoclonal antibodies. *Proc Natl Acad Sci U S A* 101:257–59.

Pasqualini, R., E. Koivunen, R. Kain, J. Lahdenranta, M. Sakamoto, A. Stryhn, R. A. Ashmun, L. H. Shapiro, W. Arap, and E. Ruoslahti. 2000. Aminopeptidase N is a receptor for tumor-homing peptides and a target for inhibiting angiogenesis. *Cancer Res* 60:722–27.

Pentz, R. D., C. B. Cohen, M. Wicclair, M. A. DeVita, A. L. Flamm, S. J. Youngner, A. B. Hamric, et al. 2005. Ethics guidelines for research with the recently dead. *Nat Med* 11:1145–49.

Pentz, R. D., A. L. Flamm, R. Pasqualini, C. J. Logothetis, and W. Arap. 2003. Revisiting ethical guidelines for research with terminal wean and brain-dead participants. *Hastings Cent Rep* 33:20–26.

Rajotte, D., W. Arap, M. Hagedorn, E. Koivunen, R. Pasqualini, and E. Ruoslahti. 1998. Molecular heterogeneity of the vascular endothelium revealed by in vivo phage display. *J Clin Invest* 102:430–37.

Rangel, R., Y. Sun, L. Guzman-Rojas, M. G. Ozawa, J. Sun, R. J. Giordano, C. S. Van Pelt, et al. 2007. Impaired angiogenesis in aminopeptidase N-null mice. *Proc Natl Acad Sci U S A* 14:4588–93.

Sergeeva, A., M. G. Kolonin, J. J. Molldrem, R. Pasqualini, and W. Arap. 2006. Display technologies: Application for the discovery of drug and gene delivery agents. *Adv Drug Deliv Rev* 58:1622–54.

Soghomonyan, S., A. Hajitou, R. Rangel, R. Pasqualini, W. Arap, J. G. Gelovani, and M. Alauddin. 2007. Molecular PET imaging of HSV1-tk reporter gene expression using ^{18}F-FEAU. *Nat Protoc* 2:416–23.

Souza, G. R., D. R. Christianson, F. I. Staquicini, M. G. Ozawa, E. Y. Snyder, R. L. Sidman, J. H. Miller, W. Arap, and R. Pasqualini. 2006a. Networks of gold nanoparticles and bacteriophage as biological sensors and cell-targeting agents. *Proc Natl Acad Sci U S A* 103:1215–20.

Souza, G. R., C. S. Levin, A. Hajitou, R. Pasqualini, W. Arap, and J. H. Miller. 2006b. In vivo detection of gold-imidazole self-assembly complexes: NIR-SERS signal reporters. *Anal Chem* 78:6232–37.

Trepel, M., W. Arap, and R. Pasqualini. 2001. Modulation of the immune response by systematic targeting of antigens to lymph nodes. *Cancer Res* 61:8110–12.

Vidal, C. I., P. J. Mintz, K. Lu, L. M. Ellis, L. Manenti, R. Giavazzi, D. M. Gershenson, et al. 2004. An HSP90-mimic peptide revealed by fingerprinting the pool of antibodies from ovarian cancer patients. *Oncogene* 23:8859–67.

Yao, V. J., M. G. Ozawa, M. Trepel, W. Arap, D. M. McDonald, and R. Pasqualini. 2005. Targeting pancreatic islets with phage display assisted by laser pressure catapult microdissection. *Am J Path* 166:625–36.

Zurita, A. J., P. Troncoso, M. Cardó-Vila, C. J. Logothetis, R. Pasqualini, and W. Arap. 2004. Combinatorial screenings in patients: The interleukin-11 receptor α as a candidate target in the progression of human prostate cancer. *Cancer Res* 64:435–39.

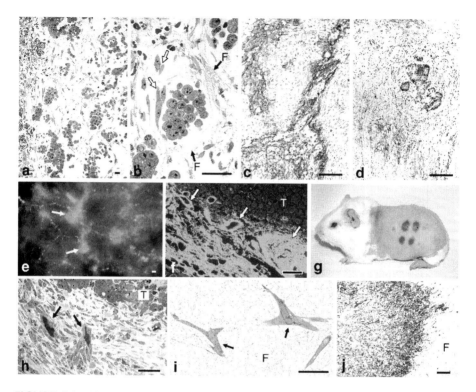

FIGURE 1.1 (a) and (b) Line 10 tumor cells 48 hours after transplant into the subcutaneous space of syngeneic strain 2 guinea pigs. Fibrin (F) forms a water-trapping gel that serves as a provisional stroma that separates tumor cells into discrete islands and that provides a favorable matrix for fibroblast (*white arrows*) and endothelial cell migration. (c) and (d) Immunohistochemical demonstration of fibrin (brown staining) in guinea pig line 1 and human colorectal adenocarcinomas, respectively. (e) and (f) Blood vessels (*arrows*) supplying line 10 guinea pig tumors are hyperpermeable to circulating macromolecular fluoresceinated dextran. (g) Miles assay illustrating permeability to Evans blue-albumin complex at sites of intradermal injections of the following: (*top row, left to right*) neutralizing anti-VEGF-A antibody, line 10 guinea pig tumor ascites fluid, mix of line 10 tumor ascites fluid and control IgG, mix of tumor ascites fluid and neutralizing anti-VEGF-A antibody; (*bottom row*) line 1 tumor ascites fluid, mix of line 1 tumor ascites fluid and control IgG, and mix of line 1 tumor ascites fluid and neutralizing anti-VEGF-A antibody. (h) Fibroblasts and blood vessels (*black arrows*) invade line 1 tumor fibrin gel, replacing it with fibrous connective tissue. (i) Fibroblasts (*arrows*) migrate through fibrin gel (F) in vitro. (j) Implanted fibrin gel (F) in subcutaneous space is replaced by ingrowth of fibroblasts and new blood vessels, creating granulation-like vascular connective tissue. Scale bars, 25 μm (b, i), 50 μm (a, c, d, h), and 100 μm (e, f, j). *Sources:* Reproduced from H. F. Dvorak, J. A. Nagy, D. Feng, L. F. Brown, and A. M. Dvorak. Vascular Permeability Factor/Vascular Endothelial Growth Factor and the Significance of Microvascular Hyperpermeability in Angiogenesis. *Curr Top Microbiol Immunol* 237 (1999): 97–132; H. F. Dvorak. Rous-Whipple Award Lecture. How Tumors Make Bad Blood Vessels and Stroma. *Amer J Path* 162 (2003): 1747–57; and H. F. Dvorak, V. S. Harvey, P. Estrella, L. F. Brown, J. McDonagh, and A. M. Dvorak. Fibrin Containing Gels Induce Angiogenesis. Implications for Tumor Stroma Generation and Wound Healing. *Lab Invest* 57 (1987): 673–86, with permission.

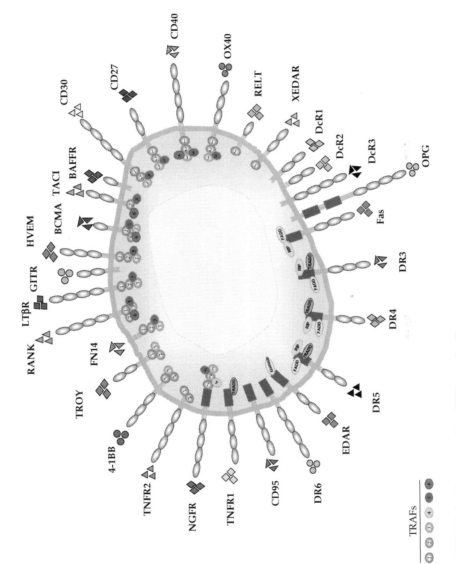

FIGURE 2.1 Receptors of the TNF superfamily.

TRAFs

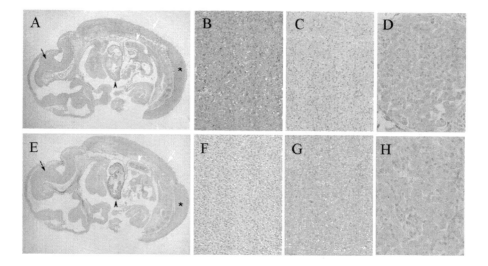

FIGURE 9.3 PrP^c and STI1 expression in mice embryos. STI1 (**A–D**) and PrP^c (**E–F**) expression were evaluated by immunohistochemistry using specific antibodies previously described (25,26). Proteins distribution (**A, E**) in E12 mice embryonic nervous system: brain (*black arrow*), spinal cord (*), and dorsal root ganglia (*white arrow*), as well as other organs such as heart (*black arrowhead*) and lung (*white arrowhead*) (37× magnification). Protein distribution in E17 mouse embryonic brain (**B, F**), medulla (**C, G**), and dorsal root ganglia (**D, H**) (300× magnification).

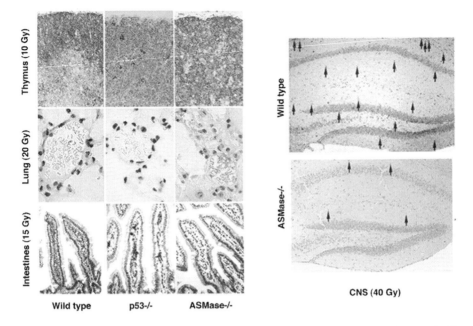

FIGURE 11.6 Radiation-induced endothelial apoptosis is ASMase-dependent and p53-independent. Tissue specimens were obtained from mice at 4–10 hours after exposure to whole-body irradiation at the indicated doses. Histologic tissue sections of 5 µm were double-stained for apoptosis by TUNEL and for endothelium using antibodies to tissue-specific endothelial surface markers. Brown nuclei are indicative of apoptosis. The residual clusters in the p53[-/-] thymus represent apoptotic microvascular endothelium. While p53 has no impact on endothelial apoptosis in the lung, gut, or brain (*arrows*), acid sphingomyelinase largely blocks this event. *Sources*: Adapted from P. Santana, L. A. Pena, A. Haimovitz-Friedman, et al. "Acid Sphingomyelinase-Deficient Human Lymphoblasts and Mice Are Defective in Radiation-Induced Apoptosis." *Cell* 86, no. 2 (1996): 189–99; F. Paris, Z. Fuks, A. Kang, et al. "Endothelial Apoptosis as the Primary Lesion Initiating Intestinal Radiation Damage in Mice." *Science* 293, no. 5528 (2001): 293–97; L. A. Pena, Z. Fuks, and R. N. Kolesnick. "Radiation-Induced Apoptosis of Endothelial Cells in the Murine Central Nervous System: Protection by Fibroblast Growth Factor and Sphingomyelinase Deficiency." *Cancer Res* 60, no. 2 (2000): 321–27.

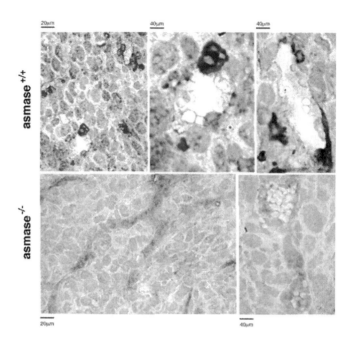

FIGURE 11.7 MCA/129 fibrosarcomas implanted into *asmase*⁻/⁻ mice display reduced baseline and radiation-induced endothelial cell apoptosis. Representative 5-μm histologic tumor sections obtained 4 hours after exposure to 15 Gy, stained for apoptosis by TUNEL and for the endothelial cell surface marker CD-34. Apoptotic endothelium manifests a red-brown TUNEL-positive nuclear signal surrounded by dark blue plasma membrane signal of CD-34 staining. *Source*: From M. Garcia-Barros, F. Paris, C. Cordon-Cardo, et al. "Tumor Response to Radiotherapy Regulated by Endothelial Cell Apoptosis." *Science* 300, no. 5622 (2003): 1155–59.

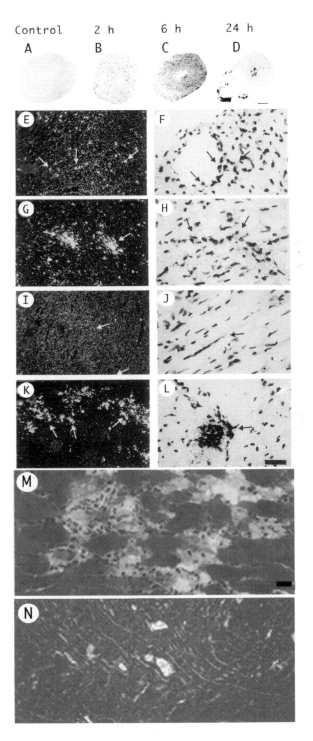

Control   2 h   6 h   24 h

FIGURE 12.1

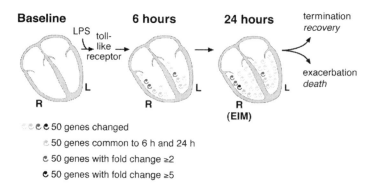

Baseline **6 hours** **24 hours** termination *recovery*

LPS toll-like receptor

exacerbation *death*

(EIM)

L — left ventricle; R — right ventricle

ⓔⓔⓔ 50 genes changed

ⓔ 50 genes common to 6 h and 24 h

ⓔ 50 genes with fold change ≥2

ⓔ 50 genes with fold change ≥5

FIGURE 12.5 Diagram summarizing EIM data. LPS acts through Toll-like receptor and initiates the heart's response to SIRS. Each spiral represents ≈50 genes or EST that are significantly changed in our microarray studies at $p \geq 0.01$. Green spirals denote the number of genes that are common to the 6- and 24-h transcriptional response; they characterize a possible stereotypic response of the innate immune system in the heart. Red and black spirals designate genes that have a fold-change of 2 or higher and 5 or higher, respectively. There is a small divergence in the response of the RV and LV at 6 h, which is intensified at 24 h. Note that the expansion of the transcriptional response in the heart at 24 h is more intense in the RV (it has more red and black spirals), the site for EIM. Hypothetically, the exacerbation of EIM would lead to death and the termination of the inflammatory process would result in recovery. R, right ventricle; L, left ventricle. Reproduced with permission from M. L. Wong, F. O'Kirwan, N. Khan, J. Hannestad, K. H. Wu, D. Elashoff, et al. "Identification, Characterization, and Gene Expression Profiling of Endotoxin-Induced Myocarditis." *Proc Natl Acad Sci U S A* 100 (2003): 14241–46, copyright by the National Academy of Sciences of the United States of America.

FIGURE 12.1 Induction of NOS2 mRNA in heart after LPS. The top image is a composite of film autoradiography showing time course of NOS2 mRNA in the heart (**a–d**) by in situ hybridization histochemistry after LPS treatment. In the heart, NOS2 mRNA induction starts at 2 h, progress at 6 h. At 24 h NOS2 levels return virtually to baseline in the LV, but are increased in the RV and a residual response is still present in the endocardium. Scale bar = 2 mm. The middle set of images shows a composite of low (*columns one and two*) and high (*column three*) magnification images of NOS2 mRNA in heart ventricles. First three rows depict the LV at 2 h (**A, B, C**); 6 h (**D, E, F**); and 24 h (**G, H, I**). The fourth column depicts the RV at 24 h (**J, K, L**). Arrows point to areas of increased NOS2 expression appearing as white dots in dark field (*first column*) and black dots in high magnification images. Note the high levels of NOS2 mRNA in the RV at 24 h (**J** and **L**). Scale bar, 378 _μm for columns one and two, and 30 _μm for column three. The bottom set of images shows bright field images of heart ventricles 24 h after induction of SIRS. This section of RV myocardium (M) is infiltrated by a moderate number of neutrophils that surround degenerated myocardiocytes. The cardiomyocytes are fragmented, pale, and vacuolated and in many areas the proteins are globular. Some globular proteins are surrounded by degenerated neutrophils characterized by fine nuclear debris, and several neutrophils contain bright eosinophilic cytoplasmic material (phagocytosis). Vacuolation varies from small to large clear, round to irregular clear spaces. Notice no signs of local inflammatory reaction in LV (N). Scale bars, 7.5 μm (M), and 31.2 μm (N). *Source*: Reproduced with permission from M. L. Wong, F. O'Kirwan, N. Khan, J. Hannestad, K. H. Wu, D. Elashoff, et al. "Identification, Characterization, and Gene Expression Profiling of Endotoxin-Induced Myocarditis." *Proc Natl Acad Sci U S A* 100 (2003): 14241–46, copyright by The National Academy of Sciences of the United States of America.

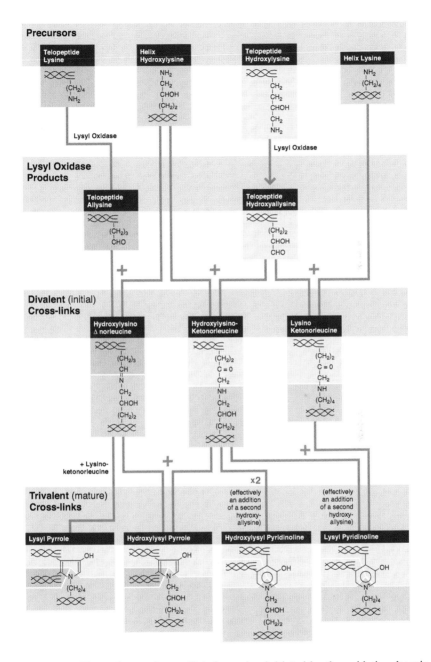

FIGURE 16.2 The pathway of cross-link formation initiated by the oxidative deamination of lysyl and hydroxylysyl side chains in collagen. Lysyl and hydroxylysyl side chains of telopeptide and triple-helical origin are tracked to mature cross-links by color. The pathways shown predominate in skeletal tissues. *Source:* Reproduced with permission from D. R. Eyre, M. A. Weiss, and J. J. Wu. "Advances in Collagen Cross-Link Analysis." *Methods* 45 (2008): 65–74.

9 Chasing Elusive Cellular Prion Protein Receptor

Vilma R. Martins, Sandro J. De Souza, and Ricardo R. Brentani

CONTENTS

9.1 FOUNDATIONS

Protein–protein interactions are major pillars of life on this planet. Although much has been devoted to their study in the last century, we still have many unanswered questions. How did protein–protein interactions arise? What are the evolutionary mechanisms responsible for the emergence of protein interacting pairs? How did the protein interaction network that sustains most of life's physiological aspects come to exist?

Like almost everything in biology, the truth probably encompasses many answers. Domain shuffling, for instance, is expected to generate a web of interactions within a given proteome. Let's suppose that domain A interacts with domain B. If both domains are shuffled to different proteins over evolutionary time, this may allow the new host proteins to bind to each other.

A second evolutionary mechanism to generate protein interaction is the coding capacity of both DNA strands in the same locus. The basis for this lies on the possible interaction between peptides coded by DNA sequences complementary to each other (1). Curiously, the genetic code is organized in a way in which there is a strong bias for complementary codons to encode amino acids with complementary hydropathy. Most of the codons coding for hydrophobic amino acids are complemented by codons coding for hydrophilic amino acids and vice versa. This led Biro (2) to suggest that peptides encoded by complementary sequences could interact. The plausibility of such interactions has been extensively discussed (3). Amino acids with opposing hydropathicities can bind one another. It has been known for more than 40 years (4) that hydrophilic amino acids with a long nonpolar lateral chain leading to a polar group can establish hydrophobic interactions with hydrophobic amino acids. Furthermore, amino–aromatic interactions have been shown to be quite common among interacting peptides (5).

Dozens of examples have been reported showing interaction of proteins encoded by complementary DNA sequences (6). Bost and coworkers (7) demonstrated a high-affinity binding of corticotrophin to synthetic peptides derived from the predicted complementary sequences. This approach was further used to study the interaction between angiotensin II and its receptor (8).

Our group has extensively used the "complementary hydropathy" approach to study protein–protein interactions in several models. First, Brentani and coworkers (9) deduced a putative binding site for the RGDS peptide in fibronectin. This peptide, WTVPTA, was able to bind fibronectin in an RGDS-dependent manner and was able to inhibit the binding of fibronectin to an osteosarcoma cell line. Antibodies raised against this antibody were able to recognize a 140-kDa fibronectin receptor. The same peptide was also effective in recognizing another RGDS-dependent receptor, GPIIbIIIa, which is present exclusively in platelets. Interestingly, sera from idiopathic thrombocytopenic purpura (ITP) patients, which have auto-antibodies against GPIIbIIIa, recognized the peptide WTVPTA (10).

We have also used the same approach to study the interaction between collagen and mammalian collagenases (11). Collagen is the most abundant component of the extracellular matrix and is cleaved by collagenase in a single position between amino acids 775 and 776. Based on the susceptible site in collagen, we deduced a peptide that should mimic the collagen-binding site in the collagenase molecule. Since collagen is a heterotrimer with two different peptide chains, we were able to deduce two complementary peptides, TKKTLRT and SSNTLRS. Interestingly, the second peptide shows sequence identity to a peptide in neutrophil collagenase, SSNPIQP. Both deduced peptides were able to bind collagen and, most important, inhibit collagenolytic activity. The same is true for antibodies raised against TKKTLRT.

One of us (RRB) has hypothesized that the use of the antisense strand in exon shuffling events would allow the donor and acceptor proteins to interact (1). In that case, interacting peptides in different proteins may have co-evolved. The argument from Gilbert (12) that exon shuffling would speed up evolution is valid for this model as well. Brentani and coworkers (9) illustrated their model by speculating that exon 7 in fibronectin, which codes for the RGDS signature, has an ORF when the antisense strand is read in the same reading frame as the sense strand. One apparent problem with this model is the splicing of the new exon. Since the antisense strand is used, the splicing of this exon is not expected to occur because the signals "GT…AG" necessary for splicing will be inverted. Therefore, for splicing to occur, new splicing signals have to be found.

In the last few years, we have used the principle of complementary hydropathy to study the interaction between the cellular prion protein and its putative receptor. Here we review the major steps taken by us to characterize this prion protein ligand.

9.2 BEGINNING

9.2.1 ANTI-PRION PROTEIN GENE AND PUTATIVE CELLULAR PRION PROTEIN RECEPTOR

Prions are the pathological agents responsible for a group of fatal neurodegenerative disorders that affect both animals and humans and can exhibit sporadic, inherited, or infectious presentations (13).

Cellular prion protein (PrP^c) expression is absolutely necessary for disease propagation. This protein is converted into an abnormal form, called PrP^{sc}, through a major conformational change. In contrast to PrP^c, the pathogenic PrP^{sc} isoform has a relatively high content of β-sheet structure, partial resistance to proteolytic digestion, insolubility in nonionic detergents, and a tendency to aggregate into amyloid-like fibrils and plaques (13). According to the protein-only hypothesis, the transmission of these diseases does not require nucleic acids, and PrP^{sc} itself is the infectious prion pathogen. It is also suggested that all properties of prions are enciphered in the conformation of PrP^{sc}, and that the great clinicopathological diversity of these diseases reflects the variety of PrP^{sc} conformers (13,14).

PrP^c is a glycosyl-phosphatidylinositol (GPI)-anchored cell surface sialoglycoprotein which is internalized and part of it recycles to the cell surface (15). It is of general agreement that PrP^c cellular trafficking plays an essential role in its pathological conversion to PrP^{sc} (16). PrP^c cellular trafficking has hardly been studied and is a field of great challenge. Some groups, including Stanley Prusiner's, presented evidence for caveolae-mediated PrP^c internalization (17). A significant number of results, however, led David Harris to propose clathrin-mediated PrP^c intracellular traffic (15). Nowadays, it is believed that more than one process may participate in PrP^c and PrP^{sc} internalization (16).

The statement from David Harris implying participation of a receptor connecting PrP^c, at the outer leaflet side of the membrane, to the clathrin machinery (18) provided us, some years ago, with a relevant question that needed to be addressed.

In 1991, Goldgaber reported the possible existence of an anti-prion protein encoded by an ORF contained in the complementary strand of the cellular prion protein gene (19). Since both ORFs are in the same phase, the anti-prion protein displays a hydropathy plot quite complementary to that of the prion protein (19). A message for this putative antisense gene was observed using strand-specific probes

(20,21), although this does not seem to come from the prion locus, since PrP null mice still express the same messenger RNA (21). Nevertheless, the anti-prion protein suggested by Goldgaber could be a likely candidate for a cellular prion receptor, since exon shuffling could account for the emergence of a new gene in a different locus, as suggested by Brentani (3).

9.3 PRPᶜ RECEPTOR CHARACTERIZATION AND IDENTIFICATION

9.3.1 FUNDAMENTAL DATA TO REVEAL PUTATIVE RECEPTOR BINDING SITE AT PRPᶜ

Based on the discussion above, we decided to evaluate the existence of the putative anti-prion protein despite its genomic localization. Two important clues helped us to propose the binding site for such a receptor at the PrPᶜ molecule. First, a domain spanning residues 117 to 135 of the chicken PrPᶜ molecule (corresponding to mouse residues 103 to 121) represents a critical region responsible for the molecule's internalization (18). Furthermore, since this domain is the most conserved among several species (22) it is reasonable to assume that it is functionally important. The second clue was provided by data showing that a PrPᶜ peptide, consisting of residues 106 to 126 of the human protein, can adopt a β-sheet conformation and reproduce, in cultured cells, the neurotoxic effects of the infectious agent (23). In fact, Forloni and coworkers were able to map an even shorter peptide from amino acids 114 to 126 that preserves the neurotoxic properties (23).

Thus, we speculated that the basis for the peptide's selective neurotoxicity was because peptide 114–126 contains the binding site recognized by the putative receptor proposed by David Harris, and thus is the only one capable of entering the cell (24).

9.3.2 COMPLEMENTARY HYDROPATHY APPROACH TO CHARACTERIZE PRPᶜ RECEPTOR

We then deduced the peptide encoded by the DNA strand complementary to the human PrPᶜ gene coding for amino acids 114 to 129. According to the basis of the complementary hydropathy approach, this peptide, HVATKAPHHGPCRSSA, which was called complementary PrPᶜ peptide (24) and later PrR (25,26), should mimic a PrPᶜ receptor binding site. The hydropathy profiles of the human PrPᶜ domain from amino acids 114 to 129 and that from the deduced complementary peptide are shown in Figure 9.1.

The complementary PrPᶜ peptide was used to immunize 10 mice and only 2 of them produced antibody titers against the peptide, probably because of the conservation among species of the protein recognized. This turned out to be true as we will further discuss. Antibodies against the complementary peptide were able to identify, in brain extracts, a single protein of about 66 kDa in brain extracts, which was localized at the cell surface although abundant expression was also present in the cytoplasm (24).

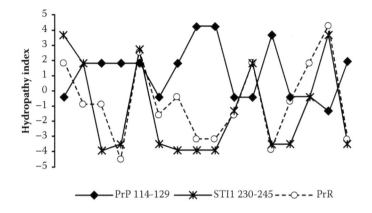

——◆—— PrP 114-129 ——✳—— STI1 230-245 - - -○- - - PrR

FIGURE 9.1 Hydropathy plot of the human PrP^c domain 114–129 (GAAAAGAVVGGL-GGYM) where we have first mapped the putative receptor binding site. The peptide codified by the complementary DNA strand correspondent to PrP^c 114–129 region was deduced (HVATKAPHHGPCRSSA) (PrR) and presented, as predicted, a complementary hydropathic profile of the former peptide. Serum raised against peptide PrR allowed us to identify STI1 as the PrP^c ligand. The PrP^c binding site at STI1 maps within domain 230 to 245 (ELGNDAYKKKDFDKAL), which has the same hydropathic profile of PrR while complementary to PrP^c 114–129. The PrR and STI1 amino acid hydrophathic profiles are shown from carboxy to amino-terminal while PrP^c is from amino to carboxy-terminal (*left to right*).

PrP^c and the 66 kDa protein from brain homogenate could be separated by differential ammonium sulfate precipitation. Isolated fractions containing each one of the proteins as well as recombinant PrP^c were used to demonstrate that PrP^c and the 66-kDa molecule bind to each other in vitro. Furthermore, using flow cytometry, we showed that the 66-kDa protein, purified from mice brain cellular membranes, was able to bind the cell surfaces of mouse neuroblastoma cells (N2A). This association was blocked either by antibodies against the complementary peptide or against PrP^c. Thus, we were able to conclude that the 66 kDa protein binds PrP^c ex vivo (24).

We further demonstrated the biological relevance of the PrP^c and the 66-kDa receptor interaction. Toxicity of the peptide 114–126 on hippocampal primary neuronal cultures was abolished by the complementary peptide. Furthermore, as predicted by the complementary hydropathy theory, complementary peptide and peptide 114–126 bind one another, leading to inactivation of the former's toxicity. It was also possible to speculate at that time (25) that the complementary peptide may also provide neuroprotective signals that minimize the toxic effects of the PrP^c 106–126 peptide. Moreover, if, as we claimed, such a peptide is neurotoxic because it contains the binding site for the receptor, allowing its internalization, an antibody raised against the receptor should inhibit peptide neurotoxicity. That was exactly the case; serum generated against the complementary peptide was able to abolish PrP^c 106–126 peptide neurotoxicty (24).

Today, looking at these data from the perspective of our more recent studies, which will be discussed later, we may add another possibility for these results. The 66-kDa protein is known to transduce neuroprotective signals through its interaction with

PrP^c (25,26). Thus, serum raised against the 66-kDa protein may work as an agonist increasing the protein interaction with PrP^c and intensifying survival signals.

9.3.3   IDENTIFICATION OF 66 KDA PRP^c RECEPTOR AS STRESS-INDUCIBLE PROTEIN 1

By the time we obtained sera against the complementary peptide, PrR, we started different approaches to decipher the identity of the 66-kDa molecule. In order to increase our chances to identify the 66-kDa, two parallel strategies using serum against PrR peptide were undertaken: screening expression libraries and isolation of specifically recognized protein spots using two-dimensional gels followed by protein sequencing using mass spectrometry. We had no success with the library screening, probably because of the low titers and modest amounts of serum available.

Before we discuss the complete characterization of our 66-kDa prion receptor, it is important to mention that along with our article in *Nature Medicine*, Rieger and coworkers published an article in the same issue proposing the 67-kDa laminin receptor as a receptor for PrP^c (27). The articles were commented on by A. Mitchell (28) who suggested that the 66-kDa identified by our group and the laminin receptor should be the same molecule involved with prion internalization. Later on, two other reports further explored the internalization of prion mediated by the 67-kDa laminin receptor (29,30).

In this context, while the experiments on the 66-kDa receptor identification were being conducted, we demonstrated that an antibody against the 67-kDa laminin receptor was unable to recognize our protein in two-dimensional gels. Furthermore, the protein recognized by our serum was unaffected by treatment with concentrated hydroxylamine, which cleaves acetyl groups and converts the 67-kDa laminin receptor into the 37-kDa precursor (24).

In two-dimensional gels, serum raised against the PrR recognized two bands, which were excised and sent to Alma Burlingame's laboratory at UCSF. Sequence obtained from both spots identified the 66-kDa protein as the mouse stress-inducible protein 1, STI1 or Hop for the human homologue (26), a co-chaperone that is able to directly interact with both Hsp90 and Hsp70 (reviewed in 31).

Rabbit serum raised against recombinant STI1 recognized the same protein identified by anti-PrR serum. Moreover, PrP^c binds to STI1 with a K_d of 1.4×10^{-7} M, and competition experiments using peptides covering the entire PrP^c molecule or PrP^c deletion mutants identified the domain within amino acids 113 to 125 of the mouse protein (114–126 in human) as the unique binding site for STI1 (26). These data confirmed our initial prediction for the receptor docking site at PrP^c (24).

On the other hand, as predicted by complementary hydropathy, the PrP^c binding site at STI1 should display a similar hydropathic profile to that of PrR. The region of the STI1 residues 230–245 was identified using the software HIDROLOG (unpublished), which provides a hydropathy index profile of amino acid sequences and searches for a domain with a similar pattern in a given protein. The hydropathic profiles of STI1 peptide 230 to 245 and the PrR peptide are very alike and complementary to the human PrP^c 114–129 domain (Fig. 9.1). A recombinant STI1 deleted for the 230–245 domain is unable to bind PrP^c, demonstrating that this is the unique binding site for PrP^c at the STI1 molecule (32).

Finally, cell surface binding and pull-down experiments demonstrated that recombinant PrP^c binds to cellular STI1 and co-immunoprecipitation assays strongly suggested that both proteins are associated in vivo (26).

9.4 AFTERMATH

9.4.1 PrP^c Interaction with STI1 Triggers Neuroprotection

We hypothesized that human PrP^c peptide 106–126 neurotoxicity was due to its interaction to the 66-kDa receptor and internalization (24). However, our data also raises the possibility that this peptide is not toxic per se, but that it impairs survival signals triggered by the 66-kDa receptor. In agreement with this idea, it has been demonstrated that introduction of truncated PrP^c gene alleles lacking either residues 32 to 121 or 32 to 134, but not those missing residues 32 to 106, into PrP^c null mice cause cerebellar neurodegeneration and ataxia (33), therefore suggesting that these deleted domains, which matched the 66-kDa receptor binding site, may be essential for the physiological functions of PrP^c and in particular neuronal survival.

In 1999 we started a collaboration with Prof. Rafael Linden from the Federal University of Rio de Janeiro. Since the identity of the 66-kDa PrP^c receptor was unknown at that time, we reasoned that the PrR peptide might be an interesting tool to evaluate the possible role of PrP^c in cell survival. Thus, we examined whether PrR peptide affected programmed cell death in an ex vivo preparation of organotypical retinal explants, a valuable model to assess cell death in the central nervous system (CNS) (34). In this model, anisomycin, an inhibitor of protein synthesis, induces cell death in recent post-mitotic cells within the neuroblastic layer.

PrR peptide was able to rescue wild-type mice neurons but not those from PrP^c ablated animals (*Prnp$^{0/0}$*) from cell death, suggesting that protection by the PrR peptide depends on PrP^c expression by retinal cells. Conversely, PrP^c peptide 106–126, while able to displace the binding between PrR and PrP^c, presented no effect over anisomycin-induced cell death. These results allowed us to conclude that rather than disrupting a neurotoxic effect mediated by PrP^c and the 66 kDa protein interaction, PrR mimicked the 66-kDa protein inducing neuroprotection (25).

Identification of the 66-kDa protein as STI1 and mapping of the PrP^c docking site at the domain 230–245 from STI1 permitted to reproduce all the results that we had previously obtained with the PrR peptide. Recombinant STI1, as well as STI1 peptide 230–245, the PrP^c docking site at the former, protected post-mitotic neurons from cell death at a concentration 10 times lower than that needed for PrR (26). A schematic representation of the peptides' and proteins' role on neuronal survival is presented in Figure 9.2.

Finally, we were also able to demonstrate that PrR, STI1, and STI1 230–245 peptide interaction with PrP^c increased PKA activity as well as the ERK 1/2 pathway, whereas only PKA activation was necessary for neuroprotection (25, unpublished results). In hippocampal neurons, the STI1-PrP^c interaction can induce neuronal

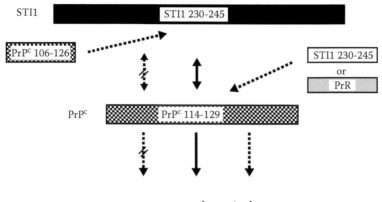

neuronal survival

FIGURE 9.2 PrP^c and STI1 interaction mediates neuronal survival. STI1, STI1 peptide 230–245, and PrR bind to PrP^c (114–129) and mediate neuronal survival. The PrP^c peptide 106–126 displaces the binding between STI1 and PrP^c and this may be one of the mechanisms related to its toxicity in hippocampal neurons. Nevertheless, in retina neurons, PrP 106–126 peptide is not toxic.

differentiation besides neuroprotection. Interestingly, the ERK 1/2 pathway was related to differentiation and PKA activation was connected to neuronal survival (32).

Murine STI1 has been described to be mainly localized in the cytoplasm (35) but its human homologue, Hop, was also found in the Golgi apparatus and small vesicles in normal cells and in the nucleoli of transformed cells (36). We demonstrated that the protein was found mainly in the cytoplasm with a small fraction at the cell surface (24,26). Data from another group (37) and from us demonstrated that STI1 can be secreted, mostly by primary astrocyte cultures (unpublished data). Therefore, these results indicate that STI1 is the natural trophic agent that leads to PrP^c activation.

It remains to be demonstrated how STI1, despite the absence of a transmembrane domain or a signal peptide for transmembrane transport (35), can be carried to the cell surface. There are other examples in the literature of proteins with such characteristics, and it has been speculated that they are either projected to the plasma membrane as part of a proteic complex or secreted by a pathway clearly distinct from the classical route through the endoplasmic reticulum and Golgi apparatus (39). Another important issue presently being experimentally addressed is how PrP^c, a GPI anchored protein, can perform signaling roles connecting extracellular signals with the intracellular milieu. Notwithstanding, PrP^c is not the only GPI protein displaying signaling capability. The urokinase type plasminogen activator receptor (uPAR), which presents complex signaling involving cell adhesion, proliferation, and migration in response to several ligands, is one example of such a protein. The signals from uPAR have to be passed through the membrane via several transmembrane adaptors, such as integrins, G-protein-coupled receptors, or caveolin (40).

An overview of our data indicates that STI1, either at the cell surface or secreted (trophic factor), binds PrP^c and is responsible for signal transduction. Although the

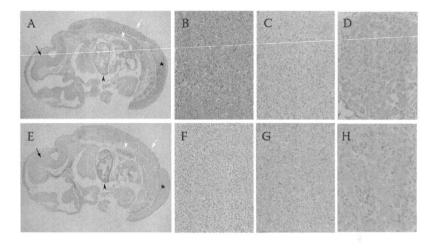

FIGURE 9.3   (See color insert following page 112.) PrP^c and STI1 expression in mice embryos. STI1 (**A–D**) and PrP^c (**E–F**) expression were evaluated by immunohistochemistry using specific antibodies previously described (25,26). Proteins distribution (**A, E**) in E12 mice embryonic nervous system: brain (*black arrow*), spinal cord (*), and dorsal root ganglia (*white arrow*), as well as other organs such as heart (*black arrowhead*) and lung (*white arrowhead*) (37× magnification). Protein distribution in E17 mouse embryonic brain (**B, F**), medulla (**C, G**), and dorsal root ganglia (**D, H**) (300× magnification).

mechanisms leading to signaling are unknown, PKA and ERK 1/2 are targets for downstream activation. Thus, it seems that PrP^c is the receptor for STI1 and not the other way around, as we had predicted. In fact, as will be discussed further, PrP^c association to other molecules from the extracellular matrix reinforces its function as a cell surface receptor.

The experimental approach described here permitted us to decipher some important cellular functions for PrP^c such as neuroprotection and differentiation. In addition, we observed that both proteins begin their expression at embryonic day 10 in mice and have a similar distribution in the nervous system: brain, spinal cord, dorsal root ganglia, as well as other organs such as heart and lung (Fig. 9.3). Thus, this indicates that their interaction may have pleiotropic functions both in and outside of the CNS.

In conclusion, the role of PrP^c–STI1 interaction in neuronal survival and differentiation will be important to understand the mechanisms involved with the neurodegeneration in prion diseases (41) and maybe with the physiopathology of other human diseases such as schizophrenia and epilepsy (42,43).

9.4.2   PrP^c Interaction with Laminin and Vitronectin and Validation of Complementary Hydropathy Theory

In 2000, while trying to identify the 66-kDa protein, we demonstrated that recombinant PrP^c was able to bind laminin, an essential extracellular matrix protein associated with neuronal pleiotropic functions such as survival, migration, and differentiation (44). In hippocampal primary cultures, neuritogenesis induced by laminin was completely blocked by anti-PrP^c antibodies. Additionally, neurons from PrP^c null mice (*Prnp*$^{0/0}$), when plated over laminin, extend 50% fewer neurites than those from wild-type animals. In order to map the PrP^c binding site at laminin, we inhibited these proteins' interaction with laminin peptides already known to be involved with neuritogenesis. The laminin γ-1 chain peptide, RDIAEIIKDI, disrupted the PrP^c–laminin interaction and, besides that, was capable of inducing neuritogenesis in wild-type neurons while being unable to stimulate *Prnp*$^{0/0}$ neurons (45). More recently, a competition approach using peptides covering the entire PrP^c molecule permitted us to map the PrP^c domain from amino acids 173–183 (mouse molecule) as the binding site for the laminin γ-1 peptide (46).

The complementary hydropathy theory has not been used in these studies; nonetheless, it is interesting to notice that hydropathy profiles of the PrP^c 173–183 and the laminin γ-1 peptide are complementary (Fig. 9.4A). If we deduce the amino acid sequence codified by the PrP^c complementary DNA strand corresponding to the 173–183 domain we obtain a peptide (GDIDAVVHQV) that has the same hydropathic profile of the laminin γ-1 peptide (Fig. 9.4A). On the other hand, we can also infer the laminin γ-1 peptide complementary strand and obtain a peptide (NILNDLCYVP) that mimics PrP^c 173 to 183 hydropathy profile (Fig. 9.4B). Thus, this turns out to be another validation for the complementary hydropathy theory.

While we were evaluating PrP^c interaction with laminin, we decided to appraise whether it could act as a broad ECM receptor, and our data showed that PrP^c binds vitronectin but not fibronectin or type IV collagen (47). As previously done, we conducted competition assays using PrP^c peptides and mapped the vitronectin binding site within amino acids 104–117 (PKTNLKHVAGAAAAG) of the murine PrP^c. We also used the software HIDROLOG, which provided the hydropathical index profile of this PrP^c region and searched for a domain with a complementary profile in the vitronectin molecule. We found two peptides with more than 70% amino acids sharing complementary hydropathy profiles to the PrP^c 104–117 domain. These peptides were synthesized and only one of them, from amino acids 307–320 (QRDSWENIFELLFW), was able to compete PrP^c–Vn interaction and to mimic the axonal growth induced by the entire molecule in wild-type mice neurons, although no effect was observed in cells from *Prnp*$^{0/0}$ animals. These data indicate that vitronectin domain 307–320 (mouse molecule) is the binding site for the PrP^c region spanning amino acids 104–117 (Fig. 9.5A).

In order to further test the complementary hydropathy theory in this model, we deduced the amino acid sequence codified by the PrP^c complementary DNA strain correspondent to the 104–117 domain and we obtained a peptide (SCRSPCDMLEVGFW) that has a complementary hydropathy profile to the former while presenting the same profile as the vitronectin 307–320 peptide (Fig. 9.5A).

The exercise can be done the other way around; a hypothetical peptide (PEEKVEDVLPRVPV) can be deduced from the complementary DNA strand of the 307–320 vitronectin domain. Again, the results are the same: the hypothetical peptide has a complementary hydropathy profile to the vitronectin 307–320 domain and the same profile of PrP^c peptide 104–117 (Fig. 9.5B).

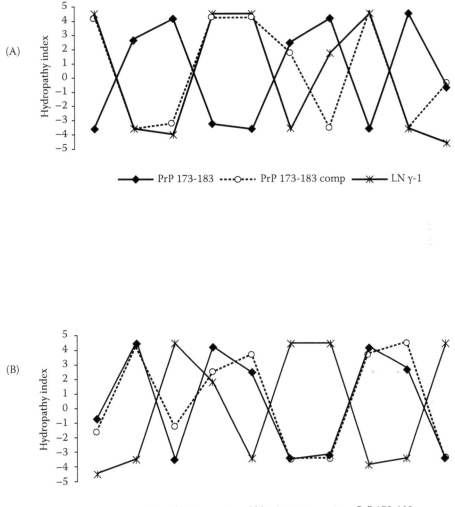

FIGURE 9.4 (A) Hydropathy plot of the mouse PrP^c domain 173–183 (NFVHDCVNIT) and its binding site at the LN γ-1 chain (RDIAEIIKDI). The peptide deduced from PrP^c 173–183 complementary DNA sequence (GDIDAVVHQV) has the same hydropathic profile of LN γ-1 chain peptide while complementary to PrP^c 173–183. The LN γ-1 chain (LN γ-1) and PrP^c 173–183 complementary peptides (PrP 173–183 comp) are shown from carboxy to amino-terminal while PrP^c 173–183 is from amino to carboxy-terminal (*left to right*). (B) The peptide deduced from LN γ-1 chain complementary DNA strand (NILNDLCYVP) has the same hydropathic profile of PrP^c 173–183 while complementary to LN γ-1 chain peptide. The LN γ-1 chain complementary peptide (LN γ-1 comp) and PrP^c 173–183 are shown from carboxy to amino-terminal while LN γ-1 chain peptide is from amino to carboxy-terminal (*left to right*).

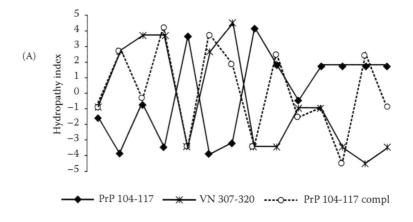

(A)

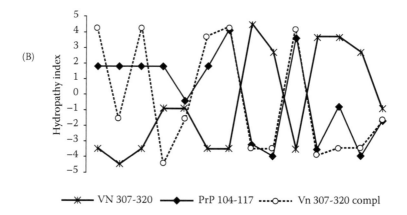

(B)

FIGURE 9.5 **(A)** Hydropathy plot of the mouse PrP^c domain 104–117 (PKTNLKHVAG-AAAA) and its binding site at vitronectin 307–320 (QRDSWENIFELLFW). The peptide deduced from PrP^c 104–117 complementary DNA sequence (SCRSPCDMLEVGFW) has the same hydropathic profile of the VN 307–320 while complementary to PrP^c 104–117. The VN 307–320 and the PrP^c 104–117 complementary peptides (PrP 104–117 comp) are shown from carboxy to amino-terminal while PrP^c 104–117 is from amino to carboxy-terminal (*left to right*). **(B)** The peptide deduced from VN 307–320 complementary DNA strand (PEEKVEDVLPRVPV) has the same profile of PrP^c 104–117 while complementary to VN 307–320 peptide. The VN 307–322 complementary peptide (VN 307–320 comp) and PrP^c 104–117 are shown from carboxy to amino-terminal while VN 307–320 peptide is from amino to carboxy-terminal (*left to right*).

9.5 CONCLUSIONS

The establishment of protein interactions is the hallmark of molecular and cellular biology. Physiological and pathological cellular processes depend not only on how and when proteins are expressed but whether they interact with their partners at a specific cellular compartment. Therefore, methodological approaches that permit the identification of cellular partners with biological active domains are of great interest. Herein, we demonstrated through different examples that the complementary hydropathy theory is a powerful tool for that matter.

ACKNOWLEDGMENTS

We thank Dr. Glaucia N. Hajj for her original contribution with the immunohistochemistry data and critical reading of the manuscript. Supported by Fundação de Amparo a Pesquisa do Estado de São Paulo (FAPESP 03-13189-2).

REFERENCES

1. R. R. Brentani. "Biological Implications of Complementary Hydropathy of Amino Acids." *J Theor Biol* 135, no. 4 (1988): 495–99.
2. J. Biro. "Comparative Analysis of Specificity in Protein-Protein Interactions. Part II: The Complementary Coding of Some Proteins as the Possible Source of Specificity in Protein-Protein Interactions." *Med Hypotheses* 7, no. 8 (1981): 981–93.
3. R. R. Brentani. "Complementary Hydropathy and the Evolution of Interactin Polypeptides." *J Mol Evol* 31 (1990): 239–43.
4. H. A. Scheraga, G. Nemethy, and I. Z. Steinberg. "The Contribution of Hydrophobic Bonds to the Thermal Stability of Protein Conformations." *J Biol Chem* 237 (1962): 2506–8.
5. S. K. Burley and G. A. Petsko. "Amino-Aromatic Interactions in Proteins." *FEBS Lett* 203, no. 2 (1986): 139–43.
6. J. R. Heal, G. W. Roberts, J. G. Raynes, A. Bhakoo, and A. D. Miller. "Specific Interactions between Sense and Complementary Peptides: The Basis for the Proteomic Code." *Chem Biochem* 3, no. 2–3 (2002): 136–51.
7. K. L. Bost, E. M. Smith, and S. E. Blalock. "Similarity between the Corticotropin (ACTH) Receptor and a Peptide Encoded by an RNA That Is Complementary to ACTH mRNA." *Proc Natl Acad Sci U S A* 82, no. 5 (1985): 1372–75.
8. T. S. Elton, S. Oparil, and J. E. Blalock. "The Use of Complementary Peptides in the Purification of an Angiotensin II Binding Protein." *J Hypertens Suppl* 6, no. 4 (1988): S404–7.
9. R. R. Brentani, S. F. Ribeiro, P. Potocnjak, R. Pasqualini, J. D. Lopes, and C. R. Nakaie. "Characterization of the Cellular Receptor for Fibronectin through a Hydropathic Complementarity Approach." *Proc Natl Acad Sci U S A* 85, no. 2 (1988): 364–67.
10. S. J. de Souza, J. Sabbaga, E. D'Amico, R. Pasqualini, and R. Brentani. "Anti-platelet Autoantibodies from ITP Patients Recognize an Epitope in GPIIb/IIIa Deduced by Complementary Hydropathy." *Immunology* 75, no. 1 (1992): 17–22.
11. S. J. de Souza and R. Brentani. "Collagen Binding Site in Collagenase Can Be Determined Using the Concept of Sense-Antisense Peptide Interactions." *J Biol Chem* 267, no. 19 (1992): 13763–67.
12. W. Gilbert. "Why Genes in Pieces?" *Nature* 271, no. 5634 (1978): 501.

13. S. B. Prusiner, M. R. Scott, S. J. DeArmond, and F. E. Cohen. "Prion Protein Biology." *Cell* 93, no. 3 (1998): 337–48.

14. J. Collinge. "Prion Diseases of Humans and Animals: Their Causes and Molecular Basis." *Ann Rev Neurosci* 24 (2001): 519–50.

15. D. A. Harris. "Trafficking, Turnover and Membrane Topology of PrP." *Br Med Bull* 66 (2003): 71–85.

16. M. A. Prado, J. Alves-Silva, A. C. Magalhaes, V. F. Prado, R. Linden, V. R. Martins, et al. "PrPc on the Road: Trafficking of the Cellular Prion Protein." *J Neurochem* 88, no. 4 (2004): 769–81.

17. P. J. Peters, A. Mironov Jr., D. Peretz, E. van Donselaar, E. Leclerc, S. Erpel, et al. "Trafficking of Prion Proteins through a Caveolae-Mediated Endosomal Pathway." *J Cell Biol* 162, no. 4 (2003): 703–17.

18. S. L. Shyng, J. E. Heuser, and D. A. Harris. "A Glycolipid-Anchored Prion Protein Is Endocytosed via Clathrin-Coated Pits." *J Cell Biol* 125, no. 6 (1994): 1239–50.

19. D. Goldgaber. "Anticipating the Anti-prion Protein?" *Nature* 351, no. 6322 (1991):106.

20. R. G. Hewinson, J. P. Lowinson, M. D. Dawson, M. J. Woodward. "Anti-prions and Other Agents." *Nature* 352, no. 6333 (1991): 291–92.

21. M. Moser, B. Oesch, and H. Bueler. "An Anti-prion Protein?" *Nature* 362, no. 6417 (1993): 213–14.

22. J. M. Gabriel, B. Oesch, H. Kretzschmar, M. Scott, and S. B. Prusiner. "Molecular Cloning of a Candidate Chicken Prion Protein." *Proc Natl Acad Sci U S A* 89, no. 19 (1992): 9097–9101.

23. G. Forloni, N. Angeretti, R. Chiesa, E. Monzani, M. Salmona, O. Bugiani, et al. "Neurotoxicity of a Prion Protein Fragment." *Nature* 362, no. 6420 (1993): 543–46.

24. V. R. Martins, E. Graner, J. Garcia-Abreu, S. J. de Souza, A. F. Mercadante, S. S. Veiga, et al. "Complementary Hydropathy Identifies a Cellular Prion Protein Receptor." *Nat Med* 3, no. 12 (1997): 1376–82.

25. L. B. Chiarini, A. R. Freitas, S. M. Zanata, R. R. Brentani, V. R. Martins, and R. Linden. "Cellular Prion Protein Transduces Neuroprotective Signals." *EMBO J* 21, no. 13 (2002): 3317–26.

26. S. M. Zanata, M. H. Lopes, A. F. Mercadante, G. N. Hajj, L. B. Chiarini, R. Nomizo, et al. "Stress-Inducible Protein 1 Is a Cell Surface Ligand for Cellular Prion That Triggers Neuroprotection. *EMBO J* 21, no. 13 (2002): 3307–16.

27. R. Rieger, F. Edenhofer, C. I. Lasmezas, and S. Weiss. "The Human 37-kDa Laminin Receptor Precursor Interacts with the Prion Protein in Eukaryotic Cells." *Nat Med* 3, no. 12 (1997): 1383–88.

28. A. Mitchell. "Puzzling Over Prion Partners." *Nat Med* 3, no. 12 (1997): 1322.

29. S. Gauczynski, J. M. Peyrin, S. Haik, C. Leucht, C. Hundt, R. Rieger, et al. "The 37-kDa/67-kDa Laminin Receptor Acts as the Cell-Surface Receptor for the Cellular Prion Protein." *EMBO J* 20, no. 21 (2001): 5863–75.

30. C. Leucht, S. Simoneau, C. Rey, K. Vana, R. Rieger, C. I. Lasmezas, et al. "The 37 kDa/67 kDa Laminin Receptor Is Required for PrP(Sc) Propagation in Scrapie-Infected Neuronal Cells." *EMBO Rep* 4, no. 3 (2003): 290–95.

31. O. O. Odunuga, V. M. Longshaw, and G. L. Blatch. "Hop: More Than an Hsp70/Hsp90 Adaptor Protein." *Bioessays* 26, no. 10 (2004): 1058–068.

32. M. H. Lopes, G. N. Hajj, A. G. Muras, G. L. Mancini, R. M. Castro, K. C. Ribeiro et al. "Interaction of cellular prion and stress-inducible protein 1 promotes neuritogenesis and neuroprotection by distinct signaling pathways." *J Neurosci.* 25 no. 49 (2005): 11330–9.

33. D. Shmerling, I. Hegyi, M. Fischer, T. Blattler, S. Brandner, J. Gotz, et al. "Expression of Amino-Terminally Truncated PrP in the Mouse Leading to Ataxia and Specific Cerebellar Lesions." *Cell* 93, no. 2 (1998): 203–14.

34. R. Linden. "The Anti-death League: Associative Control of Apoptosis in Developing Retinal Tissue." *Brain Res Brain Res Rev* 32, no. 1 (2000): 146–58.

35. M. Lassle, G. L. Blatch, V. Kundra, T. Takatori, and B. R. Zetter. "Stress-Inducible, Murine Protein mSTI1. Characterization of Binding Domains for Heat Shock Proteins and In Vitro Phosphorylation by Different Kinases." *J Biol Chem* 272, no. 3 (1997): 1876–84.

36. B. Honore, H. Leffers, P. Madsen, H. H. Rasmussen, J. Vandekerckhove, and J. E. Celis. "Molecular Cloning and Expression of a Transformation-Sensitive Human Protein Containing the TPR Motif and Sharing Identity to the Stress-Inducible Yeast Protein STI1." *J Biol Chem* 267, no. 12 (1992): 8485–91.

37. B. K. Eustace and D. G. Jay. "Extracellular Roles for the Molecular Chaperone, hsp90." *Cell Cycle* 3, no. 9 (2004).

38. F. R. Lima, C. P. Arantes, A. G. Muras, R. Nomizo, R. R. Bretani, V. R. Marins. "Cellular prion protein expression in astrocytes modulates neuronal survival and differentiation." *J Neurochem.* 103 no. 6 (2007): 2164–76.

39. W. Nickel. "The Mystery of Nonclassical Protein Secretion. A Current View on Cargo Proteins and Potential Export Routes." *Eur J Biochem* 270, no. 10 (2003): 2109–19.

40. J. S. Aguirre Ghiso, K. Kovalski, and L. Ossowski. "Tumor Dormancy Induced by Downregulation of Urokinase Receptor in Human Carcinoma Involves Integrin and MAPK Signaling." *J Cell Biol* 147, no. 1 (1999): 89–104.

41. C. Hetz, K. Maundrell, and C. Soto. "Is Loss of Function of the Prion Protein the Cause of Prion Disorders?" *Trends Mol Med* 9, no. 6 (2003): 237–43.

42. H. B. Samaia, J. J. Mari, H. P. Vallada, R. P. Moura, A. J. Simpson, and R. R. Brentani. "A Prion-Linked Psychiatric Disorder." *Nature* 390, no. 6657 (1997): 241.

43. R. Walz, R. M. Castro, M. C. Landemberger, T. R. Velasco, V. C. Terra-Bustamante, A. C. Bastos, et al. "Cortical Malformations Are Associated with a Rare Polymorphism of Cellular Prion Protein." *Neurology* 63, no. 3 (2004): 557–60.

44. L. Luckenbill-Edds. "Laminin and the Mechanism of Neuronal Outgrowth." *Brain Res Brain Res Rev* 23, no. 1–2 (1997): 1–27.

45. E. Graner, A. F. Mercadante, S. M. Zanata, O. V. Forlenza, A. L. Cabral, S. S. Veiga, et al. "Cellular Prion Protein Binds Laminin and Mediates Neuritogenesis." *Brain Res Mol Brain Res* 76, no. 1 (2000): 85–92.

46. A. S. Coitinho, M. H. Lopes, G. N. Hajj, J. I. Rossato, A. R. Freitas, C. C. Castro, M. et al. "Short-term memory formation and long-term memory consolidation are enhanced by cellular prion association to stress-inducible protein 1." *Neurobiol Dis.* 26, no. 1 (2007): 282–90.

47. G. N. Hajj, M. H. Lopes, A. F. Mercadante, S. S. Veiga, R. B. da Silveira, T. G. Santos, et al. "Cellular prion protein interaction with vitronectin supports axonal growth and is compensated by integrins." *J Cell Sci.* 120 no. 11: 915–26.

10 Discovery of Molecular Chaperone GRP78 and GRP94 Inducible by Endoplasmic Reticulum Stress

Amy S. Lee

CONTENTS

10.1 VIRAL TRANSFORMATION REVEALS GRP INDUCTION

The story of the discovery of the glucose-regulated proteins GRP78 and GRP94 began in 1974 when virologists first noticed changes in the protein profile when they transformed chick embryo fibroblasts and NRK cells with avian sarcoma viruses (1,2). In particular, the synthesis of the two proteins with molecular masses of 75 and 90 kilodaltons (kDa) were greatly enhanced. These proteins were originally thought to be transformation-specific proteins. As such, they could serve very important functions in converting a normal cell into the cancerous state. However, further investigation into the nature of induction of these proteins by the I. Pastan laboratory at the National Cancer Institute revealed that they were not encoded by the retroviruses; rather, they were cellular proteins synthesized constitutively in varying amounts under tissue culture conditions or in whole organs (3). Moreover, the accumulation of these same proteins could occur by simply culturing the cells in medium deprived of glucose in the absence of viral transformation. Thus, reinterpretation of the original observations would suggest that accumulation of these proteins was not directly due to the onset of transformation; rather, their induction was a consequence of glucose deprivation from the culture medium by the rapidly growing transformed cells (3). Because the synthesis of these proteins was regulated by the concentration of glucose in the culture medium, the 78- and 94-kDa proteins were named glucose regulated proteins GRP78 and GRP94, respectively.

Why would a lack of glucose in the culture medium lead to increased synthesis of GRP78 and GRP94? The first clue was provided by induction of these same proteins in a 3T3 fibroblast mutant AD6 that was defective in glycoprotein synthesis (4). Feeding the cells with N-acetylglucosamine, a metabolite that bypassed the metabolic block at the acetylation step of glycoproteins, restored the amounts of these proteins to levels found in normal cells. Alternatively, two glucose derivatives, glucosamine and 2-deoxyglucose, which interfered with the glycosylation process, specifically induced GRP78 and GRP94 in a variety of mammalian cells. These initial studies established that accumulation of underglycosylated proteins resulted in induction of GRP78 and GRP94. Thus, while glucose starvation could exert pleiotrophic effects on cellular metabolism, such as those affecting the pentose cycle and synthesis of nuclear precursors, a block in protein glycosylation could be one consequence that leads to the induction of GRP78 and GRP94.

10.2   K12 MUTANT LEADS TO MOLECULAR CLONING OF GRP78 AND GRP94

Halfway around the world, investigators at the Imperial Cancer Research Fund in London, UK, were searching for temperature-sensitive mutants derived from Chinese hamster cells that were blocked in cell cycle progression, as these mutants could provide important clues on how cancer cells escaped the brake on cell proliferation (5). K12 was one such mutant that could grow at 35°C but not 40.5°C due to G1 arrest (6). In analyzing the biochemical characteristics of K12, Melero and Smith (7) discovered that K12 synthesized large amounts of three specific proteins of molecular mass 94, 78, and 58 kDa when the cells were incubated at the nonpermissive temperature and this new synthesis required transcription. Nonetheless, the fact that these same proteins did not accumulate in other Chinese hamster G1-arrested mutants argued against their function in G1 arrest. It took another eight years before it was determined through genetic complementation that the defect in the K12 mutant was at the step of transfer of oligosaccharide core to polypeptides essentially a block in protein glycosylation (8).

At the same time that the K12 mutant was reported in the literature, the A. S. Lee laboratory at the University of Southern California was searching for a model system in mammalian cells to study the mechanism(s) whereby genes that were not physically linked could be transcribed in a coordinate manner. The K12 mutant with simultaneous induction of the three proteins in a manner that was dependent on new transcription offered such a model. Since the molecular size of the proteins overproduced in K12 cells at the nonpermissive temperature was nearly identical to GRP78 and GRP94, their identity was confirmed through biochemical characterization and glycosylation inhibition studies (9). In the same year, two proteins with molecular masses of 100 and 80 kDa were identified whose synthesis was specifically stimulated by the calcium ionophore A23187 (10). Direct comparisons of the calcium ionophore-inducible proteins with the glucose-regulated proteins revealed them to be identical (11,12).

Taking advantage of the K12 mutant that showed 5- to 10-fold higher induction of GRP78 as compared to that obtained by glucose starvation, a cDNA library was constructed using RNA extracted from the hamster K12 cells incubated at 40.5°C (13). Clones that hybridized preferentially with cDNA made from RNA at 40.5°C were selected. By using the hybrid-selection technique, followed by in vitro translation, a full-length cDNA clone containing a 2250-nucleotide insert encoding hamster GRP78 was isolated in 1981. Further, in vitro translation analysis revealed that GRP78 was synthesized as a slightly larger species, possessing a slightly more basic charge than the in vivo protein (9). This provided the first hint that GRP78 was first synthesized and then a short peptide was cleaved away to yield the mature GRP78. Using the same technique, a partial cDNA clone encoding the carboxyl half of hamster GRP94 was isolated in 1983 (14).

These cDNA clones were used as molecular probes to provide the first evidence that Grp78 and Grp94 were induced at the transcriptional level by glucose starvation, and each was encoded by a single copy gene in the hamster genome (14,15). The cDNA clones encoding GRP78 and GRP94 had provided valuable reagents to uncover transcription activation of Grp78 and Grp94 by a wide variety of physiological and environmental stress conditions (16–19). Furthermore, the cDNA clones were used to isolate the genomic sequences encoding for Grp78 and Grp94 gene from various mammalian species including human (20–22). The isolation of the ER stress-inducible Grp78 and Grp94 promoter opened a new research area toward understanding the transcription factor complexes that regulate this class of genes (22–25). This led to the unraveling of novel ER signal transducers that could sense ER stress and transmit the signal to the nucleus (26–28). New transgenic mouse models utilizing the Grp78 and Grp94 promoters driving reporter genes also led to discoveries on how these genes were activated in vivo in the context of the whole animal (29–31).

10.3 MATCHING GRP78 WITH BiP, THE IMMUNOGLOBULIN HEAVY CHAIN BINDING PROTEIN

To better understand the structure and function of GRP78 and GRP94, it is necessary to obtain the amino acid sequences of the two proteins. Taking advantage of the K12 mutant cell line, which produced high abundance of both GRP78 and GRP94 proteins when incubated at the nonpermissive temperature, the A. S. Lee laboratory purified GRP78 and GRP94 by two-dimensional gel electrophoresis and subjected them to amino acid sequence analysis (12). This approach generated the first amino-terminal sequence for the mature GRP78 and GRP94 proteins and established that they were distinct from the previously identified heat shock proteins. These partial amino acid sequences, first reported in 1984, provided key information for future analysis of the localization, structure, and function of both proteins.

Meanwhile, the H. R. Pelham laboratory at the Medical Research Council in the United Kingdom sought to identify HSP70-related transcripts by screening a rat liver library with a *Drosophila* hsp70 probe at low stringency. This was to test the idea that the function of HSP70 and their cohorts that existed in nonstressed cells was to solubilize aggregated, heat-damaged proteins in an ATP-dependent manner (32). This would imply that members of HSP70 protein family that expressed constitutively in nonstressed cells could serve important metabolic function. During their search, they isolated a cDNA clone referred to as p72 with features that had not previously been found in other HSP70 or related proteins. Its amino acid sequence as deduced from the cDNA clone was longer at the N-terminus than other HSP70s, and the first 18 amino acids of this extension had typical features of a secretory leader peptide and functioned as a leader peptide targeting the protein to the endoplasmic reticulum (33). Strikingly, the amino acid sequence directly following the putative leader sequence was highly similar to the N-terminal sequence reported previously for hamster GRP78 (12). Furthermore, the N-terminus of the mature hamster GRP78 corresponded exactly to the predicted cleavage site of p72, strongly suggesting that p72 is GRP78.

How then was GRP78 linked to BiP, the immunoglobulin heavy chain binding protein? BiP was originally discovered by I. G. Haas and her coworkers in Germany as a binding protein for the nonsecreted immunoglobulin heavy chains of pre-B cells that did not make light chains (34). It also bound to a small fraction of the intracellular heavy chains in B cells and plasma cells. The discovery that GRP78 was identical to BiP was made by the H. R. Pelham laboratory where they noticed that BiP and p72 shared similar properties. The definitive proof came from direct comparison of their peptide maps that turned out to be identical and cross-reactivity of BiP with a monoclonal antibody specifically recognizing GRP78 (33). The same study revealed that BiP was not restricted to B cells; in fact, the level of BiP in pre-B and B cells was comparable to that found in fibroblasts and its level was greatly elevated in antibody secreting plasma cells (33). At about the same time, studies by Hendershot and her coworkers at the University of Alabama showed that BiP served to retain incompletely assembled Ig intermediates (35). This important observation led to the first designation of GRP78/BiP as the first ER molecular chaperone. Moreover, the generation of a monospecific antibody against BiP and species-specific anti-BiP antisera contributed significantly to the identification of the function, location, and protein interactions of GRP78 in a large number of subsequent studies (36,37). For these historical reasons, GRP78 is also commonly referred to as BiP. However, it should be noted that GRP78 is the first name reported in the literature for this 78-kDa protein and, as an abundant ER chaperone protein, GRP78 has a wider repertoire of interacting client proteins than just the immunoglobulin heavy chain as implied by the name BiP.

10.4   RELATIONSHIP OF GRP78 AND GRP94 AND HEAT SHOCK PROTEINS

The glucose-regulated and heat shock proteins are two subsets of eukaryotic stress proteins that can be induced differentially, simultaneously, and reciprocally (16,38,39). However, a common confusion that exists in literature concerning GRP78 and GRP94 is whether they are heat shock proteins. From amino acid sequence comparisons, GRP78 shares about 60% homology with HSP70, including the ATP binding domain required for their shared chaperone function to facilitate protein folding (40). However, GRP78 differs from HSP70 in two important aspects. First, GRP78 has a signal peptide sequence that targets it to ER, whereas HSP70 does not contain such a sequence and is cytosolic and relocalizes to the nucleus upon stress. Second, the synthesis of GRP78 is not significantly affected by heat shock conditions that greatly elevate the level of HSP70 and other members of the heat shock protein family. In contrast, the most potent chemical inducers for GRP78 are thapsigargin, which inhibits the ER calcium ATPase pump, the calcium ionophore A23187, and tunicamycin, which blocks N-linked protein glycosylation (19,41,42). As expected, treatment of cells with azetidine, a proline analog that leads to the formation of malfolded proteins both in the cytoplasm and ER, transcriptionally activates both Hsp70 and Grp78 (18,43).

GRP94 was independently identified by several laboratories shortly after the first report of the amino acid sequence of the hamster GRP94 in 1984 (12). At the Medical Research Council, United Kingdom, while studying the role of calcium-binding proteins in the ER, G. Koch and his coworkers identified a 100-kDa protein, referred to as endoplasmin, through binding to concanavalin A and determined that it was a major calcium-binding glycoprotein in the ER (44). Determination of the amino-terminal sequence of endoplasmin revealed that it was identical to GRP94. At around the same time, the laboratory of M. Green at the Saint Louis University independently discovered an abundant ER glycoprotein referred to as ERp99 while studying the sorting mechanism for cellular proteins that were targeted to the endoplasmic reticulum (45). They isolated a full-length murine cDNA clone encoding for ERp99 from which the entire coding sequence was deduced. By aligning this sequence with the amino acid of the hamster GRP94, ERp99 was shown to be a protein with a signal peptide and was identical to GRP94. This same study also uncovered extensive homology between GRP94 and HSP90, although they were distinct proteins (45,46). The 50% sequence identity between GRP94 and HSP90 was further confirmed when the cDNA encoding for the hamster GRP94 was sequenced (47). In another study, a full-length chicken cDNA clone encoding for a protein referred to hsp108 with a signal peptide and considerable homology to HSP90 was isolated in B. W. O'Malley's laboratory at Baylor University (48). This turned out to be the homolog of GRP94 in chicken.

One main distinction between the heat shock proteins HSP70 and HSP90, and GRP78 and GRP94 was that the latter are localized in the ER. The question then arose as to how these ER lumenal proteins were retained in the ER. This mystery was partly solved through sequencing of the hamster cDNA clones for GRP94 and GRP78. It was discovered that GRP94 shared the same C-terminal tretrapeptide (KDEL) as GRP78 and protein disulfide isomerase, a third ER lumenal protein (47). Subsequently, the KDEL sequence was proved to be part of a signal for retention of proteins in the lumen of the ER.

10.5 LINK OF GRP78 TO CANCER

While it has been well accepted that GRP78 is inducible by glucose starvation independent of viral transformation as proposed in 1977, a study in 1988 revisited the issue of the link between GRP78 induction and viral transformation (49). The results showed that GRP78 could be induced in Rous sarcoma virus-transformed cells independently of glucose deprivation (49). Using mutants defective in transformation, it was further demonstrated that GRP78 was induced by p60v-src in the absence of glucose deprivation. This was consistent with the observation made around the same time that while chemically and virally transformed cells spontaneously induced GRP78 under normal culture conditions, transformed cells induced even higher levels of GRP78 under glucose starvation conditions (50). Collectively, these studies suggest that non-glucose factors might interact with glucose starvation additively or synergistically to induce Grp78 expression in transformed cells (50). A likely explanation was the altered glucose intracellular stores and an increased rate of glucose utilization by transformed and cancerous cells.

While there was ample evidence showing overexpression of GRP78 and GRP94 in a wide variety of human cancers (19,29,30,51), the requirement of GRP78 in tumor progression had only been demonstrated in a mouse fibrosarcoma xenograft model where the level of GRP78 was suppressed by antisense (52). Nonetheless, evidence is emerging that GRP78 and GRP94 are anti-apoptotic proteins that can confer a protective effect in cancer cells (53,54). This suggests that downregulation of GRP78 by small molecules such as genistein could offer a new therapeutic approach to cancer (43). Recently, the laboratory of Kazuo Shin-ya at the University of Tokyo, Japan, has isolated a novel macrocyclic compound known as versipelostatin that specifically inhibited the induction of the Grp78 promoter by glucose deprivation and disrupted other components of the unfolded protein response. This compound effectively killed glucose-deprived cancer cells, and in combination with cisplatin, it inhibited tumor growth in xenografts (55). Another exciting advance is that the laboratory of R. Pasqualini and M. Arap at the M. D. Anderson Cancer Center identified GRP78 as a tumor antigen through epitope mapping of the humoral immune responses from cancer patients (56). This suggests that GRP78 could serve as a relevant molecular target for therapeutic intervention. This idea was directly tested by the use of synthetic chimeric peptides composed of GRP78 binding peptide motifs fused to a programmed cell death-inducing sequence (57). Suppression of tumor growth was observed in xenograft and isogenic mouse models of prostate and breast cancer, thus providing further validation that targeting GRP78 in cancer may prove useful for translation into clinical applications to eliminate primary and metastatic cancer.

Although GRP78 could confer survival advantage to cancer cells such that its suppression would enhance tumor cell death and drug sensitivity, on the other hand the anti-apoptotic property of GRP78 could be beneficial to critical organs such as the brain, heart, and kidney subjected to stress (19,58,59). In support of this, GRP78 was found to protect neurons against excitotoxicity and apoptosis by suppression of oxidative stress and stabilization of calcium homeostasis (60). In another example, overexpression of GRP78 in neuroblastoma cells bearing presenilin-1 mutations almost completely restored resistance to ER stress (61). These observations strongly suggest that upregulation of GRP78 level may offer a novel therapy for neurodegenerative diseases associated with accumulation of malfolded proteins.

10.6 LINK OF GRP94 TO CANCER IMMUNOGENICITY

The link of GRP94 to cancer immunology has been an evolving story with unexpected twists and turns, cumulating to a happy ending that GRP94, also known as gp96, is currently being tested in cancer clinical trials. Back in the late 1980s, the laboratory of P. K. Srivastava at the Memorial Sloan-Kettering Cancer Center pioneered the search for cellular moieties that mediated individually distinct immunogenicity of tumors and that led to the identification of a 96-kDa surface glycoprotein referred to as gp96 (62). The original hypothesis was that there were mutations in the gp96 genes of tumors and that these mutations differed from one tumor to another. Surprisingly, sequencing of gp96 cDNA from mouse spleen (normal tissue) and fibrosarcoma cell lines (antigenic tumor) did not reveal any tumor-specific, individually distinct mutations; rather, comparison of the gp96 sequence to known sequences revealed identity to GRP94 (63). To explain tumor-specific antigenicity of gp96, it was proposed that gp96 was not immunogenic per se, but acted as an acceptor for peptides transported to the ER and enabled peptide loading of MHC class I (64). The rationale was that although GRP94 is normally intracellular, necrotic cells release GRP94-peptide complexes, which were taken up by scavenging antigen-presenting cells. Presentation of the peptides on the surfaces of these cells would lead to stimulation of T lymphocytes and a pro-inflammatory response. Immunization of Grp94-peptide preparations in mice inhibited tumor growth and no autoimmunity was observed, suggesting that the immune response was targeted against the peptides and not against GRP94 (65,66). Based on promising results with tumor inhibition in murine models, clinical trials have been initiated testing the efficacy of immunization of cancer patients with autologous cancer-derived gp96 preparations (67). The clinical trials will directly test whether this approach could provide a patient-specific, chaperone-based vaccine for immunotherapy of human cancer. On the other hand, C. V. Nicchitta and his coworkers at Duke University reported that vaccination with irradiated cells, secreting either the full length GRP94 or the GRP94 amino-terminal domain lacking the canonical peptide-binding motifs, yielded similar suppression of tumor growth and metastatic progression. These unexpected findings supported a new hypothesis that GRP94-elicited tumor suppression can occur independent of the GRP94 tissue of origin and suggest a primary role for GRP94 natural adjuvant function in antitumor immune responses (68). Thus, further experimentation is required to resolve the immunogenicity issue of GRP94 in cancer therapy.

10.7 CONCLUSION

The multiple implications of ER stress in health and disease highlight the importance of ER stress targets that act to protect cells from death. A large volume of studies have established that induction of GRP78 is a marker for ER stress and that GRP78 is a central regulator for ER stress due to its role as a major ER chaperone with anti-apoptotic properties, as well as its ability to control the activation of transmembrane ER stress sensors. GRP94 is equally important as a major ER calcium-binding protein with anti-apoptotic and immunogenic properties. Thus, the discovery of GRP78 and GRP94 has opened several new research areas. The future may produce the most exciting and fruitful research on how these proteins impact human physiology in vivo and how they can be utilized in medicine as markers as well as therapeutic targets.

ACKNOWLEDGMENTS

We thank Dr. Lawrence Hightower for the critical review of this manuscript. This work was supported in part by a grant from the National Cancer Institute CA027607.

REFERENCES

1. K. R. Stone, R. E. Smith, and W. K. Joklik. "Changes in Membrane Polypeptides That Occur When Chick Embryo Fibroblasts and NRK Cells are Transformed with Avian Sarcoma Viruses." *Virology* 58 (1974): 86–100.
2. T. Isaka, M. Yoshida, M. Owada, and K. Toyoshima. "Alterations in Membrane Polypeptides of Chick Embryo Fibroblasts Induced by Transformation with Avian Sarcoma Viruses." *Virology* 65 (1975): 226–37.
3. R. P. Shiu, J. Pouysségur, and I. Pastan. "Glucose Depletion Accounts for the Induction of Two Transformation-Sensitive Membrane Proteins in Rous Sarcoma Virus-Transformed Chick Embryo Fibroblasts." *Proc Natl Acad Sci USA* 74 (1977): 3840–44.
4. J. Pouysségur, R. P. Shiu, I. Pastan. "Induction of Two Transformation-Sensitive Membrane Polypeptides in Normal Fibroblasts by a Block in Glycoprotein Synthesis or Glucose Deprivation." *Cell* 11 (1977): 941–47.
5. D. H. Roscoe, M. Read, and H. Robinson. "Isolation of Temperature Sensitive Mammalian Cells by Selective Detachment." *J Cell Physiol* 82 (1973): 325–31.
6. B. J. Smith and N. M. Wigglesworth. "A Temperature-Sensitive Function in a Chinese Hamster Line Affecting DNA Synthesis." *J Cell Physiol* 82 (1973): 339–47.
7. J. A. Melero and A. E. Smith. "Possible Transcriptional Control of Three Polypeptides which Accumulate in a Temperature-Sensitive Mammalian Cell Line." *Nature* 272 (1978): 725–27.
8. A. S. Lee, S. Wells, K. S. Kim, and I. E. Scheffler. "Enhanced Synthesis of the Glucose/ Calcium-Regulated Proteins in a Hamster Cell Mutant Deficient in Transfer of Oligosaccharide Core to Polypeptides." *J Cell Physiol* 129 (1986): 277–82.
9. A. S. Lee. "The Accumulation of Three Specific Proteins Related to Glucose-Regulated Proteins in a Temperature-Sensitive Hamster Mutant Cell Line K12." *J Cell Physiol* 106 (1981): 119–125.
10. F. S. Wu, Y. C. Park, D. Roufa, and A. Martonosi. "Selective Stimulation of the Synthesis of an 80,000-Dalton Protein by Calcium Ionophores." *J Biol Chem* 256 (1981): 5309–12.
11. W. J. Welch, J. I. Garrels, G. P. Thomas, J. J. Lin, and J. R. Feramisco. "Biochemical Characterization of the Mammalian Stress Proteins and Identification of Two Stress Proteins as Glucose- and Ca^{2+}-Ionophore-Regulated Proteins." *J Biol Chem* 258 (1983): 7102–11.
12. A. S. Lee, J. Bell, and J. Ting. "Biochemical Characterization of the 94- and 78-Kilodalton Glucose-Regulated Proteins in Hamster Fibroblasts." *J Biol Chem* 259 (1984): 4616–21.
13. A. S. Lee, A. Delegeane, and D. Scharff. "Highly Conserved Glucose-Regulated Protein in Hamster and Chicken Cells: Preliminary Characterization of Its cDNA Clone." *Proc Natl Acad Sci USA* 78 (1981): 4922–25.

14. A. S. Lee, A. M. Delegeane, V. Baker, and P. C. Chow. "Transcriptional Regulation of Two Genes Specifically Induced by Glucose Starvation in a Hamster Mutant Fibroblast Cell Line." *J Biol Chem* 258 (1983): 597–603.

15. A. Y. Lin and A. S. Lee. "Induction of Two Genes by Glucose Starvation in Hamster Fibroblasts." *Proc Natl Acad Sci USA* 81 (1984): 988–92.

16. A. S. Lee. "Coordinated Regulation of a Set of Genes by Glucose and Calcium Ionophores in Mammalian Cells." *Trends Biochem Sci* 12 (1987): 20–23.

17. A. S. Lee. "Mammalian Stress Response: Induction of the Glucose-Regulated Protein Family." *Curr Opin Cell Biol* 4 (1992): 267–73.

18. S. S. Watowich and R. I. Morimoto. "Complex Regulation of Heat Shock- and Glucose-Responsive Genes in Human Cells." *Mol Cell Biol* 8 (1988): 393–405.

19. A. S. Lee. "The Glucose-Regulated Proteins: Stress Induction and Clinical Applications." *Trends Biochem Sci* 26 (2001): 504–10.

20. J. W. Attenello and A. S. Lee. "Regulation of a Hybrid Gene by Glucose and Temperature in Hamster Fibroblasts." *Science* 226 (1984): 187–90.

21. J. Ting and A. S. Lee. "Human Gene Encoding the 78,000-Dalton Glucose-Regulated Protein and Its Pseudogene: Structure, Conservation, and Regulation." *DNA* 7 (1988): 275–86.

22. S. C. Chang, A. E. Erwin, and A. S. Lee. "Glucose-Regulated Protein (GRP94 and GRP78) Genes Share Common Regulatory Domains and Are Coordinately Regulated by Common Trans-acting Factors." *Mol Cell Biol* 9 (1989): 2153–62.

23. K. Haze, H. Yoshida, H. Yanagi, T. Yura, and K. Mori. "Mammalian Transcription Factor ATF6 Is Synthesized as a Transmembrane Protein and Activated by Proteolysis in Response to Endoplasmic Reticulum Stress." *Mol Biol Cell* 10 (1999): 3787–99.

24. B. Roy and A. S. Lee. "The Mammalian Endoplasmic Reticulum Stress Response Element Consists of an Evolutionarily Conserved Tripartite Structure and Interacts with a Novel Stress-Inducible Complex." *Nucleic Acids Res* 27 (1999): 1437–43.

25. R. Parker, T. Phan, P. Baumeister, B. Roy, V. Cheriyath, A. L. Roy, and A. S. Lee. "Identification of TFII-I as the Endoplasmic Reticulum Stress Response Element Binding Factor ERSF: Its Autoregulation by Stress and Interaction with ATF6." *Mol Cell Biol* 21 (2001): 3220–33.

26. R. J. Kaufman. "Stress Signaling from the Lumen of the Endoplasmic Reticulum: Coordination of Gene Transcriptional and Translational Controls." *Genes Dev* 13 (1999): 1211–33.

27. K. Mori. "Tripartite Management of Unfolded Proteins in the Endoplasmic Reticulum." *Cell* 101 (2000): 451–54.

28. H. P. Harding, M. Calfon, F. Urano, I. Novoa, and D. Ron. "Transcriptional and Translational Control in the Mammalian Unfolded Protein Response." *Annu Rev Cell Dev Biol* 18 (2002): 575–99.

29. R. K. Reddy, L. Dubeau, H. Kleiner, T. Parr, P. Nichols, B. Ko, D. Dong, et al. "Cancer-Inducible Transgene Expression by the Grp94 Promoter: Spontaneous Activation in Tumors of Various Origins and Cancer-Associated Macrophages." *Cancer Res* 62 (2002): 7207–12.

30. D. Dong, L. Dubeau, J. Bading, K. Nguyen, M. Luna, H. Yu, G. Gazit-Bornstein, et al. "Spontaneous and Controllable Activation of Suicide Gene Expression Driven by the Stress-Inducible grp78 Promoter Resulting in Eradication of Sizable Human Tumors." *Hum Gene Ther* 15 (2004): 553–61.

31. C. Mao, D. Dong, E. Little, S. Luo, and A. S. Lee. "Transgenic Mouse Models for Monitoring Endoplasmic Reticulum Stress In Vivo." *Nat Med* 10 (2004): 1013–14.

32. M. J. Lewis and H. R. Pelham. "Involvement of ATP in the Nuclear and Nucleolar Functions of the 70 kd Heat Shock Protein." *EMBO J* 4 (1985): 3137–43.

33. S. Munro and H. R. Pelham. "An Hsp70-Like Protein in the ER: Identity with the 78 kd Glucose-Regulated Protein and Immunoglobulin Heavy Chain Binding Protein." *Cell* 46 (1986): 291–300.

34. I. G. Haas and M. Wabl. "Immunoglobulin Heavy Chain Binding Protein." *Nature* 306 (1983): 387–89.

35. L. M. Hendershot and J. F. Kearney. "A Role for Human Heavy Chain Binding Protein in the Developmental Regulation of Immunoglobin Transport." *Mol Immunol* 25 (1988): 585–95.

36. D. G. Bole, L. M. Hendershot, and J. F. Kearney. "Posttranslational Association of Immunoglobulin Heavy Chain Binding Protein with Nascent Heavy Chains in Nonsecreting and Secreting Hybridomas." *J Cell Biol* 102 (1986): 1558–66.

37. L. M. Hendershot, J. Y. Wei, J. R. Gaut, B. Lawson, P. J. Freiden, and K. G. Murti. "In Vivo Expression of Mammalian BiP ATPase Mutants Causes Disruption of the Endoplasmic Reticulum." *Mol Biol Cell* 6 (1995): 283–96.

38. S. A. Whelan and L. E. Hightower. "Differential Induction of Glucose-Regulated and Heat Shock Proteins: Effects of pH and Sulfhydryl-Reducing Agents on Chicken Embryo Cells." *J Cell Physiol* 125 (1985): 251–58.

39. L. E. Hightower. "Heat Shock, Stress Proteins, Chaperones, and Proteotoxicity." *Cell* 66 (1991): 191–97.

40. S. K. Wooden and A. S. Lee. "Comparison of the Genomic Organizations of the Rat grp78 and hsc73 Gene and Their Evolutionary Implications." *DNA Seq* 3 (1992): 41–48.

41. Y. K. Kim, K. S. Kim, and A. S. Lee. "Regulation of the Glucose-Regulated Protein Genes by Beta-Mercaptoethanol Requires De Novo Protein Synthesis and Correlates with Inhibition of Protein Glycosylation." *J Cell Physiol* 133 (1987): 553–59.

42. W. W. Li, S. Alexandre, X. Cao, and A. S. Lee. "Transactivation of the grp78 Promoter by Ca^{2+} Depletion. A Comparative Analysis with A23187 and the Endoplasmic Reticulum Ca^{2+}-ATPase Inhibitor Thapsigargin." *J Biol Chem* 268 (1993): 12003–9.

43. Y. Zhou and A. S. Lee. "Mechanism for the Suppression of the Mammalian Stress Response by Genistein, an Anticancer Phytoestrogen from Soy." *J Natl Cancer Inst* 90 (1998): 381–88.

44. G. Koch, M. Smith, D. Macer, P. Webster, and R. Mortara. "Endoplasmic Reticulum Contains a Common, Abundant Calcium-Binding Glycoprotein, Endoplasmin." *J Cell Sci* 86 (1986): 217–32.

45. R. A. Mazzarella and M. Green. "ERp99, an Abundant, Conserved Glycoprotein of the Endoplasmic Reticulum, is Homologous to the 90-kDa Heat Shock Protein (hsp90) and the 94-kDa Glucose Regulated Protein (GRP94)." *J Biol Chem* 262 (1987): 8875–83.

46. P. Csermely, T. Schnaider, C. Soti, Z. Prohaszka, and G. Nardai. "The 90-kDa Molecular Chaperone Family: Structure, Function, and Clinical Applications. A Comprehensive Review." *Pharmacol Ther* 79 (1998): 129–68.

47. P. K. Sorger and H. R. Pelham. "The Glucose-Regulated Protein grp94 Is Related to Heat Shock Protein hsp90." *J Mol Biol* 194 (1987): 341–44.

48. D. R. Sargan, M. J. Tsai, and B. W. O'Malley. "hsp108, a Novel Heat Shock Inducible Protein of Chicken." *Biochemistry* 25 (1986): 6252–58.

49. M. Y. Stoeckle, S. Sugano, A. Hampe, A. Vashistha, D. Pellman, and H. Hanafusa. "78-Kilodalton Glucose-Regulated Protein Is Induced in Rous Sarcoma Virus-Transformed Cells Independently of Glucose Deprivation." *Mol Cell Biol* 8 (1988): 2675–80.

50. S. R. Patierno, J. M. Tuscano, K. S. Kim, J. R. Landolph, and A. S. Lee. "Increased Expression of the Glucose-Regulated Gene Encoding the Mr 78,000 Glucose-Regulated Protein in Chemically and Radiation-Transformed C3H 10T1/2 Mouse Embryo Cells." *Cancer Res* 47 (1987): 6220–24.

51. E. Little, M. Ramakrishnan, B. Roy, G. Gazit, and A. S. Lee. "The Glucose-Regulated Proteins (GRP78 and GRP94): Functions, Gene Regulation, and Applications." *Crit Rev Eukaryot Gene Expr* 4 (1994): 1–18.
52. C. Jamora, G. Dennert, and A. S. Lee. "Inhibition of Tumor Progression by Suppression of Stress Protein GRP78/BiP Induction in Fibrosarcoma B/C10ME." *Proc Natl Acad Sci USA* 93 (1996): 7690–94.
53. R. K. Reddy, J. Lu, and A. S. Lee. "The Endoplasmic Reticulum Chaperone Glycoprotein GRP94 with Ca^{2+}-Binding and Antiapoptotic Properties Is a Novel Proteolytic Target of Calpain During Etoposide-Induced Apoptosis." *J Biol Chem* 274 (1999): 28476–83.
54. R. K. Reddy, C. Mao, P. Baumeister, R. C. Austin, R. J. Kaufman, and A. S. Lee. "Endoplasmic Reticulum Chaperone Protein GRP78 Protects Cells from Apoptosis Induced by Topoisomerase Inhibitors: Role of ATP Binding Site in Suppression of Caspase-7 Activation." *J Biol Chem* 278 (2003): 20915–24.
55. H. R. Park, A. Tomida, S. Sato, Y. Tsukumo, J. Yun, T. Yamori, Y. Hayakawa, T. Tsuruo, and K. Shin-ya. "Effect on Tumor Cells of Blocking Survival Response to Glucose Deprivation." *J Natl Cancer Inst* 96 (2004): 1300–10.
56. P. J. Mintz, J. Kim, K. A. Do, X. Wang, R. G. Zinner, M. Cristofanilli, M. A. Arap, et al. "Fingerprinting the Circulating Repertoire of Antibodies from Cancer Patients." *Nat Biotechnol* 21 (2003): 57–63.
57. M. A. Arap, J. Lahdenranta, P. J. Mintz, A. Hajitou, A. S. Sarkis, W. Arap, and R. Pasqualini. "Cell Surface Expression of the Stress Response Chaperone GRP78 Enables Tumor Targeting by Circulating Ligands." *Cancer Cell* 6 (2004): 275–84.
58. R. J. Kaufman, D. Scheuner, M. Schroder, X. Shen, K. Lee, C. Y. Liu, and S. M. Arnold. "The Unfolded Protein Response in Nutrient Sensing and Differentiation." *Nat Rev Mol Cell Biol* 3 (2002): 411–21.
59. R. J. Kaufman. "Orchestrating the Unfolded Protein Response in Health and Disease." *J Clin Invest* 110 (2002): 1389–98.
60. Z. Yu, H. Luo, W. Fu, and M. P. Mattson. "The Endoplasmic Reticulum Stress-Responsive Protein GRP78 Protects Neurons against Excitotoxicity and Apoptosis: Suppression of Oxidative Stress and Stabilization of Calcium Homeostasis." *Exp Neurol* 155 (1999): 302–14.
61. T. Katayama, K. Imaizumi, N. Sato, K. Miyoshi, T. Kudo, J. Hitomi, T. Morihara, et al. "Presenilin-1 Mutations Downregulate the Signalling Pathway of the Unfolded-Protein Response." *Nat Cell Biol* 1 (1999): 479–85.
62. P. K. Srivastava and L. J. Old. "Identification of a Human Homologue of the Murine Tumor Rejection Antigen GP96." *Cancer Res* 49 (1989): 1341–43.
63. P. K. Srivastava and R. G. Maki. "Stress-Induced Proteins in Immune Response to Cancer." *Curr Top Microbiol Immunol* 167 (1991): 109–23.
64. Z. Li and P. K. Srivastava. "Tumor Rejection Antigen gp96/grp94 Is an ATPase: Implications for Protein Folding and Antigen Presentation." *EMBO J* 12 (1993): 3143–51.
65. N. E. Blachere, Z. Li, R. Y. Chandawarkar, R. Suto, N. S. Jaikaria, S. Basu, H. Udono, and P. K. Srivastava. "Heat Shock Protein-Peptide Complexes, Reconstituted In Vitro, Elicit Peptide-Specific Cytotoxic T Lymphocyte Response and Tumor Immunity." *J Exp Med* 186 (1997): 1315–22.
66. S. Basu and P. K. Srivastava. "Heat Shock Proteins: The Fountainhead of Innate and Adaptive Immune Responses." *Cell Stress Chaperones* 5 (2000): 443–51.
67. S. Janetzki, D. Palla, V. Rosenhauer, H. Lochs, J. J. Lewis, and P. K. Srivastava. "Immunization of Cancer Patients with Autologous Cancer-Derived Heat Shock Protein gp96 Preparations: A Pilot Study." *Int J Cancer* 88 (2000): 232–38.
68. J. C. Baker-LePain, M. Sarzotti, T. A. Fields, C. Y. Li, and C. V. Nicchitta. "GRP94 (gp96) and GRP94 N-Terminal Geldanamycin Binding Domain Elicit Tissue Nonrestricted Tumor Suppression." *J Exp Med* 196 (2002): 1447–59.

11 Origin of Sphingomyelin Pathway

Richard Kolesnick

CONTENTS

11.1 INTRODUCTION

My journey to the laboratory and into the realm of lipid biochemistry was unconventional. In fact, I had no intention of becoming a scientist, and certainly not of studying lipids in an academic setting. After medical school at the University of Chicago, I returned to my beloved city, New York, and specifically to the borough in which I was born, the Bronx. My internship and residency in medicine at Montefiore Hospital were simultaneously exhilarating and wearying. A unique aspect of the training in the Montefiore program was the requirement for a senior "thesis" talk. Although this was for the most part a pro forma obligation, I never shunned the challenge of a good intellectual workout and this was no different. At that time, there was a well-known clinician at Montefiore named Harold Rifkin, a diabetes specialist who had written a clinical tome of great merit. He was well respected as a "great old man of medicine," although he probably wasn't that old—I was just pretty young (26 years old). I asked him if he would serve as my adviser, to which he agreed, on a project involving new insulins that were just becoming available at that time. The older insulins, and this was before molecular biology took hold, were purified either from cows or pigs, and invariably had some levels of contaminating proteins in the mix (1). Some patients responded to these insulins with either local or systemic allergic reactions, and hence the newer, purer monocomponent insulins were causing a stir (2,3). I was fascinated by the problem and began contacting the clinician scientists around the country who had published the primary work on these new insulins, most of whom were more than happy to send me reprints, unpublished experimental data, and even a few samples of the new products.

My presentation went over swimmingly and thereafter Dr. Rifkin invited me to his office for a chat. He told me how pleased he was with my performance and asked what I was doing after my residency was over. I told him that I thought I would take an assistant attending position at a hospital in the Bronx, to which he replied, "Oh, no you're not." He then said, "Come back in a week and talk with me again. I am going to arrange for you to have an endocrinology fellowship." Now, up until that point I had never considered doing a fellowship and had missed the deadline for all fellowship applications for the upcoming year. Nonetheless, I had great respect for Dr. Rifkin and mulled over his statement over the next week. When the time came for me to meet with him, I had decided against such a course of action, and was prepared to disappoint him when we met.

Dr. Rifkin said that he had arranged for me to stay at Montefiore and do a research fellowship with him and Dr. Roslyn Yalow. I was stunned. I couldn't believe that I was offered a position in the laboratory of a Nobel Prize winner, and I can still see myself saying, "Now that is an offer I cannot refuse." Having accepted the position, I met with Dr. Yalow to understand the offer. Further, I began to seriously consider the notion of doing an endocrine fellowship and began exploring the other programs in the city. In fact, I found that the research in quite a few programs was very interesting and I began calling around to see what was available. Eventually, I decided to go to New York University to work with Marvin Gershengorn, an upcoming thyroid researcher, who was only a few years older than I was.

11.2 POSTDOCTORAL TRAINING IN LIPID SIGNALING

The program at NYU was unusual in that the incoming fellows spent their first year in the laboratory and the second on the wards, the reverse of most programs. As I had never really been in a laboratory before (my minor degree in college was in art history), it was all so new and exciting to me. At that time the notion that phosphoinositide breakdown products might serve as second messengers was a new concept, pioneered by Bob Michel, and there was a great ongoing debate as to whether this was legitimate. Marvin Gershengorn was a stimulus-secretion coupling expert who studied calcium metabolism, and the debate fit right into his program regarding the mechanism of hormone-induced secretion and, in particular, thyrotropin-releasing hormone (TRH)-induced thyroid-stimulating hormone (TSH) and prolactin secretion from pituitary cells in culture. The lipid work was spearheaded in the group by Mario Rebecchi, a technician and eventually a graduate student under Marv.

When I arrived in the laboratory in the summer of 1981, they were well on their way to isolating inositol phosphate breakdown products of phosphoinositides via phospholipase C, and I was assigned the other metabolite, the 1,2-diacylglycerol (DAG) backbone. Little did I realize at the time that lipids were not highly esteemed in the biologic sciences, nor that there was limited information about how to isolate DAG or quantify it. There was one investigator, Arnis Kuksis, who had written a rather extensive review on the topic (4), which I used to guide my attempts. Eventually, I settled on the idea that I would label the DAG backbone by incubating cells with [³H]glycerol and the side chains with [¹⁴C]stearic acid and [³H]arachidonic acid until isotopic equilibrium appeared to be met, and then assume incorporated radiolabel reflected mass. This labeling procedure worked pretty well and I then isolated DAG by thin layer chromatography, identified the lipid by iodine staining, and quantified the results by liquid scintillation counting. Indeed, I was able to show that concomitant with TRH-induced inositol phosphate generation, DAG was accumulating over the first few minutes of TRH stimulation. This finding took on added importance when Yasutomi Nishizuka and colleagues began reporting that DAG might serve second messenger function by activating protein kinase C (PKC) (5,6). We eventually submitted two manuscripts to the *Journal of Biological Chemistry*, one on inositol phosphate and the other on DAG generation, and were told that if we combined the two they would take them as a single manuscript. This left me as second author (7), a disappointment, but nonetheless enthused by biologic research in general and lipids more specifically.

11.3   HOW STUDYING PHOSPHATIDYLCHOLINE LED TO CONCEPT OF SPHINGOMYELIN PATHWAY

At the end of my combined clinical and basic science endocrine fellowship, which lasted four years, I took a position as an assistant professor in the Division of Endocrinology in the Department of Medicine at Memorial Sloan-Kettering Cancer Center (MSKCC). Martin Sonenberg was department chair and a well-established thyroidologist and researcher. Martin was very generous and offered me space in his program. By today's standards my offer was quite modest, only 250 ft² of space, a single technician slot, and about $50,000 in start-up funds, but I didn't care; I had protected time and my own laboratory. In fact, for the first five years of my tenure at MSKCC I had no graduate students or postdoctoral fellows, doing all of the work myself with one, and eventually a second, technician. This probably served me well in the long run, as I was pretty undertrained in the laboratory upon arrival at MSKCC compared to my contemporaries.

Initially, I wanted to continue working on phosphoinositides but Marvin convinced me that I should branch out if I wanted to get respect as an independent investigator, so I elected to work on phosphatidylcholine as a secondary source of DAG for prolonged signaling through cell surface receptors. Although in the early 1980s the concept of "one receptor, one signaling pathway" predominated, by the time I opened my laboratory the notion that second messenger systems were activated in waves began to take hold. This concept conceived of signaling as a series of tandem second messenger modules with the originating module activated directly via the ligated receptor, whereas some of the subsequent messenger modules were transactivated via the primary signals. In the case of phosphatidylcholine, it was argued that phosphoinositides served as the immediate source of two independent second messengers, inositol phosphates for calcium mobilization and DAG for PKC activation (8). Thereafter, phosphatidylcholine served as a source for prolonged DAG activation as phosphatidylcholine represented the largest phospholipid pool in mammalian cells and therefore could act as a reservoir for DAG (9). In this paradigm,

DAG was generated from phosphatidylcholine in response to calcium mobilization and/or PKC activation either by the action of a phospholipase C or by coordinate activation of a phospholipase D followed by phosphatase action. This concept brought up ancillary issues, one of which interested me, which was the inability to reconcile the fact that there were hundreds if not thousands of cell surface receptors but only a handful of signaling pathways. I wrote an RO1 on this topic, which was well received at the National Institutes of Health (NIH). The tools for assessing phosphatidylcholine were still primitive at that time and again I resorted to radiolabeling, this time with [³H]choline, which would selectively label the head groups of only two major lipids in mammalian cells, phosphatidylcholine and sphingomyelin. I thought this to be ideal as it was generally accepted at the time that sphingomyelin was an inert structural lipid concentrated on the outer plasma membrane that served barrier function relative to the environment (10). Hence sphingomyelin would serve as an internal control for any changes I might find in phosphatidylcholine levels upon cellular stimulation.

I elected to compare DAG action to that of phorbol esters, as in the few years prior to my independence it had become ensconced in the literature that phorbol esters were DAG analogs capable of activating PKC (11,12). I was intrigued by the notion that a tumor promoter might act as a mimic of a physiologic lipid second messenger. In a first set of studies, I compared the action of the DAG, 1,2-dioctanlyl-glycerol (diC$_8$) to the phorbol ester, 12-O-tetradecanoylphorbol-13-acetate (TPA) on phosphatidylcholine synthesis and degradation in GH$_3$ rat pituitary cells. While both TPA and diC$_8$ induced phosphatidylcholine synthesis, only diC$_8$ induced its degradation. I was surprised at this difference and in the title of the manuscript published in the *Journal of Biological Chemistry* in July 1987 (13), stated that this represented evidence for at least some separate mechanisms of action of these lipids. At the time this seemed to make sense to me as I reasoned that TPA must act differently from DAG, at least in some aspect, as it was a tumor promoter and there was no evidence that DAG served such a function.

Follow-up studies examined the mechanism of phosphatidylcholine degradation using sphingomyelin as the internal control. Although TPA had no effect on sphingomyelin degradation, every time I stimulated cells with diC$_8$ sphingomyelin levels plummeted within minutes, an effect significantly larger than the effect on phosphatidylcholine (Fig. 11.1). DiC$_8$-induced reduction in sphingomyelin levels was accompanied by a concomitant and quantitatively equivalent increase in cellular ceramide, the sine qua non of sphingomyelinase activation. In vitro sphingomyelinase assays were set up, which generated another surprise: diC$_8$ directly activated the acid sphingomyelinase rather than the neutral form using lysates of GH$_3$ cells (Table 11.1). Acid sphingomyelinase was considered at that time to be exclusively lysosomal and involved in remodeling membranes. So why would it be activated by a lipid second messenger molecule? Only many years later as a result of the work by Ira Tabas would we come to understand that the acid sphingomyelinase gene product can be processed into a lysosomal form directed via mannose 6-phosphorylation and into a non-lysosomal secretory form upon carbohydrate trimming, and that this latter form was likely the signaling form of the enzyme (14,15).

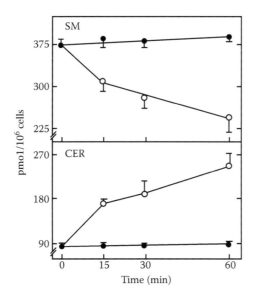

FIGURE 11.1 Time course of the effect of diC_8 on the levels of sphingomyelin and cer-amide. *Upper panel:* sphingomyelin (SM). *Lower panel:* ceramide (CER). Cells, labeled for 48 hours with [^{14}C]serine (0.1 μCi/mL) were resuspended in medium without radiola-bel containing diC_8 (200 μg/mL, *open circles*) or diluent (*closed circles*). At the indicated times, cells were separated from medium by centrifugation and extracted with 1 mL of chloroform:methanol:HCl (100:100:1, v/v/v) and 0.3 mL of balanced salt solution with 10 mM EDTA. Glycerophospholipids were saponified in methanolic KOH (0.1 M for 1 h at 37°C) prior to isolation by two-dimensional TLC or sequential one-dimensional TLC utiliz-ing chloroform:benzene:ethanol (80:40:75) followed by chloroform:methanol:28% ammonia (65:25:5) as solvents. The specific activity of serine was used to quantitate sphingolipids. The levels of [^{14}C]sphingomyelin and [^{14}C]ceramide in control incubations were 380 dpm/10^6 cells and 80 dpm/10^6 cells, respectively. The data (mean ± SE) represent duplicate determinations from three experiments. *Source*: From R. N. Kolesnick. "1,2-Diacylglycerols but Not Phorbol Esters Stimulate Sphingomyelin Hydrolysis in GH3 Pituitary Cells." *J Biol Chem* 262, no. 35 (1987): 16759–62.

Although these observations were interesting from a purely lipid perspective, they had no obvious biologic relevance. About this time Bob Bell, Yusuf Hannun, and colleagues began to publish that they thought that the sphingoid base, sphingosine, and a variety of other lysosphingolipids, might possess biologic properties (16–19). Specifically, they argued that these molecules might act as endogenous inhibitors of PKC. The biologic foundation for these initial studies was based on the ability of these sphingolipids to inactivate PKC enzymatic activity in a mixed micellar assay. This hypothesis was supported by cellular studies that showed that addition of sphingoid bases to intact cells resulted in inhibition of various events known to be mediated by PKC including phorbol ester- and formyl-methionyleucinephenyalanine (FMLP)-induced oxidative burst in human neutrophils (16), phorbol ester-induced differentiation of human promyelocytic leukemic (HL-60) cells into macrophages (17), and agonist-induced secretion from and activation of human platelets (18). They

TABLE 11.1

Effect of diC$_8$ on Acid and Neutral Sphingomyelinase Activities in Homogenates of GH$_3$ Cells

	Acid (pH 5.0)	Neutral (pH 7.4)
	nmol SM Hydrolyzed $\cdot$ mg Protein^{-1} $\cdot$ h^{-1}	
Basal	0.16 ± 0.06	0.41 ± 0.08
diC$_8$	0.60 ± 0.08[a]	0.54 ± 0.06
Mg^{2+}	0.20 ± 0.04	0.47 ± 0.11
Mg^{2+} + diC$_8$	0.69 ± 0.06[a]	0.36 ± 0.06
Triton X-100	0.84 ± 0.19[a]	0.83 ± 0.04[a]
Triton X-100 + diC$_8$	1.16 ± 0.06[a,b]	0.84 ± 0.04[a]
Triton X-100 + Mg^{2+}	0.75 ± 0.24[a]	1.33 ± 0.08[a,c]
Triton X-100 + Mg^{2+} + diC$_8$	1.33 ± 0.02[a,c]	1.41 ± 0.06[a,c]

Note: Cells, labeled for 48 hours with [^{3}H]choline (1.0 μCi/mL), were homogenized in 25 mM Tris-HC1 (pH 7.5) at 4°C. Portions of the homogenate were used in assays for acid and neutral sphingomyelinase activities. Assays were adjusted to the appropriate pH with Tris-HCl and acetate buffers (0.1 M), and diC$_8$ (150 μg/mL), Mg^{2+} (5 mM), and Triton X-100 (0.5 mg/mlL) were added. Incubations were stopped after 1 hour at 37°C and phospholipids extracted and isolated by TLC. Incubations contained 1.2×10^4 dpm of [^{3}H]sphingomyelin (SM) (0.3 nmol) and 0.16 mg of protein. All data (mean ± SE) were compared to a boiled blank control and represent duplicate determinations from four experiments for the acid activity and three experiments for the neutral activity.

[a] Significantly different from basal *(p < 0.005).*

[b] Significantly different from Triton X-100 *(p < 0.05).*

[c] Significantly different from Triton X-100 *(p < 0.005).*

Source: R. N. Kolesnick. "1,2-Diacylglycerols but Not Phorbol Esters Stimulate Sphingomyelin Hydrolysis in GH3 Pituitary Cells." *J Biol Chem* 262, no. 35 (1987): 16759–62.

conceived of the importance of this observation as relating to the pathophysiology of the sphingolipidoses (19,20), a group of inherited disorders of sphingolipid metabolism that were accompanied by marked elevation in lysosphingolipid levels (21,22). However, the PKC inhibition observed in these initial studies turned out not to be a physiologic effect but rather related to the charge on the free amine of the sphingoid base backbone (23), and eventually many molecules with free or tertiary amines were shown to be capable of inhibiting PKC in this assay (24–26). Nonetheless, these studies engendered real interest in the field.

While I was beginning to explore the differences in phorbol esters and DAGs on choline lipid metabolism I missed their papers. One day one of Marty Sonenberg's postdoctoral fellows, Dennis Mynarcik, who was aware of our findings regarding sphingomyelin hydrolysis, left a copy of *Science* on my desk with a note that said there was an article in it that I might be interested in. I was busy at that time, not having much assistance in the lab, and put the journal aside. A few months later, I came back to the journal and glanced at the article and immediately understood that my finding of stimulated sphingomyelin hydrolysis to ceramide upon DAG activation of GH_3 cells fit with what Bell and Hannun were implicating regarding sphingoid bases and lysosphingolipids as potential endogenous PKC inhibitors. While their initial studies were entirely pharmacologic, based on addition of sphingoid bases to intact cells and the observation of inhibition of PKC-mediated biologic events, there was no evidence that a signaling mechanism existed that would generate endogenous sphingoid bases that could serve such a purpose. I went ahead and showed that sphingomyelinase activation in response to DAG not only resulted in ceramide generation but also increased levels of free sphingoid bases in GH_3 cells. Further, I concluded, using cells in which PKC was down-regulated, that the effect of DAG on sphingomyelinase action was independent of PKC altogether. In the discussion of this original article I questioned whether differences in the action of the two classes of PKC activators might not have to do with the capability of DAGs, but not phorbol esters, to activate a potential inhibitory pathway for PKC. This information was published as a communication in the *Journal of Biological Chemistry* in December of 1987 (27).

Now I was really excited because I thought that I might be discovering a new signal transduction pathway. In 1987–1988, Suzanne Clegg, a technician in the lab, and I investigated this potential at the biochemical level. We showed that diC_8 induced only transient translocation of cytosolic "inactive" PKC into the membrane where it became "active," while TPA induced permanent membrane translocation followed by the eventual degradation of PKC (Fig. 11.2). I considered this consistent with the hypothesis that DAGs, but not phorbol esters, activated a potential endogenous inhibitory pathway for PKC involving sphingomyelin degradation to ceramides with subsequent de-acylation to sphingoid bases. In support of this notion, we showed that if exogenous sphingomyelinase was added to the external surface of the plasma membrane of GH_3 cells after TPA-induced redistribution of PKC into that compartment was complete, the PKC came off the membrane and redistributed back into the cytosol. Hence, the pattern of TPA plus exogenous sphingomyelinase mimicked the pattern after diC_8 (Fig. 11.3). Further, either diC_8 or exogenous sphingomyelinase increased sphingoid base levels 1.5- to 2.0-fold, and exogenous sphingosine alone was capable of redistributing membrane-bound PKC back into the cytoplasm, preventing chronic PKC activation and down-regulation by proteolytic cleavage. This paper (28), published in May 1988 in the *Journal of Biological Chemistry*, was titled "1,2-Diacylglycerols, but not Phorbol Esters, Activate a Potential Inhibitory Pathway for Protein Kinase C," and discussed the notion that activation of a sphingomyelinase by DAGs might initiate a legitimate physiologic feedback signaling pathway.

Now it was time to put the question to the test: Was the phenomenon we were observing pharmacology or biology? In other words, was there an intracellular biochemical pathway initiated by sphingomyelinase that, once engaged, was capable of generating sphingoid bases to inactivate PKC, preventing a well-defined biologic process? To address this question we examined phorbol ester-induced differentiation of HL-60 promyelocytic leukemia cells into macrophages. This event resulted in global alterations in cellular function and permitted multiple different types of readouts (growth, adherence, morphology, and enzymatic) for assessment. It had also been used extensively in the studies examining the inhibitory effects of

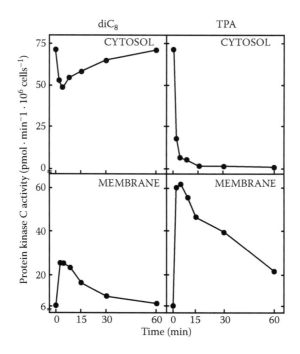

FIGURE 11.2   The effect of diC_8 and TPA on the distribution of protein kinase C activity. For these studies (n = 4), GH_3 cells were resuspended in balanced salt solution (3×10^6 cells/mL) containing diC_8 (200 μg/mL), TPA (10^{-6} M), or diluent (1% dimethyl sulfoxide). Incubations were stopped at the indicated times, and the cells were homogenized at 4°C. Protein kinase C activities in the cytosolic and particulate compartments were eluted from DEAE-cellulose columns with buffer containing 0.1 M NaCl. Protein kinase C activity was measured by the transfer of ^{32}P from [γ-^{32}P]ATP to lysine-rich histones using a standard reaction mixture. Ca^{2+}- and phospholipid-independent activity was measured in the absence of Ca^{2+}/phosphatidyl serine/diolein and usually represented less than 5% of the total activity. Incubations were terminated after 2 min and ^{32}P-labeled histones (3×10^4 dpm/assay) were quantitated by transfer to phosphocellulose paper and liquid scintillation counting. *Source*: From R. N. Kolesnick and S. Clegg. "1,2-Diacylglycerols, but Not Phorbol Esters, Activate a Potential Inhibitory Pathway for Protein Kinase C in GH3 Pituitary Cells. Evidence for Involvement of a Sphingomyelinase." *J Biol Chem* 263, no. 14 (1988): 6534–37.

exogenous sphingoid bases on PKC-mediated biologic events. As in GH_3 cells, stimulated sphingomyelin degradation by exogenous sphingomyelinase in HL-60 cells resulted in ceramide generation followed by sphingoid base production and antagonism of phorbol ester activation of PKC. Inhibition of PKC occurred by induced redistribution of the membrane-bound form into the cytoplasm, an effect mimicked by exogenous sphingoid bases. Further, sphingomyelinase-initiated PKC inhibition prevented macrophage differentiation by all criteria (see Fig. 11.4 for morphology). Degradation of other plasma membrane glycerophospholipids neither inactivated PKC nor blocked differentiation. Based on these data, we wrote in the publication, which appeared in *Journal of Biological Chemistry* in May of 1989 (29), "In summary, these studies suggest that the action of a sphingomyelinase, perhaps via the

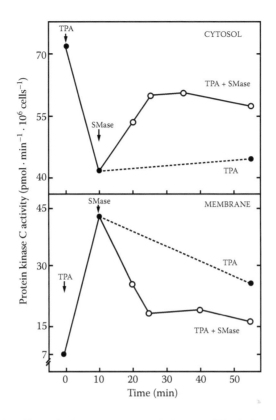

FIGURE 11.3 The effect of sphingomyelinase (SMase) on TPA-induced redistribution of protein kinase C activity. These studies (n = 5) were performed as described in Figure 11.2 except sphingomyelinase derived from *Staphylococcus aureus* (0.38 unit/mL, Sigma) was added after TPA (10^{-8} M) at the time indicated. *Source*: From R. N. Kolesnick and S. Clegg. "1,2-Diacylglycerols, but Not Phorbol Esters, Activate a Potential Inhibitory Pathway for Protein Kinase C in GH3 Pituitary Cells. Evidence for Involvement of a Sphingomyelinase." *J Biol Chem* 263, no. 14 (1988): 6534–37.

generation of free sphingoid bases, may be sufficient to comprise a physiologically relevant inhibitory pathway for protein kinase C."

A troubling aspect of these initial studies was the relatively small quantity of sphingoid bases generated despite large increases in free ceramide. This prompted a search for other ceramide metabolites. Schneider and Kennedy had reported that *Escherichia coli* diacylglycerol kinase was a promiscuous enzyme also capable of phosphorylating ceramide (30). Using this enzyme, we synthesized authentic ceramide 1-phosphate and used this compound to isolate endogenous ceramide 1-phosphate for the first time from mammalian cells (31). Further, we showed that ceramide 1-phosphate levels increase when ceramide was generated from sphingomyelin but not other sphingolipids, indicating there was a specific pathway activated by sphingomyelinase action. Late in 1990 we reported the existence of a ceramide kinase distinct from mammalian diacylglycerol kinase capable of generating ceramide 1-phosphate (32). These studies were assisted by the prior publication by Bajjalieh et al. (33) of the existence

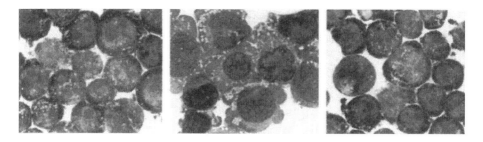

FIGURE 11.4 The effect of sphingomyelinase and TPA on the morphology of HL-60 cells. *Left-hand panel*: control; *middle panel*: TPA (0.5 nM); *right-hand panel*: TPA (0.5 nM) and sphingomyelinase (7.6×10^{-5} unit/mL). Portions of the cells, stimulated for 24 hours, were cytocentrifuged onto microscopic slides, and stained with McNeal's tetrachrome stain. Unstimulated HL-60 cells have a typical appearance of promyelocytes. These cells have a large nuclear to cytoplasmic ratio, a uniform round appearance, a central round nucleus, and a comparatively small size. TPA induces macrophage differentiation as evidenced by larger cells displaying more cytoplasm, variations in shape, cytoplasmic projections, and eccentric invaginated nuclei. These data are representative of three similar studies. (Magnification 300×.) *Source*: From R. N. Kolesnick. "Sphingomyelinase Action Inhibits Phorbol Ester-Induced Differentiation of Human Promyelocytic Leukemic (HL-60) Cells." *J Biol Chem* 264, no. 13 (1989): 7617–23.

of a mammalian ceramide kinase activity in synaptic vesicles. The fact that hydrolysis of a single plasma membrane substrate led within minutes to the synthesis of specific metabolites, at least one of which I believed was a second messenger, was evocative of the recently established phosphoinositide pathway. In fact, in the paper describing the ceramide kinase activity a figure compared the two systems (Fig. 11.5). In the accompanying text the question was raised as to whether ceramide or ceramide 1-phosphate might have signaling properties.

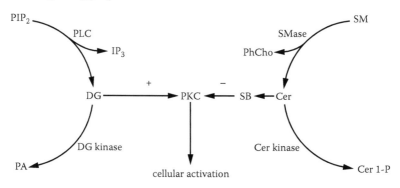

FIGURE 11.5 Analogy between the phosphoinositide pathway and a proposed sphingomyelin pathway. Abbreviations used are: PIP_2, phosphatidylinositol 4,5-bisphosphate; PLC, phospholipase C; IP_3, inositol trisphosphate; DG, diacylglycerol; PA, phosphatidic acid; PKC, protein kinase C; SB, sphingoid base; Cer, ceramide; SM, sphingomyelin; PhCho, phosphocholine; Cer 1-P, ceramide 1-phosphate. *Source*: From R. N. Kolesnick and M. R. Hemer. "Characterization of a Ceramide Kinase Activity from Human Leukemia (HL-60) Cells. Separation from Diacylglycerol Kinase Activity." *J Biol Chem* 265, no. 31 (1990): 18803–8.

11.4   EVIDENCE FOR RECEPTOR-MEDIATED SPHINGOMYELINASE ACTIVATION

Late in 1989, Hannun and coworkers published their first manuscript implicating receptor-mediated sphingomyelinase activation via vitamin D and provided evidence, using a synthetic short-chain ceramide analog, that ceramide might serve a second messenger function for HL-60 cell differentiation (34). Subsequent studies by these investigators showed this pathway was important for tumor necrosis factor-α and interferon-γ signaling (35,36). In late 1991, we published the existence of a ceramide-activated kinase (37), which we now know is Kinase Suppressor of Ras (KSR) (38). Hannun and colleagues and my group worked steadily on this system for the next few years, after which other investigators joined the process.

By the early to mid 1990s the sphingomyelin pathway appeared to be on solid footing. However, there remained a great deal of doubt within the lipid community regarding the role of an acid sphingomyelinase in signal transduction. The generation of a knockout mouse for acid sphingomyelinase by Ed Schuchman and coworkers at Mt. Sinai Hospital in New York greatly aided in dispelling much of that doubt (39). We showed in collaboration with Ed's group that this mouse, which developed Niemann Pick disease well into adulthood, displayed numerous defects in apoptosis that occurred well before the onset of a sphingomyelin storage defect, and that the phenotype was partially recapitulated in acid sphingomyelinase heterozygotes, which are analogous to the human carrier state and do not manifest Niemann Pick disease. In particular, the acid sphingomyelinase knockout mouse displays a near complete defect in endothelial cell apoptosis in response to ionizing radiation (Fig. 11.6) (39–41). Apoptosis is the primary mode of endothelial cell death in the lung, gut, and brain in response to single high-dose radiation. This apoptotic event occurs in a wave in these tissues usually within the first 1–6 hours post-irradiation with as much as 50% of the endothelium dying in this time frame. These apoptotic endothelia are then cleared from these tissues most likely by the action of tissue macrophages. Hence, this event had been previously missed as histologic tissue damage does not usually occur for days in these models. In the case of the small intestine, loss of endothelial cell apoptosis in the acid sphingomyelinase knockout mouse had a profound effect on overall response to irradiation, as these mice were protected from the lethal gastrointestinal syndrome (40), a well-defined toxicity of radiation and chemotherapy on the gut.

A similar wave of endothelial cell apoptosis occurred when MCA/129 fibrosarcomas and B16 melanomas were grown in wild-type but not knockout mice (Fig. 11.7), and as observed in the gut, the tumors were resistant to radiation in the absence of this event (42). These data prompted my colleague Zvi Fuks and I to postulate that single-dose irradiation might act differently than the more traditional fractionated radiation approach, as the threshold for endothelial cell apoptosis (>8 Gy) was well above the fractions (1.8–2.0 Gy) used in standard radiation schemes (43). When this information was viewed in the context of the work of Ira Tabas and associates showing that acid sphingomyelinase, the product of a single gene in mammalian cells, was alternatively processed into a form that trafficked either to the lysosome or into a secretory pool, it became apparent how an enzyme, previously considered to

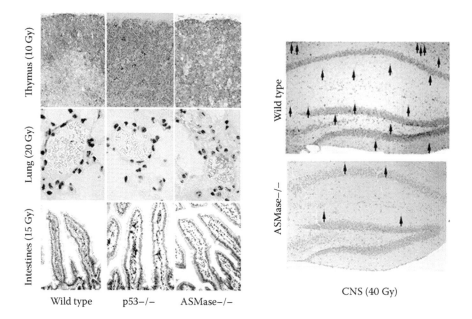

FIGURE 11.6 (See color insert following page 112.) Radiation-induced endothelial apoptosis is ASMase-dependent and p53-independent. Tissue specimens were obtained from mice at 4–10 hours after exposure to whole-body irradiation at the indicated doses. Histologic tissue sections of 5 μm were double-stained for apoptosis by TUNEL and for endothelium using antibodies to tissue-specific endothelial surface markers. Brown nuclei are indicative of apoptosis. The residual clusters in the p53$^{-/-}$ thymus represent apoptotic microvascular endothelium. While p53 has no impact on endothelial apoptosis in the lung, gut, or brain (*arrows*), acid sphingomyelinase largely blocks this event. *Sources*: Adapted from P. Santana, L. A. Pena, A. Haimovitz-Friedman, et al. "Acid Sphingomyelinase-Deficient Human Lymphoblasts and Mice Are Defective in Radiation-Induced Apoptosis." *Cell* 86, no. 2 (1996): 189–99; F. Paris, Z. Fuks, A. Kang, et al. "Endothelial Apoptosis as the Primary Lesion Initiating Intestinal Radiation Damage in Mice." *Science* 293, no. 5528 (2001): 293–97; L. A. Pena, Z. Fuks, and R. N. Kolesnick. "Radiation-Induced Apoptosis of Endothelial Cells in the Murine Central Nervous System: Protection by Fibroblast Growth Factor and Sphingomyelinase Deficiency." *Cancer Res* 60, no. 2 (2000): 321–27.

act solely for membrane turnover in the lysosome, might act as a signaling molecule in the plasma membrane.

11.5   PERSPECTIVE

After reviewing nearly 12,000 publications in the field, I look back and am struck by the role of serendipity in discovery. It still seems ironic that I had no interest in becoming a medical specialist, nor any interest in basic science, and yet I have spent a career in academic basic science. I think it is a testimony to the notion that the action of a single mentor can profoundly impact one's life. Further, I am struck by the fact that a simple internal control used for a set of lipid studies has directed the course of my career.

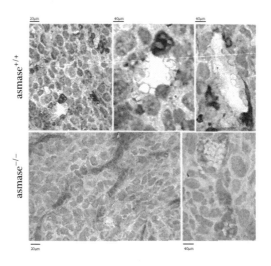

FIGURE 11.7 (See color insert following page 112.) MCA/129 fibrosarcomas implanted into *asmase⁻/⁻* mice display reduced baseline and radiation-induced endothelial cell apoptosis. Representative 5-μm histologic tumor sections obtained 4 hours after exposure to 15 Gy, stained for apoptosis by TUNEL and for the endothelial cell surface marker CD-34. Apoptotic endothelium manifests a red-brown TUNEL-positive nuclear signal surrounded by dark blue plasma membrane signal of CD-34 staining. *Source*: From M. Garcia-Barros, F. Paris, C. Cordon-Cardo, et al. "Tumor Response to Radiotherapy Regulated by Endothelial Cell Apoptosis." *Science* 300, no. 5622 (2003): 1155–59.

REFERENCES

1. C. V. Tompkins, M. C. Srivastava, P. H. Sonksen, and J. D. Nabarro. "A Comparative Study of the Distribution and Metabolism of Monocomponent Human Insulin and Porcine Proinsulin in Man." *Biochem J* 125, no. 3 (1971): 64P.
2. D. Q. Borsey and D. N. Malone. "Local Cutaneous Allergy to Monocomponent Insulin." *Postgrad Med J* 55, no. 641 (1979): 199–200.
3. A. Teuscher. "Treatment of Insulin Lipoatrophy with Monocomponent Insulin." *Diabetologia* 10, no. 3 (1974): 211–14.
4. B. J. Holub and A. Kuksis. "Metabolism of Molecular Species of Diacylglycerophospholipids," in *Advances in Lipid Research*, vol 16, eds. R. Paoletti and D. Kritchevsky, 1–125. New York: Academic Press, 1978.
5. Y. Takai, A. Kishimoto, Y. Kikkawa, T. Mori, and Y. Nishizuka. "Unsaturated Diacylglycerol as a Possible Messenger for the Activation of Calcium-Activated, Phospholipid-Dependent Protein Kinase System." *Biochem Biophys Res Commun* 91, no. 4 (1979): 1218–24.
6. Y. Kawahara, Y. Takai, R. Minakuchi, K. Sano, and Y. Nishizuka. "Possible Involvement of Ca2+-Activated, Phospholipid-Dependent Protein Kinase in Platelet Activation." *J Biochem (Tokyo)* 88, no. 3 (1980): 913–16.
7. M. J. Rebecchi, R. N. Kolesnick, and M. C. Gershengorn. "Thyrotropin-Releasing Hormone Stimulates Rapid Loss of Phosphatidylinositol and Its Conversion to 1,2-Diacylglycerol and Phosphatidic Acid in Rat Mammotropic Pituitary Cells. Association with Calcium Mobilization and Prolactin Secretion." *J Biol Chem* 258, no. 1 (1983): 227–34.

8. Y. Nishizuka. "The Albert Lasker Medical Awards. The Family of Protein Kinase C for Signal Transduction." *JAMA* 262, no. 13 (1989): 1826–33.

9. J. M. Besterman, V. Duronio, and P. Cuatrecasas. "Rapid Formation of Diacylglycerol from Phosphatidylcholine: A Pathway for Generation of a Second Messenger." *Proc Natl Acad Sci U S A* 83, no. 18 (1986): 6785–89.

10. R. N. Kolesnick. "Sphingomyelin and Derivatives as Cellular Signals." *Prog Lipid Res* 30, no. 1 (1991): 1–38.

11. M. Castagna, Y. Takai, K. Kaibuchi, K. Sano, U. Kikkawa, and Y. Nishizuka. "Direct Activation of Calcium-Activated, Phospholipid-Dependent Protein Kinase by Tumor-Promoting Phorbol Esters." *J Biol Chem* 257, no. 13 (1982): 7847–51.

12. N. A. Sharkey, K. L. Leach, and P. M. Blumberg. "Competitive Inhibition by Diacylglycerol of Specific Phorbol Ester Binding." *Proc Natl Acad Sci U S A* 81, no. 2 (1984): 607–10.

13. R. N. Kolesnick and A. E. Paley. "1,2-Diacylglycerols and Phorbol Esters Stimulate Phosphatidylcholine Metabolism in GH3 Pituitary Cells. Evidence for Separate Mechanisms of Action. *J Biol Chem* 262, no. 19 (1987): 9204–10.

14. S. L. Schissel, E. H. Schuchman, K. J. Williams, and I. Tabas." Zn2+-Stimulated Sphingomyelinase Is Secreted by Many Cell Types and Is a Product of the Acid Sphingomyelinase Gene." *J Biol Chem* 271, no. 31 (1996): 18431–36.

15. S. L. Schissel, X. Jiang, J. Tweedie-Hardman, et al. "Secretory Sphingomyelinase, a Product of the Acid Sphingomyelinase Gene Can Hydrolyze Atherogenic Lipoproteins at Neutral pH. Implications for Atherosclerotic Lesion Development." *J Biol Chem* 273, no. 5 (1998): 2738–46.

16. E. Wilson, M. C. Olcott, R. M. Bell, A. H. Merrill Jr., and J. D. Lambeth. "Inhibition of the Oxidative Burst in Human Neutrophils by Sphingoid Long C Chain Bases. Role of Protein Kinase C in Activation of the Burst." *J Biol Chem* 261, no. 27 (1986): 12616–23.

17. A. H. Merrill Jr., A. M. Sereni, V. L. Stevens, Y. A. Hannun, R. M. Bell, and J. M. Kinkade Jr. "Inhibition of Phorbol Ester-Dependent Differentiation of Human Promyelocytic Leukemic (HL-60) Cells by Sphinganine and Other Long-Chain Bases." *J Biol Chem* 261, no. 27 (1986): 12610–15.

18. Y. A. Hannun, C. S. Greenberg, and R. M. Bell. "Sphingosine Inhibition of Agonist-Dependent Secretion and Activation of Human Platelets Implies That Protein Kinase C Is a Necessary and Common Event of the Signal Transduction Pathways." *J Biol Chem* 262, no. 28 (1987): 13620–26.

19. Y. A. Hannun and R. M. Bell. "Functions of Sphingolipids and Sphingolipid Breakdown Products in Cellular Regulation." *Science* 243, no. 4890 (1989): 500–7.

20. Y. A. Hannun and R. M. Bell. "Lysosphingolipids Inhibit Protein Kinase C: Implications for the Sphingolipidoses." *Science* 235, no. 4789 (1987): 670–74.

21. K. Suzuki. "Enzymatic Diagnosis of Sphingolipidoses." *Methods Enzymol* 138 (1987): 727–62.

22. R. O. Brady. "Sphingolipidoses." *Annu Rev Biochem* 47 (1978): 687–713.

23. R. Bottega, R. M. Epand, and E. H. Ball. "Inhibition of Protein Kinase C by Sphingosine Correlates with the Presence of Positive Charge." *Biochem Biophys Res Commun* 164, no. 1 (1989): 102–7.

24. M. Moruzzi, B. Barbiroli, M. G. Monti, B. Tadolini, G. Hakim, and G. Mezzetti. "Inhibitory Action of Polyamines on Protein Kinase C Association to Membranes." *Biochem J* 247, no. 1 (1987): 175–80.

25. J. M. Besterman, W. S. May Jr., H. LeVine 3rd, E. J. Cragoe Jr., and P. Cuatrecasas. "Amiloride Inhibits Phorbol Ester-Stimulated Na+/H+ Exchange and Protein Kinase C. An Amiloride Analog Selectively Inhibits Na+/H+ Exchange." *J Biol Chem* 260, no. 2 (1985): 1155–59.

26. C. A. O'Brian, R. M. Liskamp, D. H. Solomon, and I. B. Weinstein. "Inhibition of Protein Kinase C by Tamoxifen." *Cancer Res* 45, no. 6 (1985): 2462–65.

27. R. N. Kolesnick. "1,2-Diacylglycerols but Not Phorbol Esters Stimulate Sphingomyelin Hydrolysis in GH3 Pituitary Cells." *J Biol Chem* 262, no. 35 (1987): 16759–62.

28. R. N. Kolesnick and S. Clegg. "1,2-Diacylglycerols, but Not Phorbol Esters, Activate a Potential Inhibitory Pathway for Protein Kinase C in GH3 Pituitary Cells. Evidence for Involvement of a Sphingomyelinase." *J Biol Chem* 263, no. 14 (1988): 6534–37.

29. R. N. Kolesnick. "Sphingomyelinase Action Inhibits Phorbol Ester-Induced Differentiation of Human Promyelocytic Leukemic (HL-60) Cells." *J Biol Chem* 264, no. 13 (1989): 7617–23.

30. E. G. Schneider and E. P. Kennedy. "Phosphorylation of Ceramide by Diglyceride Kinase Preparations from *Escherichia coli*. *J Biol Chem* 248, no. 10 (1973): 3739–41.

31. K. A. Dressler and R. N. Kolesnick. "Ceramide 1-Phosphate, a Novel Phospholipid in Human Leukemia (HL-60) Cells. Synthesis via Ceramide from Sphingomyelin." *J Biol Chem* 265, no. 25 (1990): 14917–21.

32. R. N. Kolesnick and M. R. Hemer. "Characterization of a Ceramide Kinase Activity from Human Leukemia (HL-60) Cells. Separation from Diacylglycerol Kinase Activity." *J Biol Chem* 265, no. 31 (1990): 18803–8.

33. S. M. Bajjalieh, T. F. Martin, and E. Floor. "Synaptic Vesicle Ceramide Kinase. A Calcium-Stimulated Lipid Kinase That Co-purifies with Brain Synaptic Vesicles." *J Biol Chem* 264, no. 24 (1989): 14354–60.

34. T. Okazaki, R. M. Bell, and Y. A. Hannun. "Sphingomyelin Turnover Induced by Vitamin D3 in HL-60 Cells. Role in Cell Differentiation." *J Biol Chem* 264, no. 32 (1989): 19076–80.

35. T. Okazaki, A. Bielawska, R. M. Bell, and Y. A. Hannun. "Role of Ceramide as a Lipid Mediator of 1 Alpha,25-Dihydroxyvitamin D3-Induced HL-60 Cell Differentiation." *J Biol Chem* 265, no. 26 (1990): 15823–31.

36. M. Y. Kim, C. Linardic, L. Obeid, and Y. Hannun. "Identification of Sphingomyelin Turnover as an Effector Mechanism for the Action of Tumor Necrosis Factor Alpha and Gamma-Interferon. Specific Role in Cell Differentiation." *J Biol Chem* 266, no. 1 (1991): 484–89.

37. S. Mathias, K. A. Dressler, and R. N. Kolesnick. "Characterization of a Ceramide-Activated Protein Kinase: Stimulation by Tumor Necrosis Factor Alpha." *Proc Natl Acad Sci U S A* 88, no. 22 (1991): 10009–13.

38. Y. Zhang, B. Yao, S. Delikat, et al. "Kinase Suppressor of Ras Is Ceramide-Activated Protein Kinase." *Cell* 89, no. 1 (1997): 63–72.

39. P. Santana, L. A. Pena, A. Haimovitz-Friedman, et al. "Acid Sphingomyelinase-Deficient Human Lymphoblasts and Mice Are Defective in Radiation-Induced Apoptosis." *Cell* 86, no. 2 (1996): 189–99.

40. F. Paris, Z. Fuks, A. Kang, et al. "Endothelial Apoptosis as the Primary Lesion Initiating Intestinal Radiation Damage in Mice." *Science* 293, no. 5528 (2001): 293–97.

41. L. A. Pena, Z. Fuks, and R. N. Kolesnick. "Radiation-Induced Apoptosis of Endothelial Cells in the Murine Central Nervous System: Protection by Fibroblast Growth Factor and Sphingomyelinase Deficiency." *Cancer Res* 60, no. 2 (2000): 321–27.

42. M. Garcia-Barros, F. Paris, C. Cordon-Cardo, et al. "Tumor Response to Radiotherapy Regulated by Endothelial Cell Apoptosis." *Science* 300, no. 5622 (2003): 1155–59.

43. Z. Fuks and R. Kolesnick. "Engaging the Vascular Component of the Tumor Response." *Cancer Cell* 8, no. 2 (2005): 89–91.

12 Endotoxin-Induced Myocarditis

A New Pathophysiological Entity in Systemic Inflammatory Response Syndrome

Ma-Li Wong and Julio Licinio

CONTENTS

12.1 INTRODUCTION

Administration of lipopolysaccharide (LPS), an endotoxin produced by gram-negative bacteria, has been used for several decades in the study of inflammation. This is an easy and simple strategy that induces dose-dependent inflammatory reaction and if given in high doses it may cause systemic inflammatory response syndrome (SIRS) and also death. We have been intrigued by the effects of systemic inflammation for many years, and have been applying this model to the study of central nervous system (CNS) response to systemic inflammation, but we have also been interested in understanding the communications between brain and periphery during systemic inflammation, as intense signs and symptoms mediated by the CNS are initiated by inflammation that originated in the periphery. Those manifestations include suppression of exploration, food intake, and sexual behavior; motor retardation; and alterations in temperature regulation and cognition (1–4); they may result from SIRS and sepsis and have been proposed to be mediated by cytokines, especially interleukin-1 beta (IL-1ß), synthesized during systemic inflammation.

In the early 1990s, while we were examining the CNS response to high doses of LPS, we documented de novo synthesis of IL-1ß, IL-RA, IL-10, and IL-13 (5) in the brain. Interestingly, patterns of IL-1ß gene induction expression were very similar to patterns of NOS2 mRNA induction (6) during SIRS. In the brain, discrete areas were induced at 2 h after LPS administration, the response peaked in most areas at around 6 h, and a residual response was found at 24 h. In fact, analyses of temporal and spatial patterns revealed that gene expression for IL-1ß, IL-1 receptor 1 (IL-1R1), and NOS2 was present in vascular and perivascular areas in the CNS (6). Those patterns have made us consider that IL-ß-NOS2 interaction in brain vasculature may be a mechanism of immune system–brain communication (7). As part of the response to systemic inflammation, vascular and nonvascular IL-1ß would bind to vascular and perivascular IL-1R1 and induce the gene expression of vascular and perivascular NOS2. This induction of NOS2 by IL-1ß could represent a mechanism of modulation for the CNS effects of peripheral inflammatory mediators. We have considered that it is likely that other sites, especially the barrier related ones, may contribute to the actions of cytokines in the brain.

Central manifestations of peripheral inflammation may be modulated by de novo IL-1ß synthesis in the context of limited cytokine counter-regulation. It has been hypothesized that in the periphery, cytokine counter-regulatory response is the predominant reaction. Interestingly, temporal and spatial patterns of gene expression have not been fully characterized in key peripheral organs involved in systemic inflammation. Systemic inflammation and sepsis are characterized by multi-organ failure, hypotension, and vascular collapse; these conditions can evolve to septic shock, and septic shock is still the major cause of death following trauma. Sepsis is a condition that continues to have a high mortality rate of up to 60% (8,9). The major mediator of the high morbidity and mortality rates characteristic of Gram-negative septic shock is LPS, which is also considered to be a major effector of hypotension, partially through increasing production of nitric oxide (NO).

Cardiac dysfunction is a key element in septic shock and it starts within the first 24 h (10,11). Clinical data lead to the clear suggestion of myocardial injury in sepsis, but this condition remains to be elucidated and the underlying pathophysiologic mechanism is unknown. We were interested in understanding the course of events in peripheral organs during systemic inflammation, and in collaboration with our colleague and long-time friend Samuel McCann, we designed a study to clarify the transcriptional events in the heart during the early phases of systemic inflammatory response syndrome induced by administration of LPS. We wanted to specifically test the following two questions: (a) Is myocardium tissue injury an element of the pathophysiology of SIRS? (b) Could the characterization of differential patterns of gene expression determine the extent of myocardium involvement during SIRS? In this chapter, we describe the discovery of a novel pathophysiological state in the heart during SIRS that we named endotoxin-induced myocarditis (EIM). This work was reported in the *Proceedings of the National Academy of Sciences* in 2003 (12). Part of this work was accomplished while we were at the Intramural Research Program, National Institutes of Health, working with Philip W. Gold, and the work was completed at the University of California, Los Angeles, to where we relocated in 1999 and accepted positions as professors in the Department of Psychiatry and Biobehavioral Sciences, Neuropsychiatric Institute.

From our journey discovering EIM, we have been led to believe that it is possible that other relevant pathophysiological states still remain to be described during SIRS. We believe that a better understanding of the body's reaction during SIRS will contribute to unraveling new aspects of the innate immune response that are important during sepsis and other inflammatory disorders, and may impart advances in pharmacotherapy of sepsis.

12.2 MYOCARDIAL DYSFUNCTION DURING SEPTIC SHOCK

Cardiac dysfunction in sepsis is multifactorial and not well understood but potentially reversible. It is often underestimated in degree and relevance. Acute septic cardiomyopathy is believed to result from hemodynamic alterations characteristic of septic shock and intrinsic cardiac dysfunction. Global myocardial ischemia is no longer favored to be a primary etiological hypothesis that could explain the intrinsic cardiac dysfunction during septic shock (13,14), because myocardial production of lactate is not increased and coronary blood flow is not decreased; it is often normal or even enhanced. Right (RV) and left ventricles (LV) are enlarged due to increased ventricular compliance. Heart dysfunction can be additionally aggravated by a secondary RV impairment caused by pulmonary hypertension during acute respiratory distress syndrome that may accompany sepsis and SIRS. Currently, it is accepted that myocardial depressant factors (MDF) cause myocyte dysfunction, which results in alterations of the intrinsic cardiac function in the LV and RV (15,16). Candidate mediators for MDF comprise many factors including NO (16–18)

Activation of constitutive NOS (NOS3) or inducible NOS (NOS2) in the myocardium may play a role in the cardiac dysfunction complicating SIRS (19). Systolic and diastolic cardiac functions in physiological states and in sepsis are thought to be mediated by NO. NO may also limit the oxidative damage to cells by scavenging reactive molecules (20).

12.3 IDENTIFICATION OF EIM

Our studies focused initially in characterizing the temporal and spatial pattern of NOS2 expression in the myocardium (12), because NOS2 is a strong MDF candidate (17). We conducted in situ hybridization studies in heart tissues that were obtained from virus- and pathogen-free male Sprague-Dawley rats treated with 5 mg of LPS. Those studies revealed that NOS2 mRNA expression peaked in the left LV at 6 h (ANOVA, $p < 0.01$) (Fig. 12.1) and returned to baseline levels 24 h after LPS administration. The distribution of NOS2 mRNA in the LV after LPS treatment was compatible with previous reports (21,22), and the temporal patterns of NOS2 induction were similar to those described in brain, pituitary, and peripheral tissues (6,23,24); therefore, in those tissues NOS2 mRNA levels that were present 24 h after SIRS induction were close to baseline levels.

Surprisingly, we found that NOS2 gene expression in the RV at 24 h after LPS administration was remarkably increased (Fig. 12.1) (ANOVA, $p < 0.01$). To our knowledge this phenomenon had not been previously reported. A residual NOS2 mRNA induction at 24 h in cardiac tissue with predominant LV contribution was

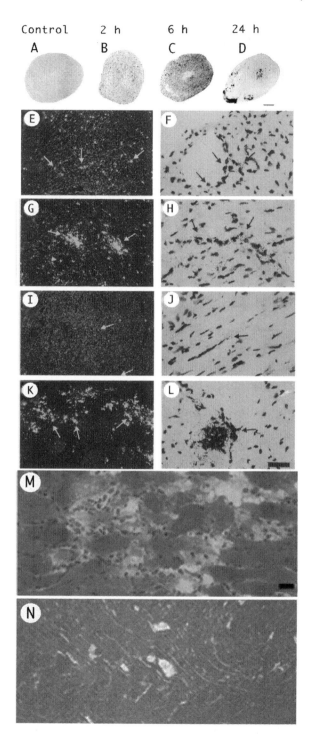

FIGURE 12.1

documented (22,24). In the RV at 24 h, we also identified a new pattern of cellular distribution, in which high levels of NOS2 mRNA were found in clusters of cells that were physically close. We observed these groups of cells in the free wall of RV but the interventricular septum was spared. Expression of NOS2 mRNA was found in cells located in the epicardium, endocardium/subendocardium, and in the myocardium layers. These cells were neutrophils predominantly, though endothelial cells and cardiomyocytes also expressed NOS2. Intense signs of inflammatory reaction were observed in the RV at 24 h. In the RV there were areas of myocardium degeneration that contained fragmented, pale, vacuolated cardiomyocytes and significant neutrophils margination in vessels and invasion of myocardium (Fig. 12.1). Myocytes that contained globular proteins were surrounded by degenerated leukocytes with phagocytic material (fine nuclear debris and bright eosinophilic cytoplasmic material). This histological description is compatible with the diagnosis of acute myocarditis with areas of focal aseptic necrosis; thus, we concluded that the RV is vulnerable to a process of EIM. Signs of local acute inflammatory reaction were not otherwise present in ventricles at 6 or 24 h. Our results indicate that pathophysiological processes are different in the RV and LV 24 h after SIRS induction and that different patterns of NOS2 gene expression characterize those differences (12).

12.4   TRANSCRIPTIONAL CHARACTERIZATION OF EIM

We debated what would be the best course of action to further advance our understanding about our histological findings in the heart. A consultation with experts

---

FIGURE 12.1   (See color insert following page 112.) Induction of NOS2 mRNA in heart after LPS. The top image is a composite of film autoradiography showing time course of NOS2 mRNA in the heart (**a–d**) by in situ hybridization histochemistry after LPS treatment. In the heart, NOS2 mRNA induction starts at 2 h, progress at 6 h. At 24 h NOS2 levels return virtually to baseline in the LV, but are increased in the RV and a residual response is still present in the endocardium. Scale bar = 2 mm. The middle set of images shows a composite of low (*columns one and two*) and high (*column three*) magnification images of NOS2 mRNA in heart ventricles. First three rows depict the LV at 2 h (**A, B, C**); 6 h (**D, E, F**); and 24 h (**G, H, I**). The fourth column depicts the RV at 24 h (**J, K, L**). Arrows point to areas of increased NOS2 expression appearing as white dots in dark field (*first column*) and black dots in high magnification images. Note the high levels of NOS2 mRNA in the RV at 24 h (**J** and **L**). Scale bar, 378 _μm for columns one and two, and 30 _μm for column three. The bottom set of images shows bright field images of heart ventricles 24 h after induction of SIRS. This section of RV myocardium (**M**) is infiltrated by a moderate number of neutrophils that surround degenerated myocardiocytes. The cardiomyocytes are fragmented, pale, and vacuolated and in many areas the proteins are globular. Some globular proteins are surrounded by degenerated neutrophils characterized by fine nuclear debris, and several neutrophils contain bright eosinophilic cytoplasmic material (phagocytosis). Vacuolation varies from small to large clear, round to irregular clear spaces. Notice no signs of local inflammatory reaction in LV (**N**). Scale bars, 7.5 μm (**M**), and 31.2 μm (**N**). *Source*: Reproduced with permission from M. L. Wong, F. O'Kirwan, N. Khan, J. Hannestad, K. H. Wu, D. Elashoff, et al. "Identification, Characterization, and Gene Expression Profiling of Endotoxin-Induced Myocarditis." *Proc Natl Acad Sci U S A* 100 (2003): 14241–46, copyright by The National Academy of Sciences of the United States of America.

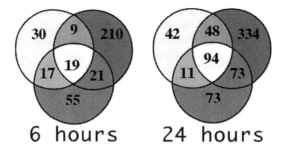

6 hours   24 hours

FIGURE 12.2 Microarray results are summarized in these two Venn diagrams. In each diagram, circles represent the main effects of our ANOVA analysis: LPS treatment (T, *top right circle*); LPS treatment by side (TxS, *top left circle*), and side (S, *bottom circle*). Numbers inside each compartment represent the number of transcripts that are significant for effect(s). Note the expansion of transcripts that are significant for treatment effect from 6 to 24 h. *Source*: Reproduced with permission from M. L. Wong, F. O'Kirwan, N. Khan, J. Hannestad, K. H. Wu, D. Elashoff, et al. "Identification, Characterization, and Gene Expression Profiling of Endotoxin-Induced Myocarditis." *Proc Natl Acad Sci U S A* 100 (2003): 14241–46, copyright by The National Academy of Sciences of the United States of America.

in heart physiology at UCLA taught us that it would be very difficult to assess RV function in rodents. Therefore, as we were using microarrays in our pharmacogenomics studies, we decided to pursue the characterization of transcriptional events related to EIM by using DNA microarrays (GeneChip Rat Genome U34A arrays, Affymetrix, Santa Clara, California, USA) to study RV and LV tissue before (6 h) and during EIM (24 h) (12). A summary of our microarray results is found in the Venn diagram (Fig. 12.2). Differently expressed genes had predominantly a treatment (LPS) effect. Unexpectedly, we found that more genes were significantly upregulated in the heart at 24 h when compared to 6 h (χ^2 analysis, $p = 0.0001$, Table 12.1), as the expression of typical pro- and anti-inflammatory genes have been reported to return to levels close to baseline 24 h after SIRS induction in brain, pituitary, and peripheral tissues (25,26).

In the RV, the number of genes that had fold-change of 2 or more was higher than in the LV at 24 h (z test for proportions, $p \leq 0.001$, Table 1) but not at 6 h (idem, $p = 0.73$). This represents a strong suggestion that EIM is accompanied by an exacerbated transcriptional response of the RV. Another fact that further supports this hypothesis comes from the analysis of the number of genes that have a significant treatment effect with a robust upregulation in transcription (fivefold or higher) levels. They increased from 0 transcripts at 6 h to 41 transcripts at 24 h (χ^2 analysis, $p < 0.0001$, Fig. 12.3, Table 12.1). Semiquantitative RT-PCR experiments confirmed our microarray findings (ANOVA, mixed effect model, $p < 0.05$) (Fig. 12.4). In these studies, fold changes reflect the net changes that occur in a heterogeneous tissue; therefore, we expected that most of our changes would be relatively small, as large fold changes would reflect very large transcription increases in a significant proportion of cells.

It has been reported that in the heart, Toll-like receptor 4 and CD14 are critical in mediating the pro-inflammatory response induced by LPS (27). It is also hypothesized that excessive production of immunomediators, such as NO and

TABLE 12.1

Chi Square Analyses for our Microarray Data: χ^2 and
Proportion Analyses of Gene Changes after LPS ip Injection

A. Upregulated Genes and Time ($p < 0.0001$)

	Upregulated Genes with Treatment Effect	Entire Dataset of Upregulated Genes
6 h	139	3903
24 h	423	3758

B. FC at 24 h and Ventricles ($p < 0.0001$)

	Treatment FC < ±2	Treatment FC ≥ ±2
Right	375	174
Left	476	73

C. FC at 6 h and Ventricles ($p = 0.045$)

	Treatment FC < ±2	Treatment FC ≥ ±2
Right	183	76
Left	203	56

D. Large FC at 24 h and Ventricles ($p < 0.0001$)

	Treatment FC < ±5	Treatment FC ≥ ±5
Right	505	44
Left	531	18

E. Large FC at 6 h and Ventricles ($p = 0.101$)

	Treatment FC < ±5	Treatment FC ≥ ±5
Right	246	13
Left	253	6

Note: In this table, we report five separate $\chi2$ analyses (**A–E**). These show associations between transcription changes and time or side of heart ventricle. The analyses revealed that significantly more genes are upregulated at 24 h than at 6 h after LPS treatment (**A**). Significantly more genes had fold change ≥2 in the RV at 24 h (**B**) and at 6 h (**C**). The proportion of genes with treatment fold changes ≥5 was significantly higher in the RV when compared to the LV at 24 h (**D**), but not at 6 h (**E**). FC, fold change.

Source: Reproduced with permission from M. L. Wong, F. O'Kirwan, N. Khan, J. Hannestad, K. H. Wu, D. Elashoff, et al. "Identification, Characterization, and Gene Expression Profiling of Endotoxin-Induced Myocarditis." *Proc Natl Acad Sci U S A* 100 (2003): 14241–46, copyright by the National Academy of Sciences of the United States of America.

pro-inflammatory cytokines, can cause cardiac decompensation (10,15), which may also contribute to the pathophysiological processes that results in sequelae of infection, ischemia, and trauma due to tissue injury. This process could be critical in the heart because cardiomyocytes are considered terminal cells, and are therefore

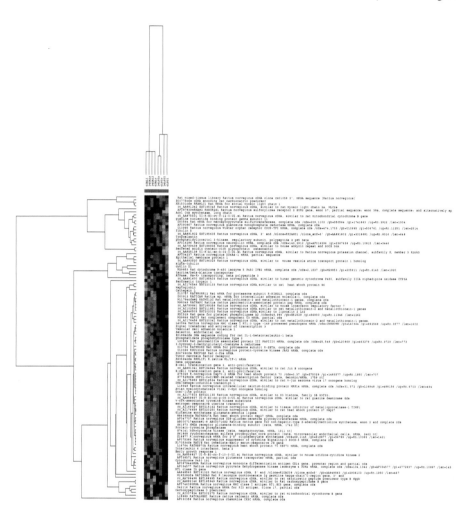

FIGURE 12.3 Cluster analysis of 103 genes and EST that produced treatment effect in cardiac ventricles at 6 and 24 h after LPS administration. These genes are likely to show sustained changes during the initial hours of an inflammatory response and they might represent the stereotypic transcriptional response pattern of the heart to inflammatory stress. *Source*: Reproduced with permission from M. L. Wong, F. O'Kirwan, N. Khan, J. Hannestad, K. H. Wu, D. Elashoff, et al. "Identification, Characterization, and Gene Expression Profiling of Endotoxin-Induced Myocarditis." *Proc Natl Acad Sci U S A* 100 (2003): 14241–46, copyright by The National Academy of Sciences of the United States of America.

unable to regenerate efficiently (28). Our findings strongly endorse the hypothesis that excessive production of immunomediators contributes to exacerbation of pathophysiological processes, as we described that large transcription inductions characterize EIM. Our data support that the following local processes are pathophysiologically relevant in EIM: increased leukocyte chemotaxis (macrophage inflammatory protein-1 alpha, macrophage inflammatory-2, and GRO); enhanced tendency to thrombophilia (plasminogen activation inhibitor); protection from foreign antigen by

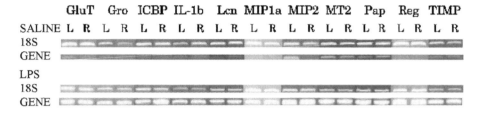

FIGURE 12.4   Semiquantitative RT-PCR of genes induced at 24 h after LPS. Our results confirm large fold-changes for 11 genes randomly selected out of genes and EST with fold-change 5 or higher at 24 h (ANOVA, $p < 0.05$). GluT, protein-glutamine γ-glutamyltransferase; Gro, growth-related oncogene; ICBP, intracellular calcium-binding protein (MRP14); IL-1b, interleukin-1 β; Lcn, lipocalin 2; MIP1a, macrophage inflammatory protein-1 α; MIP2, macrophage inflammatory protein-2 precursor; MT2, metallothionein-2; Pap, pancreatitis-associated protein precursor; Reg, regenerating gene; TIMP, tissue inhibitor of metalloproteinase-1. *Source*: Reproduced with permission from M. L. Wong, F. O'Kirwan, N. Khan, J. Hannestad, K. H. Wu, D. Elashoff, et al. "Identification, Characterization, and Gene Expression Profiling of Endotoxin-Induced Myocarditis." *Proc Natl Acad Sci U S A* 100 (2003): 14241–46, copyright by The National Academy of Sciences of the United States of America.

removal of antigen–antibody complexes (increased Fc gamma receptor); increased immunoactivation (MAK4 and DORA); increased cardiac remodeling and cellular turnover (sialoprotein, lysyl oxidase, p21 protein cip), as genes involved in those processes had robust gene expression induction (5 or higher fold changes) in the RV at 24 h after LPS (Fig. 12.5).

12.5   PATTERNS OF TRANSCRIPTIONAL RESPONSE IN HEART DURING INFLAMMATION

During SIRS, we found that a core of 102 genes (12), which represent one third of the response found at 6 h, maintained activation until 24 h (Fig. 12.5). It is likely that this pattern of activation characterizes a stereotypical response of the innate immune system in the heart during the early phase of an acute inflammatory response; thus, the identification of the critical elements of this standard response could be relevant in controlling the inflammatory process in this organ. As time elapses, different genes may be involved in the response. Therefore, analysis of gene expression patterns may help estimate not only the duration of such an acute inflammatory process, but also what changes would be required to manage the inflammatory response as different genes are involved in that response. Several cardiac pathophysiologic conditions that have been associated with pro-inflammatory cytokine expression could benefit from this research approach (29). Our microarray results are consistent with several reports of transcriptional changes relevant to the action of LPS or cytokines.

12.6   DISCUSSION

We identified endotoxin-induced myocarditis by trying to understand differential patterns of inducible NO synthase mRNA induction in cardiac ventricles during the

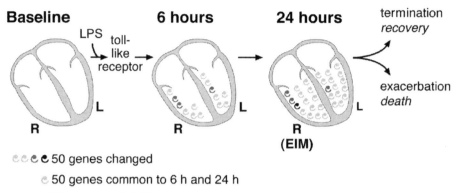

Baseline   **6 hours**   **24 hours**   termination
                                                              recovery

LPS   toll-
     like
     receptor

                                                              exacerbation
                                                              death

     L   L   L

R   R   R
                                              (EIM)

🌀🌀🌀🌀 50 genes changed

🌀 50 genes common to 6 h and 24 h

🌀 50 genes with fold change ≥2

🌀 50 genes with fold change ≥5

FIGURE 12.5 (See color insert following page 112.) Diagram summarizing EIM data. LPS acts through Toll-like receptor and initiates the heart's response to SIRS. Each spiral represents ≈50 genes or EST that are significantly changed in our microarray studies at $p \geq 0.01$. Green spirals denote the number of genes that are common to the 6- and 24-h transcriptional response; they characterize a possible stereotypic response of the innate immune system in the heart. Red and black spirals designate genes that have a fold-change of 2 or higher and 5 or higher, respectively. There is a small divergence in the response of the RV and LV at 6 h, which is intensified at 24 h. Note that the expansion of the transcriptional response in the heart at 24 h is more intense in the RV (it has more red and black spirals), the site for EIM. Hypothetically, the exacerbation of EIM would lead to death and the termination of the inflammatory process would result in recovery. R, right ventricle; L, left ventricle. Reproduced with permission from M. L. Wong, F. O'Kirwan, N. Khan, J. Hannestad, K. H. Wu, D. Elashoff, et al. "Identification, Characterization, and Gene Expression Profiling of Endotoxin-Induced Myocarditis." *Proc Natl Acad Sci U S A* 100 (2003): 14241–46, copyright by the National Academy of Sciences of the United States of America.

SIRS. EIM is characterized by the presence of myocarditis with focal areas of aseptic necrosis in the right ventricle 24 h after induction of systemic inflammation by LPS.

Our studies support that an expansion of the transcriptional response occurs in the heart from 6 to 24 h after SIRS induction. This inflammatory cascade has distinct characteristics in RV and LV, and it includes an intense inflammatory response in vivo at the site and time of EIM. Quantitative and qualitative expansions occurred in both ventricles but were more intense in the RV, which supports that excessive activation of the host inflammatory response occurred in EIM. After a single intraperitoneal (ip) injection, LPS that enter the portal vein can be eliminated completely by the liver; therefore, most LPS located in the systemic circulation enter the body through the lymphatic system. LPS plasma levels then peak at 6 h; those levels decrease to approximately 10 times lower at 24 h (30). ^{125}I-LPS administered intravenously (iv) is initially recovered predominantly in liver and blood (31). Several hours after that, migration of LPS-positive macrophages into the lungs starts and intensifies with time (32). Thus, the lungs serve as a later route of LPS excretion, remaining LPS-free for several hours. Therefore, the concentrations of LPS that reach either heart ventricle are likely to be comparable in our experiments, as the lung is not a site of rapid LPS uptake (33). Interestingly, there is a dose-dependent correlation between the uptake of LPS and tissue injury (30,32); therefore, disparity in ventricular uptake of LPS could account for modifications in the progression of the inflammatory response, and LPS internalization in cardiomyocytes promotes the activation of an endotoxin-dependent signal transduction (34). But anatomical, histological, physiological, or functional ventricular divergences could also contribute to transcriptional differences in ventricular response. Our results clearly illustrate the heterogeneity of myocardium tissues located in RV and LV.

The existence of myocardial cell injury in sepsis and SIRS is a logical assumption that derives from the finding of a high percentage of elevated levels of troponin T and I in intensive care unit patients with sepsis but without underlying coronary syndromes (35–37). Our finding supporting that cardiac tissue is susceptible to endotoxin-induced cytotoxicity validated the assumptions of those studies. Postmortem data in sepsis focused on the LV and confirm that there are no obvious signs of tissue injury in the LV in the absence of infectious endocarditis and acute myocardial infarction. But the pathology of RV tissue has not been explored, though significant right pump failure has been recognized in fatal outcomes. More pronounced RV dysfunction soon after the onset of septic shock, decreased RV performance due to decreased contractibility, and a marked increase in end-systolic volume without significant change in pulmonary artery pressure have been associated with nonsurvivors (13). RV dysfunction may also be more severe than LV dysfunction in early septic shock (24). Thus, it is plausible that RV myocardial cell injury exists during the early stages of SIRS but it has remained largely unrecognized. Our findings suggest that the development of stepwise myocardium-specific, multi-targeted strategies would prevent the amplification of the inflammatory reaction, and prioritized RV care early in septic shock would be defensible in treatment of sepsis and SIRS, particularly taking into consideration evidences of compelling and alarming unrecognized myocardial injury in sepsis (35).

Our data illustrate that the complexity and magnitude of transcriptional responses during the initial hours of SIRS induction may intensify in a cascade-like phenomenon in the heart ventricles and result in EIM, which has been characterized in the RV by a complex gene transcription pattern (Fig. 12.5). It is striking to contrast the extent of the amplification process we described with our current immunotherapy strategies, which target primarily one element of this cascade, and realize why strategies based on a single element of this cascade phenomenon would not be robust enough to considerably curb downstream events and impact disease outcome.

Exacerbated inflammatory response during SIRS and sepsis is accompanied by an insufficient host endogenous anti-inflammatory counter-response. Recently, clinical trials addressing this inadequate response have reduced the mortality in sepsis and septic shock (38,39). Low-dose glucocorticoids improve survival; though no single mediator-specific anti-inflammatory agent has been shown to improve survival significantly, they have a small beneficial effect on survival. Activated protein C, an anticoagulant that has substantial anti-inflammatory properties, has also demonstrated improved survival in clinical trials. The identification of EIM as a distinct pathophysiological entity underscores the complexity and heterogeneity of systemic inflammation and the importance of reassessing our treatment strategy in sepsis and SIRS.

REFERENCES

1. N. J. Rothwell and S. J. Hopkins. "Cytokines and the Nervous System II: Actions and Mechanisms of Action." *Trends Neurosci* 18 (1995): 130–36.
2. R. C. Bone. "Toward an Epidemiology and Natural History of SIRS (Systemic Inflammatory Response Syndrome)." *JAMA* 268 (1992): 345–55.
3. R. Dennhardt, H. J. Gramm, K. Meinhold, and K. Voigt. "Patterns of Endocrine Secretion during Sepsis." *Prog Clin Biol Res* 308 (1989): 751–56.
4. C. A. Dinarello and S. M. Wolff. "The Role of Interleukin-1 in Disease." *N Engl J Med* 328 (1993): 106–13.
5. M. L. Wong, P. B. Bongiorno, A. al-Shekhlee, A. Esposito, P. Khatri, and J. Licinio. "IL-1 Beta, IL-1 Receptor Type I and iNOS Gene Expression in Rat Brain Vasculature and Perivascular Areas." *Neuroreport* (1996): 2445–48.
6. M. L. Wong, V. Rettori, A. al-Shekhlee, P. B. Bongiorno, G. Canteros, S. M. McCann, et al. "Inducible Nitric Oxide Synthase Gene Expression in the Brain during Systemic Inflammation." *Nat Med* 2 (1996): 581–84.
7. M. L. Wong. "Cytokines in the Brain—Molecular Responses to Systemic Illness and Impact on Sickness Behaviour," in *Genes, Behaviour and Health*, eds. C. L. Bolis and J. Licinio J, 61–73 (Geneva: World Health Organization, 1999).
8. Control CfD. "Increase in National Hospital Discharge Survey Rates for Septicemia: United States, 1979–1987." *MMWR Morb Mortal Wkly Rep* 39 (1990): 31–34.
9. D. C. Angus, W. T. Linde-Zwirble, J. Lidicker, G. Clermont, J. Carcillo, and M. R. Pinsky. "Epidemiology of Severe Sepsis in the United States: Analysis of Incidence, Outcome, and Associated Costs of Care." *Crit Care Med* 29 (2001): 1303–10.
10. M. M. Parker, J. H. Shelhamer, S. L. Bacharach, M. V. Green, C. Natanson, T. M. Frederick, et al. "Profound but Reversible Myocardial Depression in Patients with Septic Shock." *Ann Intern Med* 100 (1984): 483–90.
11. A. G. Ellrodt, M. S. Riedinger, A. Kimchi, D. S. Berman, J. Maddahi, H. J. Swan, et al. "Left Ventricular Performance in Septic Shock: Reversible Segmental and Global Abnormalities." *Am Heart J* 110 (1985): 402–9.
12. M. L. Wong, F. O'Kirwan, N. Khan, J. Hannestad, K. H. Wu, D. Elashoff, et al. "Identification, Characterization, and Gene Expression Profiling of Endotoxin-Induced Myocarditis." *Proc Natl Acad Sci U S A* 100 (2003): 14241–46.
13. J. F. Dhainaut, M. F. Huyghebaert, J. F. Monsallier, G. Lefevre, J. Dall'Ava-Santucci, F. Brunet, et al. "Coronary Hemodynamics and Myocardial Metabolism of Lactate, Free Fatty Acids, Glucose, and Ketones in Patients with Septic Shock." *Circulation* 75 (1987): 533–41.
14. R. E. Cunnion, G. L. Schaer, M. M. Parker, C. Natanson, and J. E. Parrillo. "The Coronary Circulation in Human Septic Shock." *Circulation* 73 (1986): 637–44.
15. M. M. Parker, K. E. McCarthy, F. P. Ognibene, and J. E. Parrillo. "Right Ventricular Dysfunction and Dilatation, Similar to Left Ventricular Changes, Characterize the Cardiac Depression of Septic Shock in Humans." *Chest* 97 (1990): 126–31.
16. A. Kumar, C. Haery, and J. E. Parrillo. "Myocardial Dysfunction in Septic Shock." *Crit Care Clin* 16 (2000): 251–87.
17. R. M. Grocott-Mason and A. M. Shah. "Cardiac Dysfunction in Sepsis: New Theories and Clinical Implications." *Intensive Care Med* 24 (1998): 286–95.
18. D. R. Meldrum. "Tumor Necrosis Factor in the Heart." *Am J Physiol* 274 (1998): R577–95.

19. D. Ungureanu-Longrois, J. L. Balligand, I. Okada, W. W. Simmons, L. Kobzik, C. J. Lowenstein, et al. "Contractile Responsiveness of Ventricular Myocytes to Isoproterenol Is Regulated by Induction of Nitric Oxide Synthase Activity in Cardiac Microvascular Endothelial Cells in Heterotypic Primary Culture." *Circ Res* 77 (1995): 486–93.

20. S. M. McCann, J. Licinio, M. L. Wong, W. H. Yu, S. Karanth, and V. Rettorri. "The Nitric Oxide Hypothesis of Aging." *Exp Gerontol 33 (*1998): 813–26.

21. L. D. Buttery, T. J. Evans, D. R. Springall, A. Carpenter, J. Cohen, and J. M. Polak. "Immunochemical Localization of Inducible Nitric Oxide Synthase in Endotoxin-Treated Rats." *Lab Invest* 71 (1994): 755–64.

22. H. Luss, S. C. Watkins, P. D. Freeswick, A. K. Imro, A. K. Nussler, T. R. Billiar, et al. "Characterization of Inducible Nitric Oxide Synthase Expression in Endotoxemic Rat Cardiac Myocytes In Vivo and Following Cytokine Exposure In Vitro." *J Mol Cell Cardiol* 27 (1995): 2015–29.

23. K. Y. Ahn, M. G. Mohaupt, K. M. Madsen, and B. C. Kone. "In Situ Hybridization Localization of mRNA Encoding Inducible Nitric Oxide Synthase in Rat Kidney." *Am J Physiol* 267 (1994): F748–57.

24. S. F. Liu, P. J. Barnes, and T. W. Evans. "Time Course and Cellular Localization of Lipopolysaccharide-Induced Inducible Nitric Oxide Synthase Messenger RNA Expression in the Rat In Vivo." *Crit Care Med* 25 (1997): 512–18.

25. M. L. Wong, P. B. Bongiorno, V. Rettori, S. M. McCann, and J. Licini. "Interleukin (IL) 1beta, IL-1 Receptor Antagonist, IL-10, and IL-13 Gene Expression in the Central Nervous System and Anterior Pituitary During Systemic Inflammation: Pathophysiological Implications." *Proc Natl Acad Sci U S A* 94 (1997): 227–32.

26. M. R. Saban, H. Hellmich, N. B. Nguyen, J. Winston, T. G. Hammond, and R. Saban. "Time Course of LPS-Induced Gene Expression in a Mouse Model of Genitourinary Inflammation." *Physiol Genomics* 5 (2001): 147–60.

27. P. Knuefermann, S. Nemoto, A. Misra, N. Nozaki, G. Defreitas, S. M. Goyert, et al. "CD14-Deficient Mice Are Protected against Lipopolysaccharide-Induced Cardiac Inflammation and Left Ventricular Dysfunction." *Circulation* 106 (2002): 2608–15.

28. H. A. Rockman, W. J. Koch, and R. J. Lefkowitz. "Seven-Transmembrane-Spanning Receptors and Heart Function." *Nature* 415 (2002): 206–12.

29. D. L. Mann. "Stress-Activated Cytokines and the Heart: From Adaptation to Maladaptation." *Annu Rev Physiol* 65 (2003): 81–101.

30. M. Yasui, A. Nakao, T. Yuuki, A. Harada, T. Nonami, and H. Takagi. "Immunohistochemical Detection of Endotoxin in Endotoxemic Rats." *Hepatogastroenterology* 42 (1995): 683–90.

31. A. E. Warner, M. M. DeCamp Jr., R. M. Molina, and J. D. Brain. "Pulmonary Removal of Circulating Endotoxin Results in Acute Lung Injury in Sheep." *Lab Invest* 59 (1988): 219–30.

32. M. Freudenberg and C. Galanos. "Metabolic Fate of Endotoxin in Rat." *Adv Exp Med Biol* 256 (1990): 499–509.

33. M. A. Freudenberg, N. Freudenberg, and C. Galanos. "Time Course of Cellular Distribution of Endotoxin in Liver, Lungs and Kidneys of Rats." *Br J Exp Pathol* 63 (1982): 56–65.

34. D. B. Cowan, S. Noria, C. Stamm, L. M. Garcia, D. N. Poutias, P. J. del Nido, et al. "Lipopolysaccharide Internalization Activates Endotoxin-Dependent Signal Transduction in Cardiomyocytes." *Circ Res* 88 (2001): 491–98.

35. T. M. Guest, A. V. Ramanathan, P. G. Tuteur, K. B. Schechtman, J. H. Ladenson, and A. S. Jaffe. "Myocardial Injury in Critically Ill Patients. A Frequently Unrecognized Complication." *JAMA* 273 (1995): 1945–49.

36. A. Turne, M. Tsamitros, and R. Bellomo. "Myocardial Cell Injury in Septic Shock." *Crit Care Med* 27 (1999): 1775–80.

37. Y. Thiru, N. Pathan, S. Bignall, P. Habibi, and M. Levin. "A Myocardial Cytotoxic Process Is Involved in the Cardiac Dysfunction of Meningococcal Septic Shock." *Crit Care Med* 28 (2000): 2979–83.
38. J. Cohen. "The Immunopathogenesis of Sepsis." *Nature* 420 (2002): 885–91.
39. K. J. Deans, M. Haley, C. Natanson, P. Q. Eichacker, and P. C. Minneci. "Novel Therapies for Sepsis: A Review." *J Trauma* 58 (2005): 867–74.

13 Discovery of DOCK180 Superfamily of Exchange Factors

Jean-François Côté and Kristiina Vuori

CONTENTS

13.1 INTRODUCTION

Mammalian DOCK180 protein and its orthologues Myoblast City (MBC) and ced-5 in *Drosophila* and *Caenorhabditis elegans*, respectively, have been known for some time to function as critical regulators of the small GTPase Rac during several fundamentally important biological processes, such as cell motility and phagocytosis. The mechanism by which DOCK180 and its orthologues regulate Rac has remained elusive until recently. Our studies to uncover this fundamental mechanism produced several unexpected novel discoveries. Among them is the identification of a novel protein domain termed DHR-2 within the DOCK180 protein that interacts with and activates GTPases. Additionally, we identified several novel homologues of DOCK180 that possess this domain, and found that many of them directly bind to and exchange GDP for GTP both in vitro and in vivo on Rho GTPases. Thus, our studies uncovered a novel signaling module, and also resulted in a discovery of an evolutionarily conserved DOCK180-related superfamily of exchange factors, opening new opportunities in biological research. Our studies, as well as those of numerous colleagues that led to these findings are discussed in this chapter.

13.2 DISCOVERY OF DOCK180 PROTEIN

Over the last few decades, it has become clear that modular protein–protein interactions provide the underlying framework through which intracellular signal transduction takes place. At the heart of this cellular organization are adaptor proteins, which serve to bring two or more proteins together by utilizing one or more different protein binding domains, while having no enzymatic properties themselves (1). In the 1980s, Bruce Mayer and other members of the laboratory of Hidesaburo Hanafusa were instrumental in isolating and characterizing the first such protein, named Crk, which was identified as an oncogenic product of the avian sarcoma virus CT10 (2). The subsequent cDNA cloning of the cellular form of Crk from several eukaryotic origins revealed the first example of a protein that was almost completely composed of Src homology 2 (SH2) and SH3 domains (Fig. 13.1) (for a review, see 3,4). A proposal by Tony Pawson that such non-catalytic domains function to regulate cellular signaling by selective modular protein interactions fundamentally changed our view on intracellular signaling transduction (5). In today's post-genomic era, several additional SH2/SH3-containing and other classes of adaptor proteins have been identified and studied in great detail (for references, see 1).

To get further insight in the cellular functions of Crk, proteins that specifically interact with the SH2 and SH3 domains of Crk were subsequently sought after in the 1990s. Thus, several tyrosine-phosphorylated proteins that could specifically interact with the SH2 domain of Crk were identified and characterized. The two major Crk SH2 binding proteins, paxillin and p130Cas (6,7) are involved in the transmission of signals originating from focal adhesions, and this suggests a role

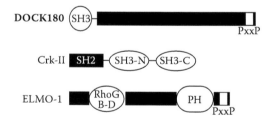

FIGURE 13.1   Signaling domains found in the DOCK180, ELMO, and CrkII proteins. At the onset of its discovery, DOCK180 was proposed to have an N-terminal SH3 domain and C-terminal proline-rich motifs. For up-to-date information on the predicted domain structure of the DOCK180 superfamily of proteins, see Figure 13.4. CrkII is composed of an N-terminal SH2 domain followed by two SH3 domains. The SH3-N of CrkII was used in the screen that led to the identification of DOCK180. ELMO-1 contains a PH domain of unknown specificity and a proline-rich motif involved in binding to the SH3 domain of DOCK180. A motif termed RhoG binding domain (RhoG-BD) involved in mediating the binding of ELMO-1 to GTP-loaded RhoG was recently reported (34).

for Crk downstream of the integrin family of cell adhesion receptors (for references, see 8–10). At the same time, Crk was found to interact with numerous proteins containing proline-rich sequence motifs via its SH3 domain. One of the two major proteins bound to the SH3 domain of Crk is C3G, which is a guanine nucleotide exchange factor for the Rap1 small GTPase, a member of the Ras superfamily (11–13). Importantly, gene knockout studies have demonstrated that C3G-dependent activation of Rap1 is required for adhesion and spreading of embryonic fibroblasts and for the early embryogenesis of mice, further supporting a role for the Crk pathway in adhesion-dependent signaling events (14). The other major Crk-SH3-binding protein, named DOCK180, was identified by Michiyuki Matsuda and coworkers in an expression library screening using the SH3 domain as a probe (15). Apart from the Crk-binding region in the C-terminus, and an SH3 domain in its N-terminus, the remaining sequence portion of this large 180-kDa protein, which is ubiquitously expressed in various tissues, did not demonstrate any homology to known functional domains (Fig. 13.1). At the onset, therefore, the function of DOCK180 remained obscure.

13.3 DOCK180: LINKING Rac GTPase TO MAMMALIAN CELLS

With the premise that a major function of Crk is to recruit its SH3 domain interacting partners to the membrane via its SH2 domain, Hasegawa et al. generated an expression vector coding for a membrane-targeted form of DOCK180 by the addition of a farnesylation signal to its C-terminus. In contrast to the wild-type DOCK180 protein, the farnesylated DOCK180 localized mainly to the plasma membrane and promoted significant morphological changes in mouse fibroblasts, resulting in a flat phenotype. Although not mechanistically understood, this was the first observation that DOCK180 could activate a signaling cascade capable of influencing cell morphology (15). Further support came from studies in which overexpression of DOCK180 together with Crk and p130Cas induced a dramatic increase in local membrane spreading reminiscent of a leading edge in polarized cells. Strikingly, coexpression of the three proteins also led to a massive localization of each component to focal adhesions. These observations suggested that this multi-protein complex transmits signals from focal adhesions upon engagement of integrins with the extracellular matrix (16).

That DOCK180 induces changes in cell morphology suggested a likely role for this protein upstream of a Rho GTPase, as members of the Rho family are known to be main regulators of cell shape and morphology via their effects on actin cytoskeleton (see section 13.6; for a review, see 17). Several groups tested this working hypothesis. Matsuda and coworkers were the first ones to notice the striking morphological similarity between fibroblasts expressing either farnesylated DOCK180 or a constitutively active form of the small GTPase Rac1, a member of the Rho family. Accordingly, they subsequently found that overexpression of DOCK180 leads to the GTP loading and activation of Rac1. Perhaps the most important experiment linking DOCK180 to Rac1 was the demonstration of a direct protein–protein interaction.

This interaction, notably, took place only when Rac1 was found in a nucleotide-free condition, suggesting that the interaction could affect the GTP-loading status of Rac1 (18). Additional independent experiments tightened the DOCK180/Rac connection. First, we were able to demonstrate that the p130Cas/Crk/DOCK180 protein complex was responsible for the integrin-induced activation of the JNK kinase cascade via Rac (19). Furthermore, Richard Klemke and colleagues reported that cell migration induced by p130Cas and Crk could be enhanced by coexpression of DOCK180 and that this phenomenon was also dependent on Rac1 (20). Additionally, mice lacking DOCK2, which is a DOCK180 homologue exclusively expressed in hematopoietic cells (21), were found to be deficient in lymphocyte migration and Rac activation in response to chemokines (22). In summary, several independent observations in mammalian cells and mammalian model systems pointed to DOCK180 as an activator of the Rac GTPase.

13.4 UNCOVERING BIOLOGICAL FUNCTIONS OF DOCK180: LESSONS FROM WORMS AND FLIES

Studies in model organisms such as *Drosophila* and *C. elegans* in the 1990s and early 2000s have been instrumental in uncovering the biological functions of DOCK180, and in supporting and guiding the cell biological studies in mammalian cells described above. Myoblast City (MBC) was first identified in a genetic screen in *Drosophila* that aimed at identifying novel genes involved in the regulation of the fusion of myoblasts during embryonic development (23). cDNA sequencing of MBC immediately revealed it to be the orthologue of human DOCK180. Subsequently, myoblast fusion and other defects found in MBC mutant flies were observed in the *Drosophila* Rac (dRac) mutant flies (24). A strong genetic connection between MBC and dRac was later established by Jeffrey Settleman's group, who demonstrated by using rough eye screens that MBC acts upstream of dRac in a signaling cascade (25).

Some time earlier, pioneering work by Bob Horvitz's laboratory aimed at the discovery of novel genes involved in the clearance of apoptotic cell corpses in the worm. These screens resulted in the identification of six genes, cell death abnormal (ced)-1, ced-2, ced-5, ced-6, ced-7, and ced-10, as components of the cellular machinery involved in phagocytosis of cells undergoing programmed cell death. Experiments involving crossing of the mutant animals revealed two different genetic pathways for these genes: ced-1, ced-6, and ced-7 constitute one pathway, while the second pathway involves ced-2, ced-5, and ced-10 (26). Recently, Hengartner and colleagues found that these two pathways converge at ced-10 to mediate corpse removal, suggesting that they do not work completely independently (27). At present, all the ced genes have been cloned and characterized to various extents. Notably, the cloning of ced-5 revealed it to be the worm orthologue of the mammalian DOCK180 protein (28). Similarly, the molecular cloning of ced-2 and ced-10 revealed homology to the human Crk and Rac1 proteins, respectively. Additional studies demonstrated that the ced-2/ced-5/ced-10 pathway also plays a crucial role in controlling the migration of the gonad distal tip cells during development (29). Both phagocytosis and cell migration are driven by membrane extension and cytoskeletal reorganization, and the studies in *C. elegans* therefore clearly demonstrated a fundamental role for ced-5 (DOCK180), as well as ced-2 (Crk) and ced-10 (Rac), in these processes. Thus, genetic data in both *Drosophila* and *C. elegans* suggested that DOCK180 and Rac reside in the same signaling pathway, DOCK180 being an upstream regulator of Rac.

13.5   ELMO: ADDITIONAL COMPONENT OF EVOLUTIONARILY CONSERVED Crk/DOCK180/Rac PATHWAY

More recently, *C. elegans* studies added an additional player, and complexity, to the signaling pathways involving the Crk, DOCK180, and Rac molecules. Thus, Chung et al. identified a novel molecule termed ced-12 that is involved in the removal of both programmed cell death and necrotic cell death corpses, as well as in the migration of the gonad distal tip cells in *C. elegans* (30). One obvious observation from this work was that the ced-12 mutants had a very similar phenotype to the one observed in the ced-2, ced-5, and ced-10 mutants. Three independent groups simultaneously reported the subsequent molecular cloning of ced-12, and its three human homologues named ELMO1 through 3 in 2001 (31–33). This novel protein family is evolutionarily conserved and is characterized by a unique N-terminal region containing a RhoG-binding module (34), a C-terminal PH-like domain, and a proline-rich region (Fig. 13.1). As anticipated, ced-12 is genetically linked to the ced-2/ced-5/ced-10 pathway and is potentially acting upstream of ced-10 (Rac) in that cascade. Work in mammalian cells suggested that ELMO/ced-12 was incapable of stimulating the GTP-loading of Rac/ced-10, however. Thus, while it was clear that the Crk/DOCK180/ELMO pathway leads to regulation of Rac, the big question was *how*, mechanistically, is that pathway capable of activating Rac? A brief overview of Rho GTPases is provided below as an introduction to the studies that uncovered this mechanism.

13.6   Rho GTPases AND THEIR REGULATORS

The Rho members are a superfamily of proteins with homology to Ras that display GTP-binding capacity in addition to a GTP hydrolase activity. These proteins have been implicated in an abundance of cellular events such as actin reorganization, gene expression control, and cell cycle regulation. Rac1, Cdc42, and RhoA have been intensively studied in the last decade and found to regulate the formation of various actin-based structures, including lamellipodia, filopodia, and stress fibers, respectively (17). When bound to GTP, these proteins are in an active conformation that enables coupling to downstream effectors. However, when bound to GDP, Rho GTPases are in an inactive conformation. In a cellular context, most of the Rho proteins are found in a GDP-loaded state. The control of the activation status of the Rho proteins is under the influence of two antagonistic families of proteins: the GTPase-activating proteins (GAPs) and the guanine exchange factors (GEFs). The GAPs interact with the active, GTP-loaded Rho proteins and stimulate their intrinsic GTPase activity, leading to the catalytic generation of inactive and GDP-loaded proteins. The catalytic activity of these negative regulators resides in the GAP domain,

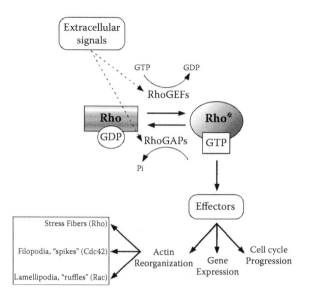

FIGURE 13.2   The Rho GTPase cycle. Rho GTPases cycle between an inactive (GDP bound) and an active (GTP bound) state. The guanine exchange factors (GEFs) promote activation while the GTPase activating proteins (GAPs) downregulate Rho GTPases. GTP-bound Rho proteins couple to their specific effectors, which mediate, among other events, the reorganization of the actin cytoskeleton.

typically around 140 amino acids in length, that is found in species ranging from yeast to mammals. On the other hand, the Rho GEFs promote the exchange of the GDP for GTP, thereby acting as activators of the small G proteins. The catalytic domain of GEFs is typically defined as a tandem Dbl homology-PH domain module. Several independent biochemical and structural studies suggest that GEFs activate Rho proteins by a two-step mechanism. First, the DH-PH domains of GEFs directly interact with Rho proteins with a concomitant release of the GDP. This transient GEF/nucleotide-free Rho GTPase complex becomes dissociated when GTP, which is in excess over GDP in cells, then incorporated in the Rho GTPase (Fig. 13.2). The GTP-loaded GTPase is then capable of activating its specific downstream target and exerting its cellular effects (for a review, see 35).

13.7   Rᴀᴄ ACTIVATION BY Cʀᴋ/ELMO/DOCK180 COMPLEX

Since all of the known GEFs for Rho GTPases contain the DH-PH module, it was widely hypothesized in the literature that DOCK180, or one of its interacting partners, was involved in recruiting a DH-PH–containing protein that would then carry out the enzymatic GTP-loading of Rac. As an alternative, we reasoned that if the powerful genetic screens in *C. elegans* and *Drosophila* systems had so far failed to identify a DH-PH GEF in the DOCK180 pathway, an intriguing possibility is that DOCK180 directly activates the Rac GTPase by catalyzing its GTP-loading via a novel mechanism. As it turns out, by testing this hypothesis, we found more than what we were looking for. In the next sections, a discussion of the combined biochemical and bioinformatics approaches that we used to identify the novel GEF domain in DOCK180 and to discover an evolutionarily conserved superfamily of DOCK180-related proteins is presented. For a detailed description, please see Cote and Vuori (36), from which the presentation is adapted.

13.8 IDENTIFICATION OF DOCK180 SUPERFAMILY OF PROTEINS

In order to get insights into the potential additional functional protein domains within DOCK180, we decided to take advantage of the ongoing genome sequencing projects, and carried out standard BLAST searches (http://www.ncbi.nlm.nih.gov/blast) using the human DOCK180 protein sequence or fragments thereof as a query. Intriguingly, database searches indicated that DOCK180 belongs to a protein superfamily, which we subsequently subgrouped based on a phylogenetic analysis (Table 13.1 and Fig. 13.3). A novel nomenclature was generated for the human gene products for presentation purposes. The eleven human DOCK180 homologues identified, named DOCK1 (=DOCK180) through DOCK11, were classified into four subfamilies that in turn were denoted DOCK-A, -B, -C, and -D. With the exception of DOCK180 and DOCK2, the various mammalian homologues were poorly characterized at the time of this discovery (see section 13.10 for an update on DOCK180 family members). In *Drosophila*, four homologues of DOCK180, one member in each subfamily, were identified. Phylogenetic analysis suggested that MBC is a member of the DOCK-A subfamily, supporting a previous report in which it was proposed that MBC is the DOCK180 orthologue in *Drosophila* (25). In *C. elegans*, three DOCK180 homologues were identified; phylogenetic analysis supported the notion that ced-5 is likely to be the orthologue of DOCK180 in *C. elegans*. Three novel homologues of DOCK180 were found in *Dictyostelium discoideum*, and the more primitive organisms *Arabidopsis thaliana* and *Saccharomyces cerevisiae* each appeared to have one DOCK180 homologue, SPIKE1 and YLR422W, respectively. Interestingly, SPIKE1 had just been identified as a gene involved in cytoskeletal organization in plant cells (37), while YLR422W remains an uncharacterized open reading frame in yeast.

As shown in Figure 13.4, several potential signaling and protein–protein interaction domains were identified in the DOCK180 superfamily of proteins, as predicted by the PFAM, SMART, and SCANSITE programs. Thus, the DOCK-A family members are characterized by the presence of an N-terminal SH3 domain, which is the predicted binding site for the ELMO family of proteins (31). Similar to DOCK180, DOCK5 (but not DOCK2) and also DOCK-B proteins contain potential binding sites for the Crk-SH3 domain in the C-terminus. As noted above, DOCK180 is known to interact with Crk and also with another adapter protein, NCKβ (38), via these sites. An N-terminal PH domain was detected in the members of the DOCK-D subfamily. PH domains are known to bind to differentially phosphorylated phosphoinositides, and they can also participate in protein–protein interactions (39). The functional significance of the PH domains within the DOCK-D proteins is currently unknown.

Two regions of high sequence homology that are conserved throughout the DOCK180 superfamily were identified by pairwise alignment, and these regions

TABLE 13.1
Classification of DOCK180 Superfamily

	Homo sapiens	D. melanogaster	C. elegans	D. discoideum	A. thaliana	S. cerevisiae
DOCK-A	**DOCK180/DOCK1** (#BAA09454) (Chr 10) **DOCK2** (KIAA0209, #BAA13200) (Chr 5) **DOCK5** (#NP_079216) (Chr 8)	**Myoblast City/ MBC** (#NP_477144)	**CED-5** (#AAC38973)	**DocA** (#AAB70856)	**SPIKE1** (#AAL74193)	**YLR422W** (#NP_013526)
DOCK-B	**DOCK3/MOCA** (KIAA0299, #BAA20759) (Chr 3) **DOCK4** (KIAA0716, #AAB83942) (Chr 7)	**CG11754** (#AAF56823)				

DOCK-C	**DOCK6** (KIAA1395, #XP_071996) (Chr 19) **DOCK7** (KIAA1771, #BAB21862) (Chr 1) **DOCK8** (#CAC22148) (Chr 9)	**CG11376** (#NP_608489)	**F46H5.4** (#NP_509219)	**Unnamed** (#AAL92252)	**SPIKE1** (#AAL74193)	**YLR422W** (#NP_013526)
DOCK-D	**DOCK9/zizimin1** (KIAA1058, #BAA83010) (Chr 13) **DOCK10** (KIAA0694, #XP_051970) (Chr 2) **DOCK11** (#XP_060056) (Chr X)	**CG6630** (#AAF52524)	**F22G12.5** (#CAB02974)	**Unnamed** (#AAM08471)		

Note: DOCK180-related proteins were identified by BLAST searches using the human DOCK180 protein sequence as a query. Classification was based on a phylogenetic analysis and BLAST identity scores. NCBI protein accession numbers are provided for all of the sequences, and chromosomal locations are indicated for the human family members.

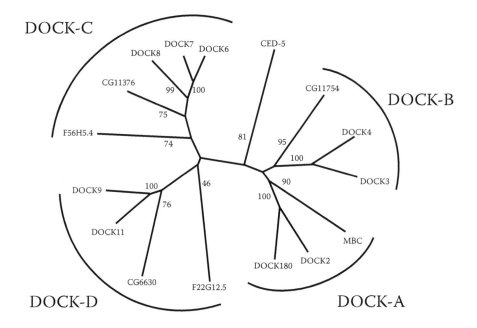

FIGURE 13.3 Phylogenetic tree of the DOCK180-related proteins. Amino acid sequences that cover the region of the DHR-2 domains of the indicated human, *Drosophila,* and *C. elegans* family members were used to generate ClustalW alignments. The tree was derived by neighbor-joining analysis applied to pairwise sequence distances calculated with the PHYLIP package using the Kimura two-parameter method to generate unrooted trees. The final output was generated in TREEVIEW and the number at each node represents the percentage of bootstrap replicates (out of 100). By utilizing in part this tree as well as BLAST search identity scores, we defined four subfamilies of DOCK180-related proteins, as indicated (DOCK-A, -B, -C, and -D).

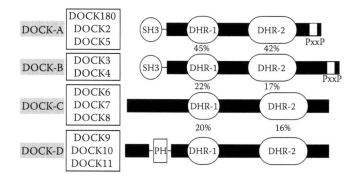

FIGURE 13.4 Structures of representative members of each subfamily of DOCK180-related proteins. Percentage identity between the DHR-1 and DHR-2 domains among the various subfamilies is indicated, in comparison to the domains in DOCK180. Most of the DOCK-A and -B family members are characterized by an N-terminal SH3 domain and a C-terminal proline-rich stretch. In addition to the DHR-1 and DHR-2 domains, the DOCK-D members harbor a PH domain of unknown function.

were named DHR-1 (*DOCK Homology Region*-1) and DHR-2 (Fig. 13.4). PFAM analysis suggested that the DHR-1 domain of DOCK180 is likely to be a C2 domain (E value: 0.47), and the DHR-1 domains of DOCK180 and DOCK2 were most similar to the C2 domain of a PI 3-kinase family member of the yeast strain *Candida albicans* (40). The C2 is a versatile signaling domain that interacts with lipids in a Ca^{2+}-dependent or -independent manner (41). While the existence of this domain would need to be experimentally verified, it could play a role in, e.g., localizing DOCK180 to the plasma membrane. In support of this hypothesis, we recently found that the DHR-1 domain of DOCK180 plays a role in targeting the DOCK180–ELMO–CrkII complex to the membrane by specifically recognizing the major lipid product of PI 3-kinases, $PtdIns(3,4,5)P_3$ (42).

No obvious domains were identified in any of the DHR-2 regions upon primary sequence analysis. Interestingly, however, threading analysis by 3D-PSSM (http://www.sbg.bio.ic.ac.uk/~3dpssm) suggested for the presence of a DH domain within the DHR-2 domain of DOCK9 that folds similar to the DH domain of β-PIX (E-value: 0.635, 50% certainty) (see 36). A tandem DH-PH domain resembling the fold found in known GEFs for Rac and Cdc42, such as SOS, intersectin, and β-PIX, could be observed in DOCK180 DHR-2 domain by threading analysis, albeit not with a significant E-value. Very intriguingly, the DHR-2 domain of DOCK180 (aas 1111-1636) overlaps with a region in DOCK180 that Matsuda and coworkers found to be necessary for DOCK180-mediated induction of Rac signaling (aas 1472-1714) (43). While structural analysis would be needed to confirm the potential presence of a DH-PH-like GEF structure in the DHR-2 domain, these preliminary data prompted us to examine the potential role of the DHR-2 domain in DOCK180-mediated Rac activation.

13.9　IDENTIFICATION OF DHR-2 DOMAIN OF DOCK180 AS NOVEL GEF FOR Rac

As noted above, the specificity of GEFs toward Rho GTPases is in part determined by their ability to directly interact with these GTPases. Following binding, GEFs catalyze nucleotide exchange by destabilizing the strong interaction between the GTPase and GDP and stabilizing the nucleotide-free state. This higher affinity transition state intermediate is then dissociated by binding of GTP. Thus, GEFs can be distinguished from other GTPase-interacting proteins by their ability to bind to the nucleotide-free state of GTPase (44). Accordingly, we examined whether the DHR-2 domain of DOCK180 could interact with nucleotide-free forms of small GTPases of the Rho-family, including RhoA, Rac1, and Cdc42. To this end, a panel of GST fusion proteins of the DOCK180 DHR-2 domain was generated (Fig. 13.5A,B). The boundaries of the highly hypothetical DH and PH domains (with no significant E-value in 3D-PSSM) are denoted in Figure 13.5A. As shown in Figure 13.5B, the DHR-2 domain of DOCK180 readily interacted with nucleotide-free Rac1 but not with Cdc42 or RhoA. None of the smaller fusion proteins corresponding to predicted subdomains of DHR-2 were able to precipitate Rac1 or any of the other GTPases tested. This would not be an unprecedented situation in case a rudimentary tandem DH-PH domain was truly present within DHR-2; regions covering both

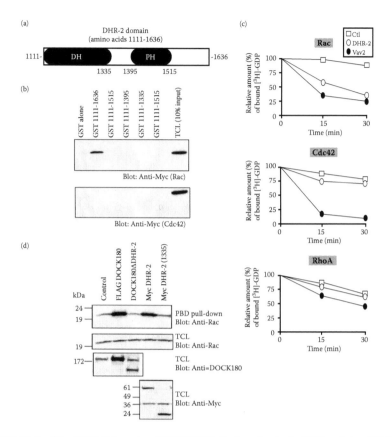

FIGURE 13.5 The DHR-2 domain of DOCK180 displays intrinsic GEF activity. (**A**) DHR-2 domain of DOCK180. Note that the presence of a DH-like and PH-like domain is hypothetical. The boundaries of the fusion proteins that were used in the GTPase binding assays are indicated. (**B**) The DHR-2 domain of DOCK180 interacts with nucleotide-free Rac1. Lysates of COS-1 cells expressing the indicated Rho GTPases were incubated with various GST fusion proteins derived from the DHR-2 domain of DOCK180 in nucleotide-free conditions (5 mM EDTA). The bound GTPases were detected by immunoblotting. (**C**) Bacterially produced and purified DHR-2 domain of DOCK180 promotes GTP-loading of Rac in vitro. GST alone (Ctl), GST DHR-2 of DOCK180, and the His-GEF domain of Vav2 were expressed in bacteria and purified. Bacterially produced GST-Rac1, GST-Cdc42, and GST-RhoA were loaded with ^{3}H-GDP and used as substrates to test for the presence of GEF activity in the DHR-2 domain of DOCK180. GST alone was used as a negative control and His-GEF domain of Vav2 was used as a positive control. All reactions were carried in a buffer containing cold GTP for 0, 15, and 30 minutes. Radiolabeled GDP bound to the GTPases was measured by using a filter-binding assay. (**D**) The DHR-2 domain of DOCK180 is both sufficient and necessary for promoting the GTP-loading of Rac in vivo. 293-T cells were transfected with an empty vector or with vectors coding for Flag-DOCK180 (wild-type), DOCK180ΔDHR-2 (deletion of amino acids 1111-1636), Myc-DHR-2 (consisting of amino acids 111-1636), or Myc-DHR-2(1335) (consisting of amino acids 1111–1335 of DOCK180). For each condition, Rac-GTP levels were measured by affinity precipitation using immobilized GST PAK PBD followed by immunoblotting. Expression levels of Rac and the various DOCK180 constructs were verified by immunoblotting of the total cell lysates (TCL).

the DH and PH domains and some of the adjacent sequences of Trio, Dbs, or Vav2 are required for these exchange factors to demonstrate maximal GTPase-binding or GEF activity (45–47).

We next investigated whether the DHR-2 domain of DOCK180 possessed GEF activity. Bacterially expressed and purified Rac1, Cdc42, and RhoA were loaded with [^{3}H]-GDP and the ability of the DHR-2 domain of DOCK180 to catalyze the exchange of the labeled GDP for cold GTP was examined. As a positive control, we used the GEF domain of Vav2 since it is known to act as a GEF for these three GTPases (48). As shown in Figure 13.5C, we found that the DHR-2 domain of DOCK180 indeed contains GEF activity for Rac1 but not Cdc42 or RhoA. As expected, the GEF domain of Vav2 readily exchanged the nucleotides for Rac1, Cdc42, and RhoA. No GEF activity was observed within the subdomains of DHR-2. Thus, the capability of the DHR-2 region to interact with nucleotide-free Rac1 correlated with its capability to catalyze nucleotide exchange for Rac1.

We next determined whether the DHR-2 domain of DOCK180 was sufficient and/or required for DOCK180-induced Rac activation in vivo. We therefore constructed a mutant form of DOCK180 that lacks the DHR-2 domain (DOCK180ΔDHR-2). In addition, we generated an expression vector for Myc-tagged DHR-2 domain alone and a Myc-tagged subfragment of the DHR-2 domain corresponding to the hypothetical DH domain. As shown in Figure 13.5D, overexpression of full-length DOCK180 resulted in a significant increase in the GTP loading of Rac, as measured by a PBD pull-down assay. DOCK180ΔDHR-2-construct in turn failed to activate Rac, suggesting that the DHR-2 domain of DOCK180 is absolutely required for DOCK180-mediated Rac activation. Exogenous expression of the isolated DHR-2 domain resulted in an increase in the GTP loading of Rac to about the same extent observed when full-length DOCK180 was expressed. Expression of the hypothetical DH domain within the DHR-2 domain alone failed to stimulate the GTP-loading of Rac.

Taken together, the findings above demonstrated that the DHR-2 domain of DOCK180 is both necessary as well as sufficient for DOCK180-mediated Rac activation, both in vitro and in vivo. As such, these studies provided a long sought-after mechanistic explanation as to how the DOCK180 signaling complex activates Rac. Additionally, we identified a novel protein domain (DHR-2) that interacts with and activates GTPases. We expect that atomic-level structural studies will yield important information on the catalytic mechanisms of this novel signaling domain. Finally, combination of bioinformatics and biochemical approaches led to the identification of an evolutionarily conserved DOCK180-related superfamily of exchange factors, opening broad vistas in biological research.

13.10   DOCK180-RELATED PROTEINS CONSTITUTE NOVEL SUPERFAMILY OF GEFS WITH DIVERSE BIOLOGICAL FUNCTIONS

Studies by us and others have continued to expand our knowledge of the DOCK180-related proteins, most if not all of which function as GEFs for Rho GTPases and contribute to a variety of biological processes. A summary of the current status in

TABLE 13.2
Activities of DOCK180-Related Proteins toward GTPases

DOCK-	Members	In Vitro Binding to Nucleotide-Free GTPases[a]	In Vitro GEF Activity on GTPases	In Vivo GEF Activity on GTPases
A	DOCK180	Rac1[b,c]	Rac1[b]	Rac1[b]
	DOCK2	Rac1[b,c]	Rac1[b]	Rac1[b]
	DOCK5	Rac2[d], Rac3[d]	?	?
B	DOCK3	Rac1[e,f,g]	?	Rac1[e,f,g]
	DOCK4	?	?	Rap1[h], Rac[i]
C	DOCK6	?	?	?
	DOCK7	?	?	?
	DOCK8	?[j]	?	?
D	DOCK9	Cdc42[b,k]	Cdc42[b]	Cdc42[k]
	DOCK10	Cdc42[l], TCL[l]	?	?
	DOCK11	Cdc42[l]	?	Cdc42[l]

[a] Denotes binding between the DHR-2 of the DOCK180-family member and the nucleotide-free GTPase, unless otherwise indicated.
[b] Côté and Vuori, *J. Cell Sci.* (115) 2002:p4901-4913.
[c] Brugnera et al., *Nat. Cell. Biol.* (4) 2002:p574-582.
[d] Our unpublished results.
[e] Binding and activation detected with full-length DOCK3, but not with the isolated DHR-2 domain.
[f] Grimsley et al., *J. Biol. Chem.* (279) 2004:p6087-6097.
[g] Namekata et al., *J. Biol. Chem.* (279) 2004:p14331-14337.
[h] Yajnik et al., *Cell* (112) 2003:p673-684.
[i] Lu et al., *Curr. Biol.* (15) 2005:p371-377.
[j] Interaction between FL DOCK8 and Rac1, Cdc42, TC10 and TCL has been detected in a yeast two-hybrid assay.
[k] Meller et al., *Nat. Cell. Biol.* (4) 2002:p639-647.
[l] Nishikimi et al., *FEBS Lett.* (579) 2005:p1039-1046.

the field can be found in Table 13.2. In our original paper (36), we demonstrated that members of the DOCK-A subfamily (at least DOCK180 and DOCK2) and DOCK-D subfamily (at least DOCK9) function as GEFs toward Rac and Cdc42, respectively. At the same time, Brugnera et al. similarly reported that a domain (which they termed DOCKER) within DOCK180 specifically recognizes nucleotide-free Rac and mediates GTP-loading of Rac in vitro (49). Importantly, this domain overlaps with the DHR-2 domain described above. Also in 2002, Meller et al. reported on purification and identification of zizimin1, which functions as a GEF for Cdc42 (50). By using biochemical approaches, these authors identified a minimal region within zizimin1 that interacts with Cdc42 and named this domain CZH2. Notably, zizimin1 is identical to DOCK9, and the CZH2 region corresponds to the DHR-2 domain.

   More recently, two groups demonstrated that the full-length DOCK3 protein (a member of the DOCK-B subfamily) binds to and activates Rac1 (51,52). DOCK3 was originally characterized as a presenilin-binding protein PBP, which is specifically expressed in neurons (53). Later on, it was renamed MOCA for "modifier of cell adhesion" due to its capability to modulate cell–substratum adhesion and amyloid-β secretion by an unknown mechanism in nerve cells (54). Cells expressing membrane-targeted, farnesylated DOCK3/MOCA show flattened morphology similar to those expressing an active mutant of Rac1. Also, DOCK3/MOCA is concentrated with actin on the growth cones in primary cultures of cortical neurons. These observations suggest that similar to DOCK180, DOCK3/MOCA may induce cytoskeletal reorganization and changes in cell morphology by regulating the activity of Rac1 (51). In addition, Ravichandran and colleagues have shown that DOCK3-mediated activation of Rac is likely to play a role in cell migration; this remains to be confirmed in the context of neuronal cells (52).

Some biomedically interesting information has emerged on the DOCK4 protein (member of the DOCK-B subfamily). Yajnik and co-workers used representational difference analysis (RDA) as a screening method for identifying homozygous deletions in genomes of various mouse tumor models. Interestingly, these authors found a deletion targeting DOCK4, and further found that DOCK4 specifically activates the Rap GTPase, enhancing the formation of adherens junctions in cells. DOCK4 mutations were also found to be present in a subset of human prostate and ovarian cancer cell lines, in which a recurrent missense mutant encodes a protein that is defective in Rap1 activation. Thus, DOCK4 encodes for a DOCK180-family member that regulates intercellular junctions and is disrupted during tumorigenesis (55). While the original data by Yajnik et al. suggested that DOCK4 exerts its effects through the activation of the Ras-related GTPase Rap1, recent data suggest that DOCK4 is a Rac activator (56). Thus, the biochemical reaction catalyzed by DOCK4 remains to be fully determined experimentally.

Less is known about the other DOCK180-related proteins. DOCK8 (DOCK-C subfamily member) is ubiquitously expressed, and is present at the cell edges in areas undergoing lamellipodia formation. In a yeast two-hybrid assay, full-length DOCK8 has affinity toward Rac1, Cdc42, TC10, and TCL (57). DOCK-D subfamily members DOCK9/zizimin1, DOCK10/zizimin3, and DOCK11/zizimin2 have been characterized to some extent. DOCK11/zizimin2 is expressed predominantly in lymphocytes, while DOCK9/zizimin1 is expressed in non-hematopoietic tissues and cells. Thus, this is reminiscent of the situation in the DOCK-A subfamily, where DOCK180 is expressed predominantly in adherent cells and DOCK2 in hematopoietic cells. DOCK-D family members demonstrate binding toward Cdc42 in vitro, and DOCK10/zizimin3 also to some extent toward TCL (50,58). The recent completion of the sequencing of the genome of *D. discoideum* revealed the existence of eight DOCK180-related genes in this organism (59). In our original report (36), we had identified three of these proteins, likely because of the poor genome sequence information available at that time. Among the eighteen Rho family members, Rac-like GTPases are abundant in *D. Discoideum*. As such, it will be interesting to determine the role of the DOCK180-related proteins in the activation of these GTPases and in the regulation of cell migration of this unicellular organism.

In sum, studies by us and others have recently led to the discovery of a family of evolutionarily conserved, DOCK180-related proteins that display GEF activity toward Rho GTPases. The DOCK180 family of proteins lack the canonical DH-PH module; instead, they rely on a novel domain, termed DHR-2, DOCKER, or CZH2, to exchange GDP for GTP on Rho targets. Multidisciplinary future studies are expected to bring forth the understanding of the detailed enzymatic mechanisms and the physiological functions of this novel superfamily of signaling molecules.

ACKNOWLEDGMENTS

The original work in the authors' laboratory has been supported by grants from NIH, Department of Defense Cancer Research Programs, California Breast Cancer Research Program, and California Tobacco-Related Disease Research Program (to K.V.), and by The Terry Fox Foundation/National Cancer Institute of Canada (to J-F.C.).

REFERENCES

1. T. Pawson. "Specificity in Signal Transduction: From Phosphotyrosine-SH2 Domain Interactions to Complex Cellular Systems." *Cell* 116 (2004): 191–203.
2. B. J. Mayer, M. Hamaguchi, and H. Hanafusa. "A Novel Viral Oncogene with Structural Similarity to Phospholipase C." *Nature* 332 (1988): 272–75.
3. M. Matsuda and T. Kurata. "Emerging Components of the Crk Oncogene Product: The First Identified Adaptor Protein." *Cell Signal* 8 (1996): 335–40.
4. S. M. Feller. "Crk Family Adaptors-Signalling Complex Formation and Biological Roles." *Oncogene* 20 (2001): 6348–71.
5. T. Pawson. "Non-catalytic Domains of Cytoplasmic Protein-Tyrosine Kinases: Regulatory Elements in Signal Transduction." *Oncogene* 3 (1988): 491–95.
6. R. B. Birge, J. E. Fajardo, C. Reichman, S. E. Shoelson, Z. Songyang, L. C. Cantley, and H. Hanafusa. "Identification and Characterization of a High-Affinity Interaction between v-Crk and Tyrosine-Phosphorylated Paxillin in CT10-Transformed Fibroblasts." *Mol Cell Biol* 13 (1993): 4648–56.
7. R. Sakai, A. Iwamatsu, N. Hirano, S. Ogawa, T. Tanaka, H. Mano, Y. Yazaki, and H. Hirai. "A Novel Signaling Molecule, p130, Forms Stable Complexes In Vivo with v-Crk and v-Src in a Tyrosine Phosphorylation-Dependent Manner." *EMBO J* 13 (1994): 3748–56.
8. M. D. Schaller. "Paxillin: A Focal Adhesion-Associated Adaptor Protein." *Oncogene* 20 (2001): 6459–72.
9. C. E. Turner. "Paxillin and Focal Adhesion Signalling." *Nat Cell Biol* 2 (2000): E231–36.
10. A. H. Bouton, R. B. Riggins, and P. J. Bruce-Staskal. "Functions of the Adapter Protein Cas: Signal Convergence and the Determination of Cellular Responses." *Oncogene* 20 (2001): 6448–58.
11. B. S. Knudsen, S. M. Feller, and H. Hanafusa. "Four Proline-Rich Sequences of the Guanine-Nucleotide Exchange Factor C3G Bind with Unique Specificity to the First Src Homology 3 Domain of Crk." *J Biol Chem* 269 (1994): 32781–87.
12. S. Tanaka, T. Morishita, Y. Hashimoto, S. Hattori, S. Nakamura, M. Shibuya, K. Matuoka, et al. "C3G, a Guanine Nucleotide-Releasing Protein Expressed Ubiquitously, Binds to the Src Homology 3 Domains of CRK and GRB2/ASH Proteins." *Proc Natl Acad Sci USA* 91 (1994): 3443–47.
13. T. Gotoh, S. Hattori, S. Nakamura, H. Kitayama, M. Noda, Y. Takai, K. Kaibuchi, et al. "Identification of Rap1 as a Target for the Crk SH3 Domain-Binding Guanine Nucleotide-Releasing Factor C3G." *Mol Cell Biol* 15 (1995): 6746–53.

14. Y. Ohba, K. Ikuta, A. Ogura, J. Matsuda, N. Mochizuki, K. Nagashima, K. Kurokawa, et al. "Requirement for C3G-Dependent Rap1 Activation for Cell Adhesion and Embryogenesis." *EMBO J* 20 (2001): 3333–41.
15. H. Hasegawa, E. Kiyokawa, S. Tanaka, K. Nagashima, N. Gotoh, M. Shibuya, T. Kurata, and M. Matsuda. "DOCK180, a Major CRK-Binding Protein, Alters Cell Morphology upon Translocation to the Cell Membrane." *Mol Cell Biol* 16 (1996): 1770–76.
16. E. Kiyokawa, Y. Hashimoto, T. Kurata, H. Sugimura, and M. Matsuda. "Evidence That DOCK180 Up-Regulates Signals from the CrkII-p130(Cas) Complex." *J Biol Chem* 273 (1998): 24479–84.
17. S. Etienne-Manneville and A. Hall. "Rho GTPases in Cell Biology." *Nature* 420 (2002): 629–35.
18. E. Kiyokawa, Y. Hashimoto, S. Kobayashi, H. Sugimura, T. Kurata, and M. Matsuda. "Activation of Rac1 by a Crk SH3-Binding Protein, DOCK180." *Genes Dev* 12 (1998): 3331–36.
19. F. Dolfi, M. Garcia-Guzman, M. Ojaniemi, H. Nakamura, M. Matsuda, and K. Vuori. "The Adaptor Protein Crk Connects Multiple Cellular Stimuli to the JNK Signaling Pathway." *Proc Natl Acad Sci USA* 95 (1998): 15394–99.
20. D. A. Cheresh, J. Leng, and R. L. Klemke. "Regulation of Cell Contraction and Membrane Ruffling by Distinct Signals in Migratory Cells." *J Cell Biol* 146 (1999): 1107–16.
21. H. Nishihara, S. Kobayashi, Y. Hashimoto, F. Ohba, N. Mochizuki, T. Kurata, K. Nagashima, and M. Matsuda. "Non-adherent Cell-Specific Expression of DOCK2, a Member of the Human CDM-Family Proteins." *Biochim Biophys Acta* 1452 (1999): 179–87.
22. Y. Fukui, O. Hashimoto, T. Sanui, T. Oono, H. Koga, M. Abe, A. Inayoshi, et al. "Haematopoietic Cell-Specific CDM Family Protein DOCK2 Is Essential for Lymphocyte Migration." *Nature* 412 (2001): 826–31.
23. M. R. Erickson, B. J. Galletta, and S. M. Abmayr. "*Drosophila* Myoblast City Encodes a Conserved Protein That Is Essential for Myoblast Fusion, Dorsal Closure, and Cytoskeletal Organization." *J Cell Biol* 138 (1997): 589–603.
24. S. Hakeda-Suzuki, J. Ng, J. Tzu, G. Dietzl, Y. Sun, M. Harms, T. Nardine, L. Luo, and B. J. Dickson. "Rac Function and Regulation during *Drosophila* Development." *Nature* 416 (2002): 438–42.
25. K. M. Nolan, K. Barrett, Y. Lu, K. Q. Hu, S. Vincent, and J. Settleman. "Myoblast City, the *Drosophila* Homolog of DOCK180/CED-5, Is Required in a Rac Signaling Pathway Utilized for Multiple Developmental Processes." *Genes Dev* 12 (1998): 3337–42.
26. R. E. Ellis, D. M. Jacobson, and H. R. Horvitz. "Genes Required for the Engulfment of Cell Corpses during Programmed Cell Death in *Caenorhabditis elegans.*" *Genetics* 129 (1991): 79–94.
27. J. M. Kinchen, J. Cabello, D. Klingele, K. Wong, R. Feichtinger, H. Schnabel, R. Schnabel, and M. O. Hengartner. "Two Pathways Converge at CED-10 to Mediate Actin Rearrangement and Corpse Removal in *C. elegans.*" *Nature* 434 (2005): 93–99.
28. Y. C. Wu and H. R. Horvitz. "*C. elegans* Phagocytosis and Cell-Migration Protein CED-5 Is Similar to Human DOCK180." *Nature* 392 (1998): 501–4.
29. P. W. Reddien and H. R. Horvitz. "CED-2/CrkII and CED-10/Rac Control Phagocytosis and Cell Migration in *Caenorhabditis elegans.*" *Nat Cell Biol* 2 (2000): 131–36.
30. S. Chung, T. L. Gumienny, M. O. Hengartner, and M. Driscoll. "A Common Set of Engulfment Genes Mediates Removal of Both Apoptotic and Necrotic Cell Corpses in *C. elegans. Nat Cell Biol* 2 (2000): 931–37.

31. T. L. Gumienny, E. Brugnera, A. C. Tosello-Trampont, J. M. Kinchen, L. B. Haney, K. Nishiwaki, S. F. Walk, et al. "CED-12/ELMO, a Novel Member of the CrkII/Dock180/Rac Pathway, Is Required for Phagocytosis and Cell Migration." *Cell* 107 (2001): 27–41.

32. Z. Zhou, E. Caron, E. Hartwieg, A. Hall, and H. R Horvitz. "The *C. elegans* PH Domain Protein CED-12 Regulates Cytoskeletal Reorganization via a Rho/Rac GTPase Signaling Pathway." *Dev Cell* 1 (2001): 477–89.

33. Y. C. Wu, M. C. Tsai, L. C. Cheng, C. J. Chou, and N. Y. Weng. "*C. elegans* CED-12 Acts in the Conserved crkII/DOCK180/Rac Pathway to Control Cell Migration and Cell Corpse Engulfment." *Dev Cell* 1 (2001): 491–502.

34. H. Katoh and M. Negishi. "RhoG Activates Rac1 by Direct Interaction with the Dock180-Binding Protein Elmo." *Nature* 424 (2003): 461–64.

35. A. Schmidt and A. Hall. "Guanine Nucleotide Exchange Factors for Rho GTPases: Turning on the Switch." *Genes Dev* 16 (2002): 1587–1609.

36. J. F. Cote and K. Vuori. "Identification of an Evolutionarily Conserved Superfamily of DOCK180-Related Proteins with Guanine Nucleotide Exchange Activity." *J Cell Sci* 115 (2002): 4901–13.

37. J. L. Qiu, R. Jilk, M. D. Marks, and D. B. Szymanski. "The Arabidopsis SPIKE1 Gene Is Required for Normal Cell Shape Control and Tissue Development." *Plant Cell* 14 (2002): 101–18.

38. Y. Tu, D. F. Kucik, and C. Wu. "Identification and Kinetic Analysis of the Interaction between Nck-2 and DOCK180." *FEBS Lett* 491 (2001): 193–99.

39. M. A. Lemmon, K. M. Ferguson, and C. S. Abrams. "Pleckstrin Homology Domains and the Cytoskeleton." *FEBS Lett* 513 (2002): 71–76.

40. E. A. Nalefski and J. J. Falke. "The C2 Domain Calcium-Binding Motif: Structural and Functional Diversity." *Protein Sci* 5 (1996): 2375–90.

41. E. Merithew and D. G. Lambright. "Calculating the Potential of C2 Domains for Membrane Binding." *Dev Cell* 2 (2002): 132–33.

42. Cote JF, Motoyama AB, Bush AB and Vuori K. A novel and evolutionarily conserved PtdIns(3,4,5)P3-binding domain is necessary for DOCK180 signaling. *Nat Cell Biol* 7 (2005): 797–807.

43. S. Kobayashi, T. Shirai, E. Kiyokawa, N. Mochizuki, M. Matsuda, and Y. Fukui. "Membrane Recruitment of DOCK180 by Binding to PtdIns(3,4,5)P3." *Biochem J* 354 (2001): 73–78.

44. J. T. Snyder, D. K. Worthylake, K. L. Rossman, L. Betts, W. M. Pruitt, D. P. Siderovski, C. J. Der, and J. Sondek. "Structural Basis for the Selective Activation of Rho GTPases by Dbl Exchange Factors." *Nat Struct Biol* 9 (2002): 468–75.

45. M. A. Booden, S. L. Campbell, and C. J. Der. "Critical but Distinct Roles for the Pleckstrin Homology and Cysteine-Rich Domains as Positive Modulators of Vav2 Signaling and Transformation." *Mol Cell Biol* 22 (2002): 2487–97.

46. S. Liu, H. Wang, M. Eberstadt, A. Schnuchel, E. T. Olejniczak, R. P. Meadows, J. M. Schkeryantz, et al. "NMR Structure and Mutagenesis of the N-Terminal Dbl Homology Domain of the Nucleotide Exchange Factor Trio." *Cell* 95 (1998): 269–77.

47. K. L. Rossman, D. K. Worthylake, J. T. Snyder, D. P. Siderovski, S. L. Campbell, and J. Sondek. "A Crystallographic View of Interactions between Dbs and Cdc42: PH Domain-Assisted Guanine Nucleotide Exchange." *EMBO J* 21 (2002): 1315–26.

48. K. Abe, K. L. Rossman, B. Liu, K. D. Ritola, D. Chiang, S. L. Campbell, K. Burridge, and C. J. Der. "Vav2 Is an Activator of Cdc42, Rac1, and RhoA." *J Biol Chem* 275 (2000): 10141–49.

49. E. Brugnera, L. Haney, C. Grimsley, M. Lu, S. F. Walk, A. C. Tosello-Trampont, I. G. Macara, H. Madhani, G. R. Fink, and K. S. Ravichandran. "Unconventional Rac-GEF Activity Is Mediated through the Dock180-ELMO Complex." *Nat Cell Biol* 4 (2002): 574–82.

50. N. Meller, M. Irani-Tehrani, W. B. Kiosses, M. A. Del Pozo, and M. A. Schwartz. "Zizimin1, a Novel Cdc42 Activator, Reveals a New GEF Domain for Rho Proteins." *Nat Cell Biol* 4 (2002): 639–47.

51. K. Namekata, Y. Enokido, K. Iwasawa, and H. Kimura. "MOCA Induces Membrane Spreading by Activating Rac1." *J Biol Chem* 279 (2004): 14331–37.

52. C. M. Grimsley, J. M. Kinchen, A. C. Tosello-Trampont, E. Brugnera, L. B. Haney, M. Lu, Q. Chen, D. Klingele, M. O. Hengartner, and K. S. Ravichandran. "Dock180 and ELMO1 Proteins Cooperate to Promote Evolutionarily Conserved Rac-Dependent Cell Migration." *J Biol Chem* 279 (2004): 6087–97.

53. A. Kashiwa, H. Yoshida, S. Lee, T. Paladino, Y. Liu, Q. Chen, R. Dargusch, D. Schubert, and H. Kimura. "Isolation and Characterization of Novel Presenilin Binding Protein." *J Neurochem* 75 (2000): 109–16.

54. Q. Chen, H. Kimura, and D. Schubert. "A Novel Mechanism for the Regulation of Amyloid Precursor Protein Metabolism." *J Cell Biol* 158 (2002): 79–89.

55. V. Yajnik, C. Paulding, R. Sordella, A. I. McClatchey, M. Saito, D. C. Wahrer, P. Reynolds, et al. "DOCK4, a GTPase Activator, Is Disrupted during Tumorigenesis." *Cell* 112 (2003): 673–84.

56. M. Lu, J. M. Kinchen, K. L. Rossman, C. Grimsley, M. Hall, J. Sondek, M. O. Hengartner, V. Yajnik, and K. S. Ravichandran. "A Steric-Inhibition Model for Regulation of Nucleotide Exchange via the Dock180 Family of GEFs." *Curr Biol* 15 (2005): 371–77.

57. A. Ruusala and P. Aspenstrom. "Isolation and Characterisation of DOCK8, a Member of the DOCK180-Related Regulators of Cell Morphology." *FEBS Lett* 572 (2004): 159–66.

58. A. Nishikimi, N. Meller, N. Uekawa, K. Isobe, M. A. Schwartz, and M. Maruyama. "Zizimin2: a Novel, DOCK180-Related Cdc42 Guanine Nucleotide Exchange Factor Expressed Predominantly in Lymphocytes." *FEBS Lett* 579 (2005): 1039–46.

59. L. Eichinger, J. A. Pachebat, G. Glockner, M. A. Rajandream, R. Sucgang, M. Berriman, J. Song, et al. "The Genome of the Social Amoeba *Dictyostelium discoideum*." *Nature* 435 (2005): 43–57.

14 Discovery of Extracellular Matrix Degradome as Novel Endogenous Regulators of Angiogenesis and Tumor Growth

Lingge Lu, Tanjore Harikrishna, and Raghu Kalluri

CONTENTS

14.1 INTRODUCTION

The extracellular matrix (ECM) is a complex structure composed of proteins, proteoglycans, and adhesive glycoproteins that provides structural and mechanical support to cells and tissues, as well as regulates a wide variety of cellular processes. The effects of ECM are mainly mediated by cell surface integrins (Giancotti and Ruoslahti 1999; Hynes 1992). At the structural level, there are two major forms of ECM: the interstitial matrix and the basement membrane (BM). While the interstitial matrix is a fibrillar network that allows cell movements through the tissues, the BM is a dense, sheet-like structure serving as a barrier to keep cells of various types separated from the surrounding stroma. The predominant components of the BM include type IV collagen, laminin, heparan-sulfate proteoglycans (HSPGs), and nidogen/entactin. Smaller amounts of type XV collagen, type XVIII collagen, fibulins, osteopontin/SPARC/BM-40, and agrin are also found in the BM. Via the self-assembly laminin and type IV collagen, BMs form super-molecular networks providing a scaffold with which nidogen, HSPG, and other BM proteins interact and adhere. The specific protein components of the BM vary from tissue to tissue, thus contributing to the tissue-specific morphology and function. The vascular basement membrane (VBM), sandwiched between blood vessel endothelial cells and mural cells (vascular smooth muscle cells and pericytes), is important for providing structural support and also regulation of angiogenesis—the formation of new blood capillaries from pre-existing vasculature (Kalluri 2003).

14.2   BASEMENT MEMBRANES AND ANGIOGENESIS

Angiogenesis plays a crucial role in physiological processes such as embryonic development, wound healing, and ovarian cycle, as well as pathological conditions including atherosclerosis, rheumatoid arthritis, diabetic retinopathy, tumor growth, and metastasis. The normal angiogenesis process, tightly regulated by both stimulators and inhibitors, involves vasodilation of existing vessels, degradation of surrounding matrix, endothelial cell proliferation and migration to form lumens, remodeling, and finally maturation by recruiting mural cells and matrix support (Carmeliet and Jain 2000; Folkman 1995; Folkman and D'Amore 1996; Folkman and Shing 1992; Hanahan and Folkman 1996; Risau 1997). It has been revealed that BM proteins such as laminin and type IV collagen modulate and orchestrate the process of angiogenesis (Form et al. 1986; Madri 1997; Madri and Pratt 1986). Endothelial cells that reside on the BM are normally quiescent, suggesting that the BM inhibits endothelial cell growth and migration. Up-regulated angiogenic stimuli via growth factors such as VEGF and FGF-2 induce degradation of BM proteins by matrix degrading enzymes, which liberates endothelial cells and facilitates their migration and proliferation. This action results in detached endothelial cells coming into direct contact with interstitial matrix components such as vitronectin, fibronectin, type I collagen, and thrombin. Pro-angiogenic growth factors also induce endothelial cells to produce matrix molecules (Eliceiri and Cheresh 2001). These observations suggest a positive role for the provisional matrix on endothelial cell proliferation. Therefore, the process of matrix remodeling provides either an inhibitory or a proliferative cue for endothelial cells. Unbalanced pro- and anti-angiogenesis signal could lead to pathologic angiogenesis.

It is now been well accepted that tumor growth and metastasis require establishment of their own blood supply—the process of tumor angiogenesis (Folkman 1971; Hanahan and Weinberg 2000). Tumor cells express various pro-angiogenic factors such as VEGF and FGF, which induce angiogenesis. However, tumor-associated blood vessels are different from the normal vasculature. They are often irregularly shaped, dilated, leaky, and hemorrhagic (Jain 1988; McDonald and Baluk 2002; McDonald and Foss 2000). Tumor blood vessels also exhibit reduced pericyte coverage (Benjamin et al. 1999; Morikawa et al., 2002) and possibly abnormal VBM. Studies suggest that the VBMs of tumor vessels have a loose association with endothelial cells and pericytes, as well as displaying broad extension from the vessel walls and multiple layers (Baluk et al. 2003).

14.3 TYPE IV COLLAGEN-DERIVED ENDOGENOUS INHIBITORS OF ANGIOGENESIS

Type IV collagen is one of the major components of VBM and is highly conserved among most organisms (Blumberg et al. 1987). Six α chains of type IV collagen, named $\alpha1-\alpha6$, have been identified, which share 50%–70% homology and form heterotrimers. The $\alpha1$ and $\alpha2$ chains are ubiquitously expressed, whereas the other four isoforms exhibit a restricted expression pattern. Each α chain is composed of an N-terminal cysteine-rich 7S domain, a central triple-helical domain, and a C-terminal non-collagenous (NC1) domain (Hudson et al. 1993). The NC1 domain is involved in the assembly of the type IV collagen network and regulation of cell behavior (Cameron et al. 1991; Chelberg et al. 1990; Furcht 1984; Herbst et al. 1988; Tsilibary et al. 1988). The functional importance of type IV collagen is further supported by identification of the genetic kidney diseases such as Alport syndrome (Barker et al. 1990; Hudson et al. 1992; Hudson et al. 1993; Kalluri et al. 1997), which is due to mutations in either $\alpha3$, $\alpha4$, or $\alpha5$ chains. Goodpasture syndrome is an autoimmune disease caused by auto-antibodies against a C-terminal region of the $\alpha3$ chain (Butkowski et al. 1987; Hudson et al. 1993; Kalluri et al. 1994).

It has been shown that blockage of BM collagen synthesis by inhibitors such as GPA 1734, D609, cis-hydroxyproline, and β-aminopropionitrile inhibit angiogenesis and tumor growth (Grant et al. 1994; Ingber and Folkman 1988; Maragoudakis et al. 1993; Oberbaumer et al. 1982). Other studies suggest that different domains of type IV collagen have distinct roles on the behavior of endothelial cells. While the triple helical fragments are nearly as active as intact type IV collagen in promoting endothelial cell adhesion and migration, the NC1 domains are less active (Furcht 1986; Herbst et al. 1988; Tsilibary et al. 1988). Tsilibary et al. demonstrated that the $\alpha1$ NC1 domain, as well as a 12-amino acid peptide within the $\alpha1$ NC1 domain, disrupted the assembly of type IV collagen. These results together with the observation that VBM proteins and provisional matrix molecules have either negative or positive influence on endothelial cells, led our laboratory to hypothesize that proteolytic degradation fragments of BM liberated during the inductive phase of angiogenesis may have the potential to regulate endothelial cell proliferation and migration (Kalluri 2002). To test the hypothesis, BMs from human placenta, amnion, and testis were isolated and subjected to digestion with various proteinases such as MMPs, elastase, cathespins, and serine proteinases, present usually in high concentrations in the tumor microenvironment (Berchem et al. 2002; Coussens et al. 2002; Egeblad and Werb 2002; Gershtein et al. 2001; Ghosh et al. 2002; Kaufmann et al. 2002; Krepela 2001; Levicar et al. 2002; Ray and Stetler-Stevenson 1994; Staack et al. 2002; Tang et al. 2001). The initial experiments with BM digestion were carried out using MMP-2, MMP-7, MMP-3, MMP-9, and MMP-13. The degradation fragments were separated by anion exchange chromatography, gel filtration, and HPLC. Several fragments were recovered and some of them confirmed as NC1 domains of type IV collagen by amino acid sequence and Western blot analysis. The studies with these NC1 domain fragments revealed that $\alpha1$, $\alpha2$, and $\alpha3$ NC1 domains inhibited endothelial cell tube formation significantly, whereas the α4 and α5 NC1 domains had insignificant effect. The α6 NC1 domain exhibited weak inhibitory effect in the original screen. The fragments of α1(IV), α2(IV), and α3(IV) NC1 were named arresten, canstatin, and tumstatin, respectively.

Arresten is a 26-kDa molecule isolated from BM of human placenta and produced as a recombinant protein in *Escherichia coli* and 293 embryonic kidney cells (Colorado et al. 2000). It inhibits FGF-2–induced endothelial cell proliferation and arrests VEGF-stimulated cell cycle progression in the S-phase. Arresten also inhibits endothelial tube formation and blood vessel formation in Matrigel assay. In vivo studies using human xenograft tumors reveal that arresten inhibits tumor growth and metastasis. The effects of arresten are potentially mediated via its binding to α1β1 integrin on proliferating endothelial cells.

The 24-kDa canstatin was also isolated from human placenta and produced as a recombinant protein in *E. coli* and 293 embryonic kidney cells (Kamphaus et al. 2000). It inhibits migration, proliferation, and tube formation associated with human and mouse endothelial cells and induces endothelial apoptosis. In human xenograft tumor models, treatment with canstatin significantly decreases tumor size, which is consistent with the decrease of CD31 positive vasculature (Kamphaus et al. 2000).

The 28-kDa tumstatin is the most extensively studied molecule among the type IV collagen-derived fragments. Tumstatin has been shown to exhibit anti-angiogenic activity through a series of in vitro and in vivo experiments. It inhibits endothelial cell proliferation and tube formation, and promotes apoptosis of endothelial cells with no effect on cancer cells. Tumstatin also blocks recruitment of capillaries in Matrigel plugs and inhibits tumor growth in mouse tumor models (Maeshima et al. 2000b). The region responsible for its angiogenic activity has been localized to amino acid 54-132, using deletion mutagenesis. The amino acid 185–203 region of α3(IV) NC1 domain, which is involved in antitumor activity (Han et al. 1997), is not active until the region is exposed by truncation.

Via binding to αvβ3 integrin in an RGD-independent manner (Maeshima et al. 2000a), tumstatin inhibits activation of focal adhesion kinase (FAK), phosphatidylinositol 3-kinase (PI 3-kinase), protein kinase B (PKB/Akt), and mammalian target of rapamycin (mTOR) and prevents the dissociation of eukaryotic initiation factor 4E protein (elF4E) from 4E-binding protein 1, thus leading to inhibition of cap-dependent protein synthesis (Maeshima et al. 2002). Tumstatin can be found in the circulation at physiological level of 330–350 ng/mL in mice, and is generated most efficiently by MMP-9 cleavage of α3(IV) collagen. The importance of MMP-9 is highlighted by experiments using MMP-9 deficient mice, which show decreased circulating tumstatin and accelerated tumor growth (Hamano et al. 2003). Mice with a genetic deletion of α3(IV) collagen exhibited increased tumor growth associated with enhanced pathological angiogenesis, which can be abolished by recombinant tumstatin provided at a normal physiological concentration. However, physiological angiogenesis associated with development and tissue repair is unaltered in the α3(IV) collagen-deficient mice. The activity of tumstatin requires the expression of αvβ3 integrin on the proliferating tumor vasculature.

14.4 TYPE XVIII COLLAGEN- AND TYPE XV COLLAGEN-DERIVED ENDOGENOUS INHIBITORS OF ANGIOGENESIS—ENDOSTATIN AND ENDOSTATIN-LIKE FRAGMENT FROM TYPE XV COLLAGEN (EFC-XV)

O'Reilly et al. found that conditioned medium from a murine hemangioendothelioma cell line, EOMA, inhibited proliferation of FGF-stimulated bovine endothelial cells. As several angiogenesis inhibitors have heparin binding activity, a heparin sepharose column was used to isolate the functional molecule, followed by gel filtration and HPLC purification. The inhibitory activity was found to be associated with a 20-kDa protein, named endostatin. Amino acid sequencing revealed that endostatin was an NC1 domain fragment of type XVIII collagen. Recombinant endostatin from *E. coli* inhibited endothelial cell proliferation, angiogenesis, and tumor growth. Systemic toxicity and drug resistance were not observed (O'Reilly et al. 1997). Due to the high homology between the NC1 domains of type XV collagen and type XVIII collagen, an endostatin-like fragment from type XV collagen was also identified as an endogenous angiogenesis inhibitor (Ramchandran et al. 1999; Sasaki et al. 2000).

Type XVIII and type XV belong to a subfamily of mutiplexin collagens, characterized by a triple helix interrupted by multiple NC domains. They have an N-terminal thrombospondin-like module and high homology in the C-terminal NC1 domain (Erickson and Couchman 2000). Through post-translational modification, heparan sulfate or chondroitin sulfate side chains can be added to type XVIII and type XV collagens, respectively (Halfter et al. 1998; Li et al. 2000). There are three mouse and two human splicing variants of type XVIII collagen (Muragaki et al. 1995; Rehn et al. 1996; Rehn and Pihlajaniemi 1995; Saarela et al. 1998). Expression of type XVIII and type XV collagens has been identified in most BMs including the VBM (Tomono et al. 2002). The function of these collagens is still unclear. Mutations of the human COL18A1 gene are associated with autosomal recessive Knobloch syndrome, causing myopia, retinal degeneration, and detachment. Mice deficient in type XVIII collagen are viable, with no apparent abnormalities. However, the mice exhibit delayed regression of hyaloid vessels and abnormal outgrowth of retinal vasculture (Fukai et al. 2002; Ylikarppa et al. 2003a). Mice lacking type XV collagen are also viable but develop muscular degeneration and cardiac defects (Eklund et al. 2001). There is no apparent overlapping function between type XV and XVIII collagens (Ylikarppa et al. 2003b).

Extensive studies have been carried out on the role of endostatin in angiogenesis. Endostatin inhibits endothelial proliferation and migration, and induces apoptosis and G1 arrest of endothelial cells. It inhibits tumor angiogenesis and tumor growth in various models (Dixelius et al. 2003). Endostatin may also target circulating endothelial progenitor cells (Capillo et al. 2003). This molecule binds to cell surface $\alpha5\beta1$, $\alpha v\beta3$, $\alpha v\beta5$ intergrins and glypicans (Karumanchi et al. 2001; Rehn et al. 2001; Sudhakar et al. 2003). It directly interacts with VEGF receptors and blocks VEGF signaling (Kim et al. 2002). Its anti-angiogenic activity has been recently found to be dependent on E-selectin (Yu et al. 2004). Although the mechanism of endostatin's action is still not completely understood, recent reports using gene array and proteomic analysis have shown that endostatin up-regulates several anti-angiogenesis factors and down-regulates a group of pro-angiogenesis factors (Abdollahi et al. 2004; Mazzanti et al. 2004). Endostatin can be found in the vessel wall, platelets, and circulation (Dixelius et al. 2003). Several enzymes are involved in the generation of endostatin. While the cleavage by cathepsin L releases 20-kDa endostatin (Felbor et al. 2000), MMPs and elastase produce a 30-kDa and a 28-kDa endostatin-containing fragment, respectively (Ferreras et al. 2000; Lin et al. 2001; Wen et al. 1999).

14.5 PERLECAN-DERIVED ENDOGENOUS INHIBITOR OF ANGIOGENESIS—ENDOREPELLIN

Perlecan is a major HSPG in BM distributed in various organs. It is highly conserved from human to *Caenorhabditis elegans*. The perlecan protein core is large, about 470 kDa, and contains five domains (I to V) from the N terminal. Heparan sulfate (HS) side chains can be added to the perlecan core protein by post-translation modification. Through the HS chains and the core protein, perlecan interacts with other BM proteins such as laminin, type IV collagen, nidogen, and fibulin, and forms the stable BM structure (Iozzo 1998). Recent studies have revealed the functional importance of perlecan during development. Targeted disruption of perlecan in mice results in embryonal death at E10–E12 or shortly after birth. The embryos exhibit severe chondrodysplasia and deterioration of well-formed BM. Furthermore, malformation of cardiac outflow tract and transposition of great vessels in embryos have also been observed, suggesting the roles of perlecan in vasculogenesis (Arikawa-Hirasawa et al. 1999; Costell et al. 2002; Costell et al. 1999). Perlecan stimulated angiogenesis in a rabbit ear model (Aviezer et al. 1994). It is also involved in tumor angiogenesis. Perlecan expression has been detected in human breast carcinoma and primary liver tumors, as well as in the blood vessel walls (Guelstein et al. 1993; Roskams et al. 1998). Down-regulation of perlecan with anti-sense perlecan blocks tumor growth and angiogenesis (Sharma et al. 1998).

In search for the proteins interacting with perlecan, Iozzo's group screened a keratinocyte cDNA library in the yeast two-hybrid system using the entire domain V of perlecan as bait (Mongiat et al. 2003). The domain V of perlecan, which was renamed endorepellin, contains three laminin G domain-like modules (LG) and four EGF-like repeats (EG). One of the strongest interacting clones encoded the C-terminal half of type XVIII collagen, which included the region for angiogenesis inhibitor endostatin. Further experiments demonstrated that endostatin interacted with the LG2 domain of endorepellin directly. Endorepellin suppresses HUVEC migration and tube formation in vitro. It also inhibits blood vessel formation in chorioallantoic membrane (CAM) assay and Matrigel plug assay. By binding to $\alpha2\beta1$ integrin, endorepellin increases cAMP and induces activation of PKA and FAK, and transient activation of p38 and heat shock protein 27, resulting in disassembly of actin stress fiber and focal adhesions (Bix et al. 2004). The anti-angiogenic activity of endorepellin attributes mostly to its LG3 domain (Bix et al. 2004), which can be cleaved from endorepellin by BMP-1/Tolloid family of metalloproteases (Gonzalez et al. 2004). The LG3 fragments can also be generated by endogenous proteolysis (Brown et al. 1997). The LG3 fragments have also been found in the urine of patients with end-stage renal failure and amniotic fluid of pregnant women with premature rupture of fetal membranes (Oda et al. 1996; Vuadens et al. 2003), suggesting that proteolytic processing of endorepellin and generation of LG3 may have biological significance.

REFERENCES

Abdollahi, A., P. Hahnfeldt, C. Maercker, H. J. Grone, J. Debus, W. Ansorge, J. Folkman, L. Hlatky, and P. E. Huber. 2004. Endostatin's antiangiogenic signaling network. *Mol Cell* 13:649–63.

Arikawa-Hirasawa, E., H. Watanabe, H. Takami, J. R. Hassell, and Y. Yamada. 1999. Perlecan is essential for cartilage and cephalic development. *Nat Genet* 23:354–58.

Aviezer, D., D. Hecht, M. Safran, M. Eisinger, G. David, and A. Yayon. 1994. Perlecan, basal lamina proteoglycan, promotes basic fibroblast growth factor-receptor binding, mitogenesis, and angiogenesis. *Cell* 79:1005–13.

Baluk, P., S. Morikawa, A. Haskell, M. Mancuso, and D. M. McDonald. 2003. Abnormalities of basement membrane on blood vessels and endothelial sprouts in tumors. *Am J Pathol* 163:1801–15.

Barker, D. F., S. L. Hostikka, J. Zhou, L. T. Chow, A. R. Oliphant, S. C. Gerken, M. C. Gregory, M. H. Skolnick, C. L. Atkin, and K. Tryggvason. 1990. Identification of mutations in the COL4A5 collagen gene in Alport syndrome. *Science* 248:1224–27.

Benjamin, L. E., D. Golijanin, A. Itin, D. Pode, and E. Keshet. 1999. Selective ablation of immature blood vessels in established human tumors follows vascular endothelial growth factor withdrawal. *J Clin Invest* 103:159–65.

Berchem, G., M. Glondu, M. Gleizes, J. P. Brouillet, F. Vignon, M. Garcia, and E. Liaudet-Coopman. 2002. Cathepsin-D affects multiple tumor progression steps in vivo: proliferation, angiogenesis and apoptosis. *Oncogene* 21:5951–55.

Bix, G., J. Fu, E. M. Gonzalez, L. Macro, A. Barker, S. Campbell, M. M. Zutter, et al. 2004. Endorepellin causes endothelial cell disassembly of actin cytoskeleton and focal adhesions through alpha2beta1 integrin. *J Cell Biol* 166:97–109.

Blumberg, B., A. J. MacKrell, P. F. Olson, M. Kurkinen, J. M. Monson, J. E. Natzle, and J. H. Fessler. 1987. Basement membrane procollagen IV and its specialized carboxyl domain are conserved in *Drosophila*, mouse, and human. *J Biol Chem* 262:5947–50.

Brown, J. C., T. Sasaki, W. Gohring, Y. Yamada, and R. Timpl. 1997. The C-terminal domain V of perlecan promotes beta1 integrin-mediated cell adhesion, binds heparin, nidogen and fibulin-2 and can be modified by glycosaminoglycans. *Eur J Biochem* 250:39–46.

Butkowski, R. J., J. P. Langeveld, J. Wieslander, J. Hamilton, and B. G. Hudson. 1987. Localization of the Goodpasture epitope to a novel chain of basement membrane collagen. *J Biol Chem* 262:7874–77.

Cameron, J. D., A. P. Skubitz, and L. T. Furcht. 1991. Type IV collagen and corneal epithelial adhesion and migration. Effects of type IV collagen fragments and synthetic peptides on rabbit corneal epithelial cell adhesion and migration in vitro. *Invest Ophthalmol Vis Sci* 32:2766–73.

Capillo, M., P. Mancuso, A. Gobbi, S. Monestiroli, G. Pruneri, C. Dell'Agnola, G. Martinelli, L. Shultz, and F. Bertolini. 2003. Continuous infusion of endostatin inhibits differentiation, mobilization, and clonogenic potential of endothelial cell progenitors. *Clin Cancer Res* 9:377–82.

Carmeliet, P. and R. K. Jain. 2000. Angiogenesis in cancer and other diseases. *Nature* 407:249–57.

Chelberg, M. K., J. B. McCarthy, A. P. Skubitz, L. T. Furcht, and E. C. Tsilibary. 1990. Characterization of a synthetic peptide from type IV collagen that promotes melanoma cell adhesion, spreading, and motility. *J Cell Biol* 111:261–70.

Colorado, P. C., A. Torre, G. Kamphaus, Y. Maeshima, H. Hopfer, K. Takahashi, R. Volk, et al. 2000. Anti-angiogenic cues from vascular basement membrane collagen. *Cancer Res* 60:2520–26.

Costell, M., R. Carmona, E. Gustafsson, M. Gonzalez-Iriarte, R. Fassler, and R. Munoz-Chapuli. 2002. Hyperplastic conotruncal endocardial cushions and transposition of great arteries in perlecan-null mice. *Circ Res* 91:158–64.

Costell, M., E. Gustafsson, A. Aszodi, M. Morgelin, W. Bloch, E. Hunziker, K. Addicks, R. Timpl, and R. Fassler. 1999. Perlecan maintains the integrity of cartilage and some basement membranes. *J Cell Biol* 147:1109–22.

Coussens, L. M., B. Fingleton, and L. M. Matrisian. 2002. Matrix metalloproteinase inhibitors and cancer: trials and tribulations. *Science* 295:2387–92.

Dixelius, J., M. J. Cross, T. Matsumoto, and L. Claesson-Welsh. 2003. Endostatin action and intracellular signaling: beta-catenin as a potential target? *Cancer Lett* 196:1–12.

Egeblad, M., and Z. Werb. 2002. New functions for the matrix metalloproteinases in cancer progression. *Nat Rev Cancer* 2:161–74.

Eklund, L., J. Piuhola, J. Komulainen, R. Sormunen, C. Ongvarrasopone, R. Fassler, A. Muona, et al. 2001. Lack of type XV collagen causes a skeletal myopathy and cardiovascular defects in mice. *Proc Natl Acad Sci U S A* 98:1194–99.

Eliceiri, B. P., and D. Cheresh. 2001. Adhesion events in angiogenesis. *Curr Opin Cell Biol* 13:563–68.

Erickson, A. C., and J. R. Couchman. 2000. Still more complexity in mammalian basement membranes. *J Histochem Cytochem* 48:1291–1306.

Felbor, U., L. Dreier, R. A. Bryant, H. L. Ploegh, B. R. Olsen, and W. Mothes. 2000. Secreted cathepsin L generates endostatin from collagen XVIII. *Embo J* 19:1187–94.

Ferreras, M., U. Felbor, T. Lenhard, B. R. Olsen, and J. Delaisse. 2000. Generation and degradation of human endostatin proteins by various proteinases. *FEBS Lett* 486:247–51.

Folkman, J. 1971. Tumor angiogenesis: therapeutic implications. *N Engl J Med* 285:1182–86.

Folkman, J. 1995. Angiogenesis in cancer, vascular, rheumatoid and other disease. *Nat Med* 1:27–31.

Folkman, J., and P. A. D'Amore. 1996. Blood vessel formation: what is its molecular basis? *Cell* 87:1153–55.

Folkman, J., and Y. Shing. 1992. Angiogenesis. *J Biol Chem* 267:10931–34.

Form, D. M., B. M. Pratt, and J. A. Madri. 1986. Endothelial cell proliferation during angiogenesis. In vitro modulation by basement membrane components. *Lab Invest* 55:521–30.

Fukai, N., L. Eklund, A. G. Marneros, S. P. Oh, D. R. Keene, L. Tamarkin, M. Niemela, et al. 2002. Lack of collagen XVIII/endostatin results in eye abnormalities. *Embo J* 21:1535–44.

Furcht, L. T. 1984. Role of cell adhesion molecules in promoting migration of normal and malignant cells. *Prog Clin Biol Res* 149:15–53.

Furcht, L. T. 1986. Critical factors controlling angiogenesis: cell products, cell matrix, and growth factors. *Lab Invest* 55:505–9.

Gershtein, E. S., S. V. Medvedeva, I. V. Babkina, N. E. Kushlinskii, and N. N. Trapeznikov. 2001. Tissue- and urokinase-type plasminogen activators and type 1 plasminogen activator inhibitor in melanomas and benign skin pigment neoplasms. *Bull Exp Biol Med* 132:670–74.

Ghosh, S., Y. Wu, and M. S. Stack. 2002. Ovarian cancer-associated proteinases. *Cancer Treat Res* 107:331–51.

Giancotti, F. G., and E. Ruoslahti. 1999. Integrin signaling. *Science* 285:1028–32.

Gonzalez, E. M., C. C. Reed, G. Bix, J. Fu, Y. Zhang, B. Gopalakrishnan, D. S. Greenspan, and R. V. Iozzo. 2004. BMP-1/tolloid-like metalloproteases process endorepellin, the angiostatic C-terminal fragment of perlecan. *J Biol Chem* 280:7080–87.

Grant, D. S., M. C. Kibbey, J. L. Kinsella, M. C. Cid, and H. K. Kleinman. 1994. The role of basement membrane in angiogenesis and tumor growth. *Pathol Res Pract* 190:854–63.

Guelstein, V. I., T. A. Tchypysheva, V. D. Ermilova, and A. V. Ljubimov. 1993. Myoepithelial and basement membrane antigens in benign and malignant human breast tumors. *Int J Cancer* 53:269–77.

Halfter, W., S. Dong, B. Schurer, and G. J. Cole. 1998. Collagen XVIII is a basement membrane heparan sulfate proteoglycan. *J Biol Chem* 273:25404–12.

Hamano, Y., M. Zeisberg, H. Sugimoto, J. C. Lively, Y. Maeshima, C. Yang, R. O. Hynes, Z. Werb, A. Sudhakar, and R. Kalluri. 2003. Physiological levels of tumstatin, a fragment of collagen IV alpha3 chain, are generated by MMP-9 proteolysis and suppress angiogenesis via alphaV beta3 integrin. *Cancer Cell* 3:589–601.

Han, J., N. Ohno, S. Pasco, J. C. Monboisse, J. P. Borel, and N. A. Kefalides. 1997. A cell binding domain from the alpha3 chain of type IV collagen inhibits proliferation of melanoma cells. *J Biol Chem* 272:20395–401.

Hanahan, D., and J. Folkman. 1996. Patterns and emerging mechanisms of the angiogenic switch during tumorigenesis. *Cell* 86:353–64.

Hanahan, D., and R. A. Weinberg. 2000. The hallmarks of cancer. *Cell* 100:57–70.

Herbst, T. J., J. B. McCarthy, E. C. Tsilibary, and L. T. Furcht. 1988. Differential effects of laminin, intact type IV collagen, and specific domains of type IV collagen on endothelial cell adhesion and migration. *J Cell Biol* 106:1365–73.

Hudson, B. G., R. Kalluri, S. Gunwar, M. Weber, F. Ballester, J. K. Hudson, M. E. Noelken, et al. 1992. The pathogenesis of Alport syndrome involves type IV collagen molecules containing the alpha 3IV chain: evidence from anti-GBM nephritis after renal transplantation. *Kidney Int* 42:179–87.

Hudson, B. G., S. T. Reeders, and K. Tryggvason. 1993. Type IV collagen: structure, gene organization, and role in human diseases. Molecular basis of Goodpasture and Alport syndromes and diffuse leiomyomatosis. *J Biol Chem* 268:26033–36.

Hynes, R. O. 1992. Integrins: versatility, modulation, and signaling in cell adhesion. *Cell* 69:11–25.

Ingber, D., and J. Folkman. 1988. Inhibition of angiogenesis through modulation of collagen metabolism. *Lab Invest* 59:44–51.

Iozzo, R. V. 1998. Matrix proteoglycans: from molecular design to cellular function. *Annu Rev Biochem* 67:609–52.

Jain, R. K. 1988. Determinants of tumor blood flow: a review. *Cancer Res* 48:2641–58.

Kalluri, R. 2002. Discovery of type IV collagen non-collagenous domains as novel integrin ligands and endogenous inhibitors of angiogenesis. *Cold Spring Harb Symp Quant Biol* 67:255–66.

Kalluri, R., V. H. Gattone 2nd, M. E. Noelken, and B. G. Hudson. 1994. The alpha 3 chain of type IV collagen induces autoimmune Goodpasture syndrome. *Proc Natl Acad Sci U S A* 91:6201–5.

Kalluri, R., C. F. Shield, P. Todd, B. G. Hudson, and E. G. Neilson. 1997. Isoform switching of type IV collagen is developmentally arrested in X-linked Alport syndrome leading to increased susceptibility of renal basement membranes to endoproteolysis. *J Clin Invest* 99:2470–78.

Kamphaus, G. D., P. C. Colorado, D. J. Panka, H. Hopfer, R. Ramchandran, A. Torre, Y. Maeshima, J. W. Mier, V. P. Sukhatme, and R. Kalluri. 2000. Canstatin, a novel matrix-derived inhibitor of angiogenesis and tumor growth. *J Biol Chem* 275:1209–15.

Karumanchi, S. A., V. Jha, R. Ramchandran, A. Karihaloo, L. Tsiokas, B. Chan, M. Dhanabal, et al. 2001. Cell surface glypicans are low-affinity endostatin receptors. *Mol Cell* 7:811–22.

Kaufmann, R., U. Junker, K. Junker, K. Nuske, C. Ranke, M. Zieger, and J. Scheele. 2002. The serine proteinase thrombin promotes migration of human renal carcinoma cells by a PKA-dependent mechanism. *Cancer Lett* 180:183–90.

Kim, Y. M., S. Hwang, B. J. Pyun, T. Y. Kim, S. T. Lee, Y. S. Gho, and Y. G. Kwon. 2002. Endostatin blocks vascular endothelial growth factor-mediated signaling via direct interaction with KDR/Flk-1. *J Biol Chem* 277:27872–79.

Krepela, E. 2001. Cysteine proteinases in tumor cell growth and apoptosis. *Neoplasma* 48:332–49.

Levicar, N., J. Kos, A. Blejec, R. Golouh, I. Vrhovec, S. Frkovic-Grazio, and T. T. Lah. 2002. Comparison of potential biological markers cathepsin B, cathepsin L, stefin A and stefin B with urokinase and plasminogen activator inhibitor-1 and clinicopathological data of breast carcinoma patients. *Cancer Detect Prev* 26:42–49.

Li, D., C. C. Clark, and J. C. Myers. 2000. Basement membrane zone type XV collagen is a disulfide-bonded chondroitin sulfate proteoglycan in human tissues and cultured cells. *J Biol Chem* 275:22339–47.

Lin, H. C., J. H. Chang, S. Jain, E. E. Gabison, T. Kure, T. Kato, N. Fukai, and D. T. Azar. 2001. Matrilysin cleavage of corneal collagen type XVIII NC1 domain and generation of a 28-kDa fragment. *Invest Ophthalmol Vis Sci* 42:2517–24.

Madri, J. A. 1997. Extracellular matrix modulation of vascular cell behaviour. *Transpl Immunol* 5:179–83.

Madri, J. A., and B. M. Pratt. 1986. Endothelial cell-matrix interactions: in vitro models of angiogenesis. *J Histochem Cytochem* 34:85–91.

Maeshima, Y., P. C. Colorado, and R. Kalluri. 2000a. Two RGD-independent alpha vbeta 3 integrin binding sites on tumstatin regulate distinct anti-tumor properties. *J Biol Chem* 275:23745–50.

Maeshima, Y., P. C. Colorado, A. Torre, K. A. Holthaus, J. A. Grunkemeyer, M. B. Ericksen, H. Hopfer, Y. Xiao, I. E. Stillman, and R. Kalluri. 2000b. Distinct antitumor properties of a type IV collagen domain derived from basement membrane. *J Biol Chem* 275:21340–48.

Maeshima, Y., A. Sudhakar, J. C. Lively, K. Ueki, S. Kharbanda, C. R. Kahn, N. Sonenberg, R. O. Hynes, and R. Kalluri. 2002. Tumstatin, an endothelial cell-specific inhibitor of protein synthesis. *Science* 295:140–43.

Maragoudakis, M. E., E. Missirlis, G. D. Karakiulakis, M. Sarmonica, M. Bastakis, and N. Tsopanoglou. 1993. Basement membrane biosynthesis as a target for developing inhibitors of angiogenesis with anti-tumor properties. *Kidney Int* 43:147–50.

Mazzanti, C. M., A. Tandle, D. Lorang, N. Costouros, D. Roberts, G. Bevilacqua, and S. K. Libutti. 2004. Early genetic mechanisms underlying the inhibitory effects of endostatin and fumagillin on human endothelial cells. *Genome Res* 14:1585–93.

McDonald, D. M., and P. Baluk. 2002. Significance of blood vessel leakiness in cancer. *Cancer Res* 62:5381–85.

McDonald, D. M., and A. J. Foss. 2000. Endothelial cells of tumor vessels: abnormal but not absent. *Cancer Metastasis Rev* 19:109–20.

Mongiat, M., S. M. Sweeney, J. D. San Antonio, J. Fu, and R. V. Iozzo. 2003. Endorepellin, a novel inhibitor of angiogenesis derived from the C terminus of perlecan. *J Biol Chem* 278:4238–49.

Morikawa, S., P. Baluk, T. Kaidoh, A. Haskell, R. K. Jain, and D. M. McDonald. 2002. Abnormalities in pericytes on blood vessels and endothelial sprouts in tumors. *Am J Pathol* 160:985–1000.

Muragaki, Y., S. Timmons, C. M. Griffith, S. P. Oh, B. Fadel, T. Quertermous, and B. R. Olsen. 1995. Mouse Col18a1 is expressed in a tissue-specific manner as three alternative variants and is localized in basement membrane zones. *Proc Natl Acad Sci U S A* 92:8763–67.

O'Reilly, M. S., T. Boehm, Y. Shing, N. Fukai, G. Vasios, W. S. Lane, E. Flynn, J. R. Birkhead, B. R. Olsen, and J. Folkman. 1997. Endostatin: an endogenous inhibitor of angiogenesis and tumor growth. *Cell* 88:277–85.

Oberbaumer, I., H. Wiedemann, R. Timpl, and K. Kuhn. 1982. Shape and assembly of type IV procollagen obtained from cell culture. *Embo J* 1:805–10.

Oda, O., T. Shinzato, K. Ohbayashi, I. Takai, M. Kunimatsu, K. Maeda, and N. Yamanaka. 1996. Purification and characterization of perlecan fragment in urine of end-stage renal failure patients. *Clin Chim Acta* 255:119–32.

Petitclerc, E., A. Boutaud, A. Prestayko, J. Xu, Y. Sado, Y. Ninomiya, M. P. Sarras Jr., B. G. Hudson, and P. C. Brooks. 2000. New functions for non-collagenous domains of human collagen type IV. Novel integrin ligands inhibiting angiogenesis and tumor growth in vivo. *J Biol Chem* 275:8051–61.

Ramchandran, R., M. Dhanabal, R. Volk, M. J. Waterman, M. Segal, H. Lu, B. Knebelmann, and V. P. Sukhatme. 1999. Antiangiogenic activity of restin, NC10 domain of human collagen XV: comparison to endostatin. *Biochem Biophys Res Commun* 255:735–39.

Ray, J. M., and W. G. Stetler-Stevenson. 1994. The role of matrix metalloproteases and their inhibitors in tumour invasion, metastasis and angiogenesis. *Eur Respir J* 7:2062–72.

Rehn, M., E. Hintikka, and T. Pihlajaniemi. 1996. Characterization of the mouse gene for the alpha 1 chain of type XVIII collagen Col18a1 reveals that the three variant N-terminal polypeptide forms are transcribed from two widely separated promoters. *Genomics* 32:436–46.

Rehn, M., and T. Pihlajaniemi. 1995. Identification of three N-terminal ends of type XVIII collagen chains and tissue-specific differences in the expression of the corresponding transcripts. The longest form contains a novel motif homologous to rat and *Drosophila* frizzled proteins. *J Biol Chem* 270:4705–11.

Rehn, M., T. Veikkola, E. Kukk-Valdre, H. Nakamura, M. Ilmonen, C. Lombardo, T. Pihlajaniemi, K. Alitalo, and K. Vuori. 2001. Interaction of endostatin with integrins implicated in angiogenesis. *Proc Natl Acad Sci U S A* 98:1024–29.

Risau, W. 1997. Mechanisms of angiogenesis. *Nature* 386:671–74.

Roskams, T., R. De Vos, G. David, B. Van Damme, and V. Desmet. 1998. Heparan sulphate proteoglycan expression in human primary liver tumours. *J Pathol* 185:290–97.

Saarela, J., R. Ylikarppa, M. Rehn, S. Purmonen, and T. Pihlajaniemi. 1998. Complete primary structure of two variant forms of human type XVIII collagen and tissue-specific differences in the expression of the corresponding transcripts. *Matrix Biol* 16:319–28.

Sasaki, T., H. Larsson, D. Tisi, L. Claesson-Welsh, E. Hohenester, and R. Timpl. 2000. Endostatins derived from collagens XV and XVIII differ in structural and binding properties, tissue distribution and anti-angiogenic activity. *J Mol Biol* 301:1179–90.

Sharma, B., M. Handler, I. Eichstetter, J. M. Whitelock, M. A. Nugent, and R. V. Iozzo. 1998. Antisense targeting of perlecan blocks tumor growth and angiogenesis in vivo. *J Clin Invest* 102:1599–1608.

Staack, A., F. Koenig, D. Daniltchenko, S. Hauptmann, S. A. Loening, D. Schnorr, and K. Jung. 2002. Cathepsins B, H, and L activities in urine of patients with transitional cell carcinoma of the bladder. *Urology* 59:308–12.

Sudhakar, A., H. Sugimoto, C. Yang, J. Lively, M. Zeisberg, and R. Kalluri. 2003. Human tumstatin and human endostatin exhibit distinct antiangiogenic activities mediated by alpha v beta 3 and alpha 5 beta 1 integrins. *Proc Natl Acad Sci U S A* 100:4766–71.

202 Protein Discovery Technologies

Tang, W. H., H. Friess, P. B. Kekis, M. E. Martignoni, A. Fukuda, A. Roggo, A. Zimmerman, and M. W. Buchler. 2001. Serine proteinase activation in esophageal cancer. *Anticancer Res* 21:2249–58.

Tomono, Y., I. Naito, K. Ando, T. Yonezawa, Y. Sado, S. Hirakawa, J. Arata, T. Okigaki, and Y. Ninomiya. 2002. Epitope-defined monoclonal antibodies against multiplexin collagens demonstrate that type XV and XVIII collagens are expressed in specialized basement membranes. *Cell Struct Funct* 27:9–20.

Tsilibary, E. C., G. G. Koliakos, A. S. Charonis, A. M. Vogel, L. A. Reger, and L. T. Furcht. 1988. Heparin type IV collagen interactions: equilibrium binding and inhibition of type IV collagen self-assembly. *J Biol Chem* 263:19112–18.

Vuadens, F., C. Benay, D. Crettaz, D. Gallot, V. Sapin, P. Schneider, W. V. Bienvenut, D., et al. 2003. Identification of biologic markers of the premature rupture of fetal membranes: proteomic approach. *Proteomics* 3:1521–25.

Wen, W., M. A. Moses, D. Wiederschain, J. L. Arbiser, and J. Folkman. 1999. The generation of endostatin is mediated by elastase. *Cancer Res* 59:6052–56.

Ylikarppa, R., L. Eklund, R. Sormunen, A. I. Kontiola, A. Utriainen, M. Maatta, N. Fukai, B. R. Olsen, and T. Pihlajaniemi. 2003a. Lack of type XVIII collagen results in anterior ocular defects. *Faseb J* 17:2257–59.

Ylikarppa, R., L. Eklund, R. Sormunen, A. Muona, N. Fukai, B. R. Olsen, and T. Pihlajaniemi. 2003b. Double knockout mice reveal a lack of major functional compensation between collagens XV and XVIII. *Matrix Biol* 22:443–48.

Yu, Y., K. S. Moulton, M. K. Khan, S. Vineberg, E. Boye, V. M. Davis, P. E. O'Donnell, J. Bischoff, and D. S. Milstone. 2004. E-selectin is required for the antiangiogenic activity of endostatin. *Proc Natl Acad Sci U S A* 101:8005–10.

15 RING Finger Proteins as E3 Ubiquitin Ligases

Claudio A. P. Joazeiro and Tony Hunter

CONTENTS

15.1 INTRODUCTION

E3 ubiquitin ligases are the components of the ubiquitin (Ub) system that confer specificity to the covalent attachment of Ub to substrates. Most RING finger (RNF)-containing proteins examined to date exhibit E3 activity. There are nearly 300 genes encoding RNF-containing proteins in the human genome, potentially comprising the largest family of E3s. In this chapter, we examine the evidence that led to the discovery of a functional role for the RNF domain in ubiquitylation, and in particular we describe the thinking processes behind our own work demonstrating that c-Cbl and other RNF-containing proteins act as E3s. For specific references, please consult the following reviews covering the topics referred to here: Thien and Langdon (Nat. Rev. Mol. Cell Biol 2:294) and Pickart (Annu. Rev. Biochem 70:503).

15.2 c-CBL AS NEGATIVE REGULATOR OF RTK SIGNALING

In the late 1990s, at a time when Francis Crick was still alive and well and often seen eating lunch at the Salk Institute's cafeteria, we were working together as a postdoctoral fellow and his advisor. At that time, we were especially interested in elucidating how receptor tyrosine kinase (RTK) signaling is negatively regulated. It was already relatively well known how RTKs elicit positive signals to promote changes in the gene expression program in the cytoskeletal architecture and in cellular metabolism. It was also appreciated that the specificity of the response

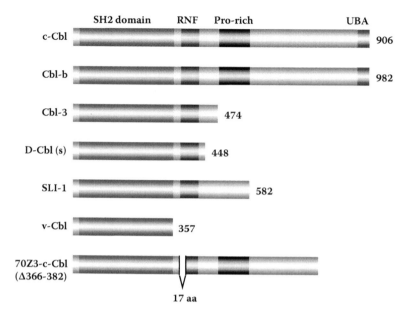

FIGURE 15.1 Cbl family members. Cbl proteins from mammals (c-Cbl, Cbl-b, and Cbl-3), *Drosophila melanogaster* (D-Cbl, only short form shown) and *Caenorhabditis elegans* (SLI-1). Motifs shown are the phosphotyrosine-binding (SH2), RING finger, proline-rich, and UBA. The oncogenic proteins v-Cbl and 70Z3-c-Cbl are also shown. *Source*: Modified from Thien, C. B., and W. Y. Langdon. 2001. Cbl: many adaptations to regulate protein tyrosine kinases. *Nat Rev Mol Cell Biol* 2:294–307.

to extracellular signals via cell surface receptors requires the precise control of both the duration and the intensity of the signals elicited by such receptors. This requirement was particularly well illustrated from studies performed in the context of T-cell receptor (TCR) signaling, and suggested that critical mechanisms should be in place not only to turn on, but also to terminate the response to growth factors and other stimuli. In fact, a few molecules with the properties of negative regulators of signaling, such as c-Cbl, had already been identified through both biochemical and genetic approaches. However, the mechanism of action of most such signaling inhibitors remained poorly understood.

c-Cbl is a ubiquitously expressed 120-kDa cytoplasmic proto-oncoprotein; its gene was originally cloned in the early 1990s by Wally Langdon at the Hanson Centre for Cancer Research in Adelaide, Australia. Langdon identified c-*Cbl* as the mammalian homologue of a retroviral oncogene (v-Cbl, for *Casitas B*-lineage *l*ymphoma; Fig. 15.1). However, only four years later would the role of the c-Cbl as a negative regulator of RTK signaling first be suggested from a genetic screen carried out in Paul Sternberg's laboratory at Caltech. Worms that exhibited a defect in vulva development due to a hypomorphic mutation in the epidermal growth factor receptor (EGFR) were screened for mutations that restored vulval induction. This led to the identification of a loss-of-function mutant of the c-Cbl ortholog, SLI-1 (Fig. 15.1), suggesting its function as a negative regulator of RTK signaling. Consistent with this finding, various groups rapidly demonstrated that mammalian c-Cbl was present in a complex with the activated EGFR and other RTKs, and that c-Cbl acted functionally as an RTK signaling inhibitor. Moreover, introduction of the equivalent loss-of-function mutation found in *sli-1* (Gly306 to Glu) into c-Cbl abrogated interaction with the EGFR. Langdon and Hamid Band (Brigham and Women's Hospital) also showed that v-Cbl and c-Cbl 70Z3, a c-Cbl mutant that had been isolated from a transformed pre–B-cell line, are both competent to bind to EGFR, but potentiate rather than inhibit EGFR signaling, thus correlating with their transforming potential.

Proteins that directly associate with activated RTKs often do so via interaction of their SH2 domains with Tyr-phosphorylated receptor peptide sequences. SH2 domain-containing proteins elicit downstream signals either by acting as enzymes (e.g., phospholipase C-γ), by playing adapter roles (e.g., Nck) or through other mechanisms. Collaborative studies in 1995 of the Schlessinger laboratory at NYU, David Bowtell at the Peter MacCallum Cancer Institute in Melbourne, and the Langdon group at the University of Western Australia showed that the novel amino terminal region of c-Cbl, which encompasses v-Cbl, could associate with the activated EGFR. It is also in this N-terminal region where the *sli-1* Gly306Glu loss-of-function mutation lies. We now know that this domain is a variant SH2 domain, according to structural evidence provided by the groups of Michael Eck and Nikola Pavletich. We also know, based on elegant experiments performed in Yossi Yarden's laboratory at the Weizmann Institute in Israel, that this variant SH2 domain in human c-Cbl specifically interacts with the EGFR via Tyr1045—one of the multiple Tyr residues that become autophosphorylated in response to receptor activation. It is interesting that SH2 domain-mediated interactions with phosphotyrosine (pTyr)–containing peptides are involved in recruiting of both positive and negative regulators of signaling to RTKs. The recruitment of c-Cbl via interaction with pTyr presents a remarkably elegant example of how cells ensure that only those receptors that are activated become targeted to degradation, as is further discussed below.

By the spring of 1998, when we began working on c-Cbl, its function as an adapter protein in signal transduction had been well characterized. Until then, there had been a clear bias, reflected in the published literature, toward the characterization of this role of c-Cbl, as opposed to understanding its mechanism of action as a signaling inhibitor. This bias may have been due in part to the earlier extensive description of p120, a protein that had been shown to become rapidly and conspicuously Tyr-phosphorylated in response to various stimuli, such as TCR activation. After Wally Langdon cloned c-Cbl and, with Larry Samelson (NIH), showed that p120 and c-Cbl were the same protein, multiple pTyr residues were quickly mapped in the C-terminal half of the protein and identified as sites for interaction with other SH2 domain-containing proteins. The C-terminal half of c-Cbl also has Pro-rich motifs, which signaling aficionados promptly recognized as sites that could mediate interaction with the SH3 domains of yet other proteins.

c-Cbl and SLI-1 were already recognized as negative regulators of RTK signaling, SLI-1 exhibited no clear homology to the C-terminal half of c-Cbl (Fig. 1), where the adapter function lies. In contrast, amino acid conservation between SLI-1 and the N-terminal half of c-Cbl was substantial. David Bowtell and Michael Czech had also pointed out that the Drosophila Cbl and EGFR interact, despite the obvious absence of adapter sites in the fly Cbl. These observations suggested that the C-terminal half of c-Cbl was not essential for its function as a negative regulator and led us to hypothesize that the negative regulatory functions of the molecule lie in the conserved N-terminal region. Because only two domains could be recognized in this region, the variant SH2 domain and an RNF, in the fall of 1998 we started working on a simple model, according to which the pTyr-binding domain of c-Cbl would be responsible for substrate targeting, while the RNF domain might mediate the negative regulatory function.

This model also helped explain how the two transforming alleles of c-Cbl might do their job; v-Cbl encodes a protein containing the entire SH2 domain but lacking all amino acids C-terminal to it, including the RNF. c-Cbl 70Z3 contains a discrete 17-amino acid deletion which overlaps with the N-terminal portion of the RNF. The deletion includes the first Cys in the RNF consensus and was predicted to disrupt the structure of this domain. Both proteins are competent to associate with activated RTKs but up-regulate rather than down-regulate their signaling, as shown by the groups of Hamid Band and Wally Langdon.

15.3   MECHANISM OF ACTION OF c-CBL

RTK level and activity had long been known to be regulated by multiple mechanisms: gene transcription, cell surface delivery, ligand binding, autophosphorylation, negative feedback phosphorylation, dephosphorylation, endocytosis, and others. The first reports of ligand-induced RTK ubiquitylation, as well as the proposal that this might be linked to their down-regulation, were published in 1992 by Seijiro Mori, Lena Claesson Welsh, and Callie Heldin, who were working with the platelet-derived growth factor receptor (PDGFR) at the Uppsala branch of the Ludwig Institute for Cancer Research. In July 1998, Hamid Band's group (Brigham and Women's Hospital) reported that the expression of c-Cbl, but not of c-Cbl 70Z3, was associated with the increased ubiquitylation and down-regulation of the PDGFR. Similar findings with the EGFR and the receptor for colony stimulating factor-1 were reported shortly afterward by, respectively, the groups of Yossi Yarden, in collaboration with Langdon, and Richard Stanley. These results were key to the understanding of how c-Cbl acts as a negative regulator of RTKs. The months subsequent to these publications must have been very busy with hypothesis-testing activities in many laboratories, as they certainly were for us.

The questions arose as to whether the ubiquitylation of RTKs in response to c-Cbl expression was a direct or indirect effect, and exactly how c-Cbl leads to increased RTK ubiquitylation. We started to pay more regular visits to the institute's library, in part to find a quiet space to dive into the Ub literature, and in part to catch the magnificent views of the Pacific Ocean and Torrey Pines State Park. At the time, the Ub field was beginning to make rapid progress, with growing evidence that Ub plays important regulatory roles in many cellular processes, breaking with the traditional view that the Ub-proteasome system mostly serves as a garbage collector for misfolded and damaged proteins.

Ub is an 8-kDa protein whose free species is not competent to be directly conjugated to its substrates. Ub must first be activated through an energy-dependent process that leads to the formation of a transient thiol-ester between its C-terminus and the side chain of an active site Cys present in the E1 Ub-activating enzyme. Ub is subsequently transferred, again in the form of a thiol-ester bond, to the catalytic Cys of one of a few E2 Ub-conjugating enzymes. Specificity of ubiquitylation is conferred by E3 Ub ligases. E3s have the dual ability to interact with E2s and with target substrates, and to mediate the stable linkage of the Ub C-terminus to the ε-amino group in a Lys residue of the substrate. Although the list of proteins specifically regulated by ubiquitylation had been growing continuously, by the late 1990s only few E3s were known, the most notorious being E6AP; Ubr1; the anaphase-promoting complex (APC); the SCF complexes, composed of Skp1, a cullin, and an F-box protein; and Mdm2. The existing list was clearly not sufficient to explain how ubiquitylation of a large variety of proteins could be accomplished with specificity. A family of proteins containing a domain *homologous to the E6AP C-terminus* (HECT) was already recognized and shown to act as E3s but consisted of fewer than 30 genes in humans. Except for the HECT domain, there were no other motifs associated with E3 activity that could be used as predictors of E3 function in the rapidly growing genomic databases.

c-Cbl was already known as an adapter protein in signaling, and therefore, a number of simple models evoking its direct participation in RTK ubiquitylation immediately came to mind. For example, the c-Cbl RNF might recruit either E3 Ub ligases or an E3-stimulatory factor to promote the ubiquitylation of activated receptors. One of the early hypotheses that we tested was whether c-Cbl might associate with an E3 of the C2-WW-HECT family. This hypothesis was based on the knowledge that various molecules containing C2 domains (e.g., phospholipase C-γ) are among those that bind to activated RTKs to mediate downstream signaling. It was conceivable that these receptors might use a similar mechanism to recruit inhibitors of signaling; it was also part of the hypothesis that such an C2-WW-HECT E3 would be recruited via the conserved c-Cbl N-terminal half, and that this binding might be stabilized by the interaction between the WW domains in the E3 and c-Cbl's Pro-rich region in the less conserved C-terminal half. In January of 1996, Richard Stanley (Albert Einstein College of Medicine) had published a paper showing that c-Cbl itself became ubiquitylated, we imagined, as a result of this hypothetical interaction. The pursuit of the elusive c-Cbl-interacting E3 was briefly addressed experimentally in the laboratory without much success. As it turns out, the Pro-rich-mediated interaction of the Cbl-family protein, Cbl-b, and the WW domain of HECT E3s can in fact occur, as recently reported by Allan Weissman's group at the NIH, but this interaction seems to be important for the ubiquitylation and negative regulation of Cbl-b itself.

15.4 RNF-CONTAINING PROTEINS

Yet another model for the mechanism of c-Cbl-mediated RTK ubiquitylation appeared even more likely as we became familiarized with other RNF proteins that were characterized at the time. Many RNF-containing proteins appeared to be involved in the regulated degradation of other proteins! Clearly, we should spend more time trying to understand the functions of this domain.

The RNF is a zinc-binding domain that was first recognized in the RING (*R*eally *I*nteresting *N*ew *G*ene 1) protein by Paul Freemont in 1993, and was immediately recognized to define a new protein family. The motif is present in thousands of proteins in databases and was initially proposed to function in protein–protein and protein–nucleic acid interactions. Most RNF domains consist of 50–100 amino acids following the consensus sequence CX2CX(9–39)CX(1–3)HX(2–3)C/HX2CX(4–48) CX2C, where the Cys and His represent zinc binding residues and X represents any amino acid. Depending on the presence of a Cys or His at the fifth coordination site, RNF domains can be further classified as RING-HC (such as c-Cbl's) or RING-H2, respectively, although the functional relevance of this difference is unknown. Unlike the tandem arrangement of zinc binding sites characteristic of zinc fingers, RING fingers exhibit two interleaved zinc-binding sites.

What seemed to be a common role of RNF-containing proteins in protein degradation was not sufficient to reveal the whole story. The involvement of Sina in the degradation of Tramtrack, of the photomorphogenesis factor COP1 in the negative regulation of Hy5, and of Hrd1p/Der3p in ER-associated protein degradation still did not explain what c-Cbl's role in ubiquitylation was. On the other hand, the appreciation that the RNF domain was a commonality among some members of the short list of non-HECT E3s was more informative. Three RNF proteins were known to participate directly in ubiquitylation: Ubr1/E3α, Mdm2 (whose RNF had been suggested to function as an RNA binding motif), and Rad18p. However, the role or even the requirement for the RNF in these proteins' function in ubiquitylation had not yet been examined. Finally, a RNF subunit (Apc11) had also been found by mass spectrometry in Kim Nasmyth's lab as a component of the APC E3 complex, and this subunit was shown to be essential for the APC function. To us, these observations promptly established a link among c-Cbl, Ubr1p, Mdm2, and the APC and led us to wonder whether c-Cbl might itself function as an E3 Ub ligase, and whether this activity was mediated by the RING domain. The hypothesis of c-Cbl's role was also conceived, through insight, after listening to presentations by Stanley, Band, Gil Lefkowitz from Yarden's laboratory, and Stan Lipkowitz (NIH) on their work describing the Cbl-regulated RTK ubiquitylation. As it turned out later (below) other investigators participating in Cbl workshop at the 10th International Conference on Second Messengers and Phosphoproteins (Jerusalem, November of 1998) were probably asking themselves the same question.

15.5 DEMONSTRATION THAT c-CBL FUNCTIONS AS RNF-DEPENDENT E3

We were able to rapidly assemble *in vitro* ubiquitylation reactions, thanks to the generosity of various people. Fumiaki Yamao at the National Institute of Genetics in Mishima, Japan, provided us with a human E1 expression construct. Mitsuyoshi Nakao at Kumamoto University in Japan provided a mouse Ubc4 expression construct. Our collaborator Simon Wing (McGill University) later provided both purified Ubc4 protein and antibodies against it. We had selected Ubc4 because this E2 had already been shown by Seijiro Mori and Yasushi Saito (Chiba University School of Medicine, Japan) to promote EGFR ubiquitylation, at least in an *in vitro* assay using a reticulocyte lysate.

For a few months we attempted to detect E3 activity with c-Cbl protein that had been purified from transiently transfected cells by immunoprecipitation. A number of reaction formats were tested, such as substrate-independent E3 activity, and PDGFR ubiquitylation *in vitro*. We presume that the lack of success of those early attempts was due to insufficient amounts of c-Cbl protein. Luckily, we later came across a protocol designed by Hideyo Yasuda and collaborators at the Tokyo University of Pharmacy and Life Science. These researchers had demonstrated the Ub ligase activity of Mdm2, using a simple reaction mixture consisting of E1, an E2 (UbcH5), GST-Mdm2, Ub, and ATP. Importantly, E3 activity under these conditions could be detected regardless of the presence of a specific substrate. We thus decided to test GST fusion proteins with c-Cbl under conditions similar to those described by Yasuda.

We were fortunate to have Yun-Cai Liu working locally, at the La Jolla Institute of Allergy and Immunology. He had been studying the c-Cbl protein for several years together with Amnon Altman, and had contributed a significant amount to the understanding of how c-Cbl acts as an adapter in T-cell signaling. We placed a telephone call to Yun-Cai to ask whether he would provide *Escherichia coli* expression constructs encoding the RNF of both c-Cbl and the 70Z3 mutant that he had described in earlier work. When we mentioned the model we wanted to test and the reasons why we were requesting those reagents, he became very excited. As he explained, he had also been interested in understanding the role of c-Cbl's RNF. In fact, he had used the c-Cbl RNF in an earlier yeast two-hybrid screen for interacting proteins and found an E2 Ub-conjugating enzyme among the hits. Moreover, he had tested and found that the 70Z3 mutant RNF failed to interact with that E2! These findings provided further support to the model that c-Cbl might act as an E3. Yun-Cai generously provided the constructs, marking the beginning of a very productive collaboration, which was extended later to other Ub ligases.

Armed with protein made using Yun-Cai's GST-c-Cbl fusion constructs, we were ready to perform ubiquitylation assays. What was especially attractive with this experimental design with respect to other types of assays was that it utilized the bacterially expressed c-Cbl. Since bacteria do not have a Ub system, we would have a strong argument against a positive result being due to an E3 co-purifying with the c-Cbl RNF.

It was one of those late evening experiments. E1, Ubc4, and the GST-RNF fusions to be tested for E3 activity were incubated with ^{125}I-labeled Ub under Yasuda's reaction conditions. The reaction products were separated by denaturing gel electrophoresis and identified by autoradiography. The results were remarkably clear as the x-ray film came out of the developing machine. A GST fusion with the wild-type RNF, but not with the RNF from c-Cbl 70Z3 or GST alone, promoted the ligation of Ub to proteins in the reaction, in a time-dependent manner (Fig. 15.2). We had found the pot of gold!

15.6   MECHANISMS UNDERLYING c-CBL'S E3 ACTIVITY

Subsequent experiments were designed to put the model to further test and to characterize mechanisms. For example, we wanted to know what Ub was ligated to in these

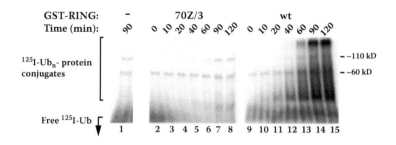

FIGURE 15.2   Intrinsic ubiquitin ligase activity of the c-Cbl RING finger.

reactions, in the absence of a c-Cbl substrate. Based on the sizes of the conjugates, it did not seem that unanchored Ub chains were abundant products. Immunoblotting with antibodies to GST revealed that at least part of the product was the ubiquitylated GST-c-Cbl RNF. This reminded us of the in vitro protein kinase assays that were run almost daily by everyone else in the lab, where the autophosphorylated kinase often was the most obvious product.

Importantly, the PDGF-Rβ could also serve as a direct substrate for ubiquitylation by c-Cbl in our assays. The substrate that we used for these reactions was PDGF-Rβ that had been immunopurified from unstimulated cells and then autophosphorylated in vitro using γ^{32}P-labeled ATP. The GST fusion with the c-Cbl RNF alone was not competent to ubiquitylate the receptor, and ubiquitylation was observed only when the fusion protein included the variant SH2 domain.

We also examined the effects of point mutations altering conserved residues on the RNF domain. Cys381 is the first Cys in the RNF consensus; Trp408 is conserved among a subset of RNF proteins (including those then known to be involved in protein degradation) predicted to be exposed on the RNF surface; thus, its mutation was not expected to disrupt the overall domain structure. Mutation of either residue to Ala led to loss of E3 activity, providing additional evidence of the direct involvement of the RNF in c-Cbl-mediated ubiquitylation.

In another line of experiments, we found that, unlike HECT domain-type E3s, there was no evidence of formation of covalent thiol-ester intermediates between c-Cbl and Ub. We thus proposed that Ub was transferred directly from a thiol-ester with c-Cbl-bound Ubc4~Ub to substrates.

How does the RNF mediate Ub ligase activity? Yun-Cai obtained a critical result for answering this question, building on his earlier yeast two-hybrid data and showing that Ubc4 and GST-c-Cbl RNF, both expressed and purified from *E. coli*, were capable of interacting directly in a "pulldown" assay. Importantly, mutation of either c-Cbl Cys381 or Trp408 abolished the interaction, correlating with the defective E3 activity of these mutants. The results with the c-Cbl Trp408Ala mutant were beautifully expanded soon afterward by the c-Cbl • UbcH7 (E2) co-crystal structure solved by Ning Zheng in Nikola Pavletich's laboratory at the Memorial Sloan-Kettering Cancer Center, which made it crystal clear that Trp408 directly contacts with the E2.

Around that time, a number of laboratories had been conducting experiments to characterize the composition of the multi-subunit E3 complex, SCF. Until then, the SCF complex was thought to consist of Skp1, a cullin and F-box proteins, and was thus seemingly unrelated to other known E3s. Yet another subunit of the complex, Rbx1/Roc1/Hrt1, was found in experiments led by Zhen-Qiang Pan (Mount Sinai School of Medicine), Yue Xiong (University of North Carolina at Chapel Hill), Dorota Skowyra (Saint Louis University), Ray Deshaies (Caltech), Joan Conaway (Oklahoma Medical Research Foundation), Steve Elledge (Baylor College), and Wade Harper (Baylor College). This new subunit had been missed in earlier biochemical purification experiments because of its small size and because the recombinant complex expressed in insect cells was unknowingly complemented by insect Rbx1. As it turned out, Rbx1 was another RNF domain-containing protein. Those investigators went on to demonstate that Rbx1 was essential for the E3 activity of SCF. However, both Rbx1 and the cullin subunit were required for E2 binding and E3 activity in *in vitro* assays, precluding one from distinguishing the role of the individual subunits. It was our work with c-Cbl and Allan Weissman's work with 8 other RING domain-containing proteins, a few months later, that established that E3 activity is intrinsic to, and not associated with, the RNF. In fact, we now know that for most E3s the RNF domain is both required and sufficient for activity.

Shortly after publication of our work in *Science* in October 1999, Yossi Yarden and Roland Baron's groups also published results leading to the similar conclusion that c-Cbl acts as an RNF-dependent E3. Less than a year later, the structure of c-Cbl bound to UBCH7 was reported by Nikola Pavletich's laboratory—the first structure solved for an E3 in complex with an E2. The structure revealed the expected contacts between c-Cbl's RNF and UBCH7 (including c-Cbl's Trp408) and, in addition, revealed an unexpected contribution of the linker alpha helix that connects the SH2 and RNF domains. By then, c-Cbl had become a paradigm for RNF-type E3 Ub protein ligases.

15.7   EVIDENCE SUPPORTING GENERAL ROLE OF RNF IN UBIQUITYLATION

The observation that several E3s have an RNF domain suggested that the latter might play a general role in ubiquitylation. This observation was greeted with general interest by other postdoctoral fellows in the laboratory. In particular, Wei Jiang, now associate professor at the Burnham Institute, initiated a number of excellent discussions. And Joel Leverson, now a scientist at Abbott Laboratories, and Han Huang, who's currently pursuing a religious career, helped set up assays for ubiquitylation and collaborated to expand on the RNF-E3 model in the context of their own studies.

As described above, we had hypothesized that the RNF protein Apc11 might be the "catalytic" subunit of APC, the multisubunit Ub ligase that mediates mitotic progression. Like SCF complexes, APC has both RNF and cullin subunits (Apc11 and Apc2, respectively). In collaboration with Joel, we provided evidence that Apc11 defines the minimal Ub ligase activity of the APC—its ability to act as an E3 was dependent on the integrity of the RNF but, unlike the Rbx1 of SCF complexes, Apc11 did not require the cullin subunit for E3 activity. Joel also took advantage of the budding yeast genetics to show that the integrity of the Apc11 RNF was essential for budding yeast cell viability.

The inhibitor of apoptosis, cIAP2, also contains an RNF. With Han, we found that the cIAP2 RNF possesses intrinsic Ub ligase activity and promotes ubiquitylation of caspases 3 and 7 but not caspase-1 in vitro. An intriguing observation made by Han, Joel, and us was that the RNF domains isolated from some proteins can recognize specific substrates, at least to some extent. As such, the isolated RNF of cIAP2, but not those of c-Cbl or Apc11, was capable of (inefficiently) promoting caspase-7 ubiquitylation; conversely, the RNF of Apc11, but not those of c-Cbl or cIAP2, mediated some cyclin B ubiquitylation. However, in the latter case, the recognition site on the substrate did not correspond to its functionally characterized "degron," or destabilizing sequence. This apparent paradox can perhaps be explained simply by the fact that RNFs are usually not the main sites of substrate recognition in E3s. Nonetheless, the data suggest that RNFs can contribute to substrate recognition. In fact, we now know that some E3s, such as Parkin, can recognize at least a subset of substrates entirely via an RNF-mediated mechanism.

Realizing the importance of these findings, we rapidly took the initiative to propose external collaborations to test the putative E3 activity of several other RNF-containing proteins. Together with Heinz Ruffner in Inder Verma's also at Salk, we showed this to be the case for BRCA1, the breast and ovarian cancer-specific tumor suppressor; with Carolyn Rankin (University of Kansas) we worked on Parkin, which is mutated in autosomal recessive juvenile parkinsonism (AR-JP); and with Randy Hampton across the street at UCSD we examined Hrd1, which is associated with endoplasmic reticulum-associated degradation (ERAD).

This work, together with that published contemporaneously by the Weismann group showing that other RFNs had E3 ligase activity, provided the evidence to suggest that E3 activity is a general function of RNF-containing proteins. The Weissman work was especially relevant because they demonstrated E3 activity of 8 random RING proteins that had not been previously implicated in protein ubiquitylation or degradation.

15.8   CONCLUDING REMARKS

The knowledge on E3s has expanded dramatically since the publication of our initial findings of c-Cbl's RNF-mediated E3 activity. The search for the "ubiquitin ligase" keyword on PubMed resulted in only 54 papers published before 1999, while the same search led to 4,643 hits as of February 13, 2005. We now know that E3 activity is not intrinsic to all RNF, although it is becoming clear that most RNF-containing proteins act as E3s themselves or as part of E3 complexes. The latter is the case, for example, of Rbx1 and of BARD1, which require heterodimerization with a cullin subunit or with BRCA1, respectively. We also know that many RNF E3s are directly involved in disease, such as BRCA1, Parkin, VHL, MID1, and Mdm2, to name a few. This observation has spurred a number of biotech and pharmaceutical companies to initiate drug-discovery programs with E3s as targets. With less than a third of all human E3s having any type of functional assignment, we expect that this class of proteins will be the subject of many historical accounts yet to come.

ACKNOWLEDGMENTS

We thank Wally Langdon for comments on the text as well as our many contributors and collaborators who have made this work possible.

REFERENCES

Joazeiro C. A., S. S. Wing, H. Huang, J. D. Leverson, T. Hunter, and Y. C. Liu. 1999. The tyrosine kinase negative regulator c-Cbl as a RING-type, E2-dependent ubiquitin-protein ligase. *Science* 286:309–12.

Thien, C. B., and W. Y. Langdon. 2001. Cbl: many adaptations to regulate protein tyrosine kinases. *Nat Rev Mol Cell Biol* 2:294–307.

Pickart, C. M. 2001. Mechanisms underlying ubiquitination. *Annu Rev Biochem* 70:503–33.

16 Discovery of Chemical Nature of Cross-Links in Collagen
A Personal Retrospective[*]

Paul Bornstein

CONTENTS

In 1963 I joined the laboratory of Karl Piez at the National Institute for Dental Research. This assignment satisfied my national service obligation as a physician during the Vietnam War. Karl was interested in the structure of the triple helical protein, type I collagen, the only collagen type known at the time (there are now at least 28 genetically distinct collagen types known in mammals). Karl had previously shown that soluble collagen, i.e., collagen that could be extracted from animal and human skin, or from some other tissues by non-denaturing solvents such as cold acetic acid, consisted of two identical chains termed α1, and a third homologous chain, α2. These chains were identified by their characteristic positions of elution during carboxymethyl (CM) cellulose chromatography, and by their amino acid compositions. However, α1/α1 and α1/α2 dimers, termed β11 and β12, respectively, could also be identified, based on their sedimentation equilibrium patterns during ultracentrifugation and by their amino acid compositions. These dimers were linked by covalent bonds, since they could not be dissociated by heat or by denaturing agents such as urea or guanidine. Furthermore, the majority of the collagen in skin, and to even a greater extent in tissues such as tendon or ligament, could not be solubilized, even with denaturing agents. The collagen in these tissues was clearly stabilized by covalent bonds, but what was the chemical nature of these bonds, and was there a relation between the inter-chain bonds in soluble β components and those in insoluble collagen?

As a physician, I was aware of the importance of collagen in connective tissues, i.e., tendons, ligaments, skin, and bone, and of the requirement for ascorbic acid (vitamin C) in the post-translational hydroxylation of peptidyl proline to form hydroxyproline, but I was not prepared to study the protein at a basic science level. Nevertheless I chose to pursue an investigation of the biosynthesis and structure

---

[*] This chapter is dedicated to the memory of Karl Piez (1924–2006), a pioneer in the field of collagen chemistry.

of collagen cross-links as my project in the Piez lab, because these questions were inherently interesting and were also the subject of a great deal of both investigative effort and disagreement at the time (1).

The vast majority of non-α-amino α-carboxy peptide covalent bonds in proteins are formed by the non-enzymatic oxidation of sulfhydryl side chains of cysteine residues to produce peptidyl disulfide cystines. Such bonds are commonly referred to as cross-links, a category that includes ε-amino γ-glutamyl iso-peptide bonds, catalyzed by the enzyme, transglutaminase. However, disulfide cross-links could not account for the insolubility of collagen, because the fibrillar protein lacks cystine, as judged by amino acid analysis. At a later time it was established that the non-triple helical COOH-terminal extensions in the biosynthetic precursor, procollagen, were linked by disulfide bonds (2).

Before we were able to pursue the question of the nature of inter-chain cross-links in collagen, a related issue required resolution. Collagen α chains are very long chains, composed of more than 1000 amino acids. Based largely on experiments involving the treatment of collagen with hydroxylamine, which was thought to cleave intra-chain ester bonds, Gallop developed a repeating subunit model of collagen α chains (3). The need for a subunit model was bolstered by the hypothetical argument, which now 40 years later seems quaint, that genes that encoded chains consisting of more than 1000 uninterrupted amino acids were unlikely to exist since the chance for random errors in transcription and translation would result in too large a fraction of defective protein (Paul Gallop, personal communication).

To provide evidence for or against the subunit hypothesis, I decided to use the then relatively new method for cleavage of protein chains at methionyl residues by cyanogen bromide (CNBr) (4). If collagen α chains were composed of repeating subunits, then chromatographic resolution of the cleavage products, followed by amino acid analysis of the CNBr-produced peptides, should provide evidence for or against the subunit model. The results of these experiments demonstrated that collagen α chains were composed of an uninterrupted string of amino acids and argued conclusively against the Gallop subunit model (5).

The first step in our effort to characterize a collagen cross-link was to isolate a fragment that contained it. The relatively small number of methionines in collagen (6–8 residues per 100,000 molecular weight) made CNBr cleavage an attractive initial approach. Our rationale was as follows. If interchain cross-links were relatively few in number per chain, specific in location, and stable to CNBr cleavage, then a chromatographic comparison of the cleavage products of an α1 and an α2 chain with those of β12 should reveal the presence of one or more fragments in the β component that contained cross-links. In our experiments we chose soluble rat skin collagen as a source of fragments and used chromatography on phosphocellulose, which provided better resolution of the smaller peptides than CM cellulose, but retained the larger peptides on the column.

The results of these experiments were successful beyond our most optimistic expectations. As shown schematically in Figure 16.1, peptides that we could later conclude were involved in the formation of an interchain cross-link are shaded. In digests of both the α1 and α2 chains, peptide CB1 exists in two forms, CB1 and

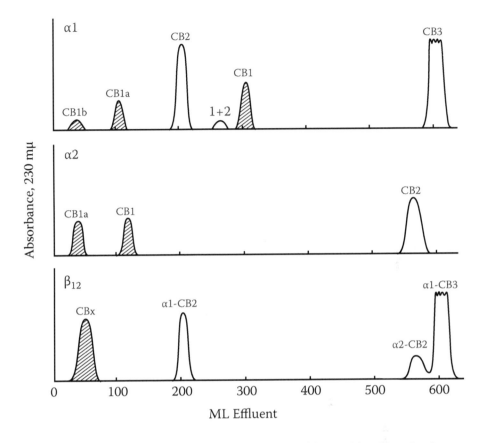

FIGURE 16.1 Phosphocellulose elution patterns of peptides resulting from the cleavage of the α1 and α2 chains and the β12 component of type I collagen with CNBr. The shaded peaks represent peptides from the α1 and α2 chains that participate in the formation of the cross-link in β12, and the peptide in β12 that contains the cross-link. *Source*: Reproduced with permission from P. Bornstein. "Covalent Cross-Links in Collagen: A Personal Account of Their Discovery." *Matrix Biol* 22 (2003): 385–91.

CB1a, which differ in amino acid composition only in that CB1a lacks a lysyl residue. The chromatogram of the α1 chain digest also shows a peak, CB1b, which is identical in amino acid composition to CB1a and results from the artifactual early elution of this peptide. Peptides CB1 and CB1a from both the α1 and α2 chains are missing from digests of the β12 dimer; instead, a new peptide, labeled CBx in Figure 16.1, is present. The amino acid composition of peptide CBx is equivalent to the sum of the compositions of peptides CB1a from the α1 and α2 chains, and the presence of two moles of homoserine per mole of peptide (methionine is converted to homoserine after digestion with CNBr) indicates that it is a double-stranded peptide. Peptides CB2 and CB3 in digests of the α1 chain and peptide CB2 in digests of the α2 chain are present in digests of β12, and would therefore appear not to be involved in cross-link formation (6).

On the basis of these findings, we could conclude tentatively that a lysyl residue on each of the $\alpha 1$ and $\alpha 2$ chains serves as a precursor of an inter-chain cross-link in collagen, which is present in $\beta 12$. This supposition was supported by experiments with ^{14}C-lysine-labeled collagen, which showed that peptides CB1a from both the $\alpha 1$ and $\alpha 2$ chains, and peptide CBx from $\beta 12$ were labeled, despite the absence of lysine on amino acid analysis of these peptides (7). A number of experiments involving limited cleavage of collagen with chymotrypsin, trypsin, and CNBr, performed with Dr. Andrew Kang, indicated that the lysyl residues that participate in cross-link formation were located near the NH_2 termini of the α chains (8).

Strong additional support for the role of peptidyl lysine in the formation of cross-links in collagen came from experiments with lathyrogens. It had been known for some time that treatment of growing animals with lathyritic agents, such as β-aminopropionitrile, resulted in bone abnormalities, weakened tendons and ligaments, and an increase in the extractability of collagen from tissues with non-denaturing solvents (9). When the CNBr-produced peptides from control and lathyritic rat skin collagen were compared, there was a marked reduction in lathyritic collagen in the ratios of peptides CB1a to CB1 in both the $\alpha 1$ and $\alpha 2$ chains (10). This finding indicated that lathyrogens inhibited the process by which lysyl residues in peptide linkage are converted to precursors of inter-chain cross-links in collagen.

If lysyl residues serve as precursors of covalent cross-links in collagen, how might the side chain of this amino acid, which terminates in an ε-amino group, be converted into a reactive moiety that could participate in the formation of a cross-link? At the time that these studies were in progress, a number of publications reported the presence of aldehydes in collagen, and several mechanisms were suggested that could implicate the participation of aldehydes in the formation of cross-links. However, lysyl side chains were either specifically excluded or not considered in these studies (11). We nevertheless postulated that conversion of the ε-amino group in the lysyl side chain to an aldehyde by an amine oxidase, thereby forming the δ-semialdehyde of α-amino adipic acid, followed by an aldol condensation reaction, could produce a covalent interchain cross-link. As a first step in a test of this hypothesis, peptide CB1a of the $\alpha 1$ chain was treated with N-methylbenzothiazolone hydrazone (MBTH), and the absorption spectrum of the derivative as a function of pH was monitored. This spectrum was consistent with the presence of a saturated aldehyde in peptide CB1a. After reduction of CB1a with potassium borohydrate, the expected product, ε-hydroxy-α-amino caproic acid, was identified by amino acid analysis. Similar treatment of peptide CB$\alpha 1$, as a control, failed to yield these compounds.

When peptide CBx from $\beta 12$ was treated with MBTH, a pH-dependent absorption spectrum, similar to that seen with $\alpha 1a$ but shifted to a lower wavelength, was observed. These findings were consistent with those observed with α,β unsaturated aldehydes (7). We therefore presumed that the initial aldol condensation product of two saturated aldehydes was subject to dehydration, resulting in the formation of a more stable α,β unsaturated aldehyde. Similar analyses were performed with peptide CBx from $\beta 11$, the peptide containing a cross-link joining two $\alpha 1$ chains, and amino acid analyses of the two peptides were consistent with their postulated origins.

Further support for the role of aldehydes in the formation of interchain cross-links in collagen was provided by kinetic studies of radioactively labeled α chains, which demonstrated that $\alpha 1^{Lys}$ and $\alpha 2^{Lys}$ chains are precursors of $\alpha 1^{Ald}$ and $\alpha 2^{Ald}$ chains, which in turn are precursors of β components (12).

It should be noted that early in our studies of the biosynthesis of collagen cross-links, we became aware of parallel studies of cross-links in elastin. Elastin is a highly insoluble protein that is cross-linked by polyfunctional amino acids, termed desmosine and isodesmosine. Desmosines were shown to be derived from peptidyl lysyl residues and, like collagen cross-links, the synthesis of these compounds was inhibited by lathyrogens, which blocked the conversion of lysyl side chains to aldehyde intermediates. These conclusions were supported by the demonstration of the incorporation of radiolabeled lysine into desmosines (13,14).

Although the elucidation of the biosynthesis and structure of the intramolecular cross-links that formed β components was a significant advance in our understanding of collagen cross-links, it was clear that intermolecular cross-links were also required to achieve the stability of collagen needed in most tissues. At the time that I left the Piez laboratory in 1967, it was unclear whether the bimolecular cross-links we described were sufficient to provide this stability, or whether more complex structures such as the desmosines in elastin or entirely different mechanisms were responsible for the formation of intermolecular cross-links. Since the cross-links we identified were limited to the NH_2-terminal non-triple helical region of the protein, related questions included (a) are inter-chain cross-links also present in the much larger triple-helical domain, and in the COOH-terminal non-triple helical region of the molecule and (b) do hydroxylysyl side chains, which predominate in the triple helical domain, also participate in cross-link formation?

It is now clear, based largely on the work of Eyre and coworkers (15) that the answer to these questions is yes. Furthermore, the product of the aldol condensation reaction of two lysyl-derived aldehydes represents only a minor fraction of the cross-links in most tissues. As shown in Figure 16.2, the initial aldehyde derivative of either lysyl or hydroxylysyl side chains, produced by lysyl oxidase, can interact with the side chains of either lysine or hydroxylysine to form divalent cross-links, which can then proceed to form mature trivalent pyrole and pyridinoline cross-links. The latter cross-links and the pathways for their formation have much in common with the synthesis of desmosines in elastin. It would appear that, as in the desmosines, only the first step in this complex reaction pathway is catalyzed enzymatically. Subsequent condensations occur spontaneously. Furthermore, the extent to which trivalent cross-links predominate depends on the function of the tissue and the extent to which stress is normally applied to it. Thus, bone contains both pyroles and pyridinolines, whereas mature cartilage contains only pyridinolines.

We were fortunate, when we started our work on collagen cross-links more than 40 years ago, that we used skin and rat tail tendon, tissues in which the more simple divalent cross-links predominate. As suggested by Figure 16.2, had we used a more highly cross-linked tissue, the task of deciphering the results with the approach that we had chosen would have been far more difficult.

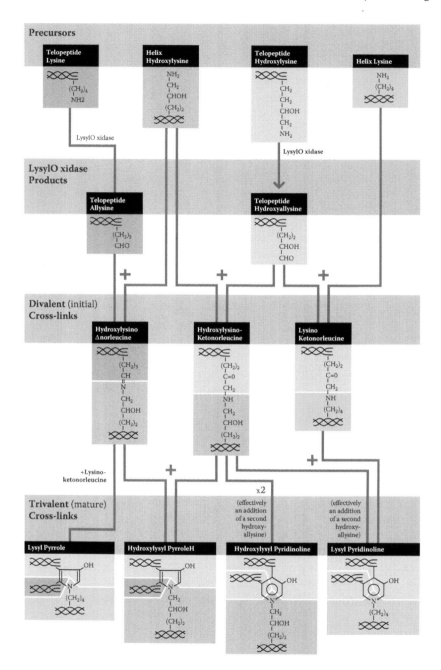

FIGURE 16.2 (See color insert following page 112.) The pathway of cross-link formation initiated by the oxidative deamination of lysyl and hydroxylysyl side chains in collagen. Lysyl and hydroxylysyl side chains of telopeptide and triple-helical origin are tracked to mature cross-links by color. The pathways shown predominate in skeletal tissues. *Source*: Reproduced with permission from D. R. Eyre, M. A. Weiss, and J. J. Wu. "Advances in Collagen Cross-Link Analysis." *Methods* 45 (2008): 65–74.

REFERENCES

1. J. J. Harding. "The Unusual Links and Cross-Links of Collagen." *Adv Protein Chem* 20 (1965): 109–90.
2. P. H. Byers, E. M. Click, E. Harper, and P. Bornstein. "Inter-chain Disulfide Bonds in Procollagen Are Located in a Large Non-Triple Helical COOH-Terminal Domain." *Proc Natl Acad Sci USA* 72 (1975): 3009–13.
3. P. M. Gallop. "3,2,1:A,B,C Sub-unit Hypothesis for the α Chains of Tropocollagen." *Nature* 209 (1966): 73–74.
4. B. Witkop. "Nonenzymatic Methods for the Preferential and Selective Cleavage and Modification of Proteins." *Adv Protein Chem* 16 (1961): 221–321.
5. P. Bornstein and K. A. Piez. "Collagen: Structural Studies Based on the Cleavage of Methionyl Bonds." *Science* 148 (1965): 1353–55.
6. P. Bornstein, A. H. Kang, and K. A. Piez. "The Nature and Location of the Intramolecular Cross-Links in Collagen." *Proc Natl Acad Sci USA* 55 (1966): 417–24.
7. P. Bornstein and K. A. Piez. "The Nature of the Intramolecular Cross-Links in Collagen. The Separation and Characterization of Peptides from the Cross-Link Region of Rat Skin Collagen." *Biochemistry* 5 (1966): 3460–73.
8. P. Bornstein, A. H. Kang, and K. A. Piez. "The Limited Cleavage of Native Collagen with Chymotrypsin, Trypsin, and Cyanogen Bromide." *Biochemistry* 5 (1966): 3803–12.
9. C. I. Levene and J. Gross. "Alterations in the State of Molecular Aggregation of Collagen Induced in Chick Embryos by β-Aminopropionitrile (Lathyrus Factor)." *J Exp Med* 110 (1959): 771–92.
10. P. Bornstein. "Cross-Linking of Collagen Chains." *Fed Proceed* May-June (1966): 1004–9.
11. O. O. Blumenfeld and P. M. Gallop. "Amino Aldehydes in Tropocollagen: The Nature of a Probable Cross-Link." *Proc Natl Acad Sci* 56 (1966): 1260–67.
12. K. A. Piez, G. R. Martin, A. H. Kang, and P. Bornstein. "Heterogeneity of the α Chains of Rat Skin Collagen and Its Relation to the Biosynthesis of Cross-Links." *Biochemistry* 5 (1966): 3813–20.
13. E. J. Miller, G. R. Martin, and K. A. Piez. "The Utilization of Lysine in the Biosynthesis of Elastin Cross-Links." *Biochem Biophys Res Commun* 117 (1964): 248–53.
14. S. M. Partridge, D. F. Elsden, J. Thomas, A. Dorfman, A. Telser, and P.-L. Ho. "Incorporation of Labelled Lysine into the Desmosine Cross-Bridges in Elastin." *Nature* 209 (1966): 299–300.
15. D. R. Eyre, M. A. Weiss, and J.-J. Wu. "Advances in Collagen Crosslink Analysis." *Methods* 45 (2008): 65–74.

17 Discovery of SPARC as Prototype for Matricellular Proteins

E. Helene Sage

CONTENTS

17.1 INTRODUCTION

It was the proverbial dark and stormy night (typical of Seattle winters) when we first saw it—an M_r 43,000 band on an SDS-polyacrylamide gel, apparently the major biosynthetic product secreted by cultured endothelial cells. Although heavily labeled with [^{3}H]-proline, the next experiment showed that this protein was not a collagen, a forsaken identity relegating it to the backburner in a laboratory focused primarily (but not entirely) on collagens. My postdoctoral mentor, Paul Bornstein, showed great intellectual generosity (and undoubtedly patience), by allowing me to work on what we termed "43k" (with a promise not to bother him with details "until the manuscript was written"). By all counts this decision was both wise and prophetic, as 43k (now known as SPARC) has provided us with puzzles, insights, and enjoyment over the last 30 years. And, as the story began to unfold, it became clear to us that we were working on far more than the biochemical characterization of a small, extracellular glycoprotein.

17.2 HISTORICAL PERSPECTIVE

"I'd prefer to work on proteins that dissolve." I made this remark to Paul Bornstein after my arrival in Seattle in the autumn of 1977, from the perspective of a long dissertation project on the evolution of the protein elastin. As the repetitive sequences of this protein "rubber" were just being elucidated by Bill Gray (my advisor) and

Larry Sandberg at the University of Utah, I was excited to see how the structure of this extremely hydrophobic protein was related to its function in blood vessels—were its passive recoil and morphological arrangement in the aorta fine tuned to the physiological requirements of animals that were homeo- or poikilothermic, or that showed a high versus low blood pressure? Naturally, I encountered the highly insoluble, cross-linked network of elastin fibers that contributed to my impatience with connective tissue proteins, but more importantly, to a PhD and an awareness of similar types of proteins, such as collagen.

Paul Bornstein's laboratory was well recognized for its work on collagen structure, but there were two new findings in this field that piqued my interest: the secretion by cells of procollagen (the soluble biosynthetic precursor of collagen) and the existence of a gene family, termed collagen types. The latter spoke to structure–function relationships for the different collagens, e.g., type I as the major fibrillar collagen and type IV as the collagen of basement membranes. This preparation was to prove useful during our discovery years with SPARC.

After a sabbatical with Jon Singer at UCSD, Paul had recently started experiments dealing with the cell surface and its associated proteins, collagen type I and fibronectin, with an overarching hypothesis that the exterior environment of a cell regulated cellular function. This direction, which included a thorough analysis of what was subsequently termed "extracellular matrix" (ECM), interested me the most, and we embarked on a long quest to characterize the secretory phenotypes of cells in culture. I focused chiefly on vascular endothelial cells, and Paul's hypothesis later became known as "dynamic reciprocity."

Our first paper on SPARC/43k was published in the *Journal of Biological Chemistry* in 1984 (1). The purified protein contained carbohydrate, bound to bovine serum albumin, and was secreted in copious amounts by most cells in vitro, especially endothelial cells. Unknown to us, John Termine's group at the National Institutes of Health (NIH) had already published a paper in *Cell* on the same protein, which they had termed "osteonectin" because of its abundance in bone. The identity between the two proteins became apparent when Brigid Hogan, then at the National Institute for Medical Research (NIMR) at Mill Hill (United Kingdom), invited me to her laboratory to collaborate on SPARC, a term she had coined from a translated cDNA that coworkers had isolated from a differential screen of mouse teratocarcinoma cells induced to differentiate into parietal endoderm.

Indeed, the extraembryonic Reichert's membrane (secreted by parietal endoderm cells in vivo) produced copious amounts of SPARC, and its sequence matched that of the N-terminal portion of bovine endothelial 43k protein. We presented our data at the Basement Membrane Gordon Conference that summer (1986) and published our findings later that year (2). Soon thereafter, other investigators described this protein in a basement membrane tumor, in platelet α-granules, and other tissues (these articles are cited in Reference 3).

17.3 PROPERTIES AND FUNCTIONS OF SPARC

Translation of the SPARC cDNA predicted a molecular mass of approximately 32,000 (excluding the signal peptide and carbohydrate modification). The M_r of 43,000 from

SDS-PAGE was therefore aberrant but characteristic of highly charged proteins (the pI of SPARC was estimated to be 4.3). Several domains were identified from the sequence: acidic, α-helical region with low Ca^{+2}-binding affinity (domain 1), a follistatin domain containing EGF repeats and two Cu^{+2}-binding sequences (domain 2), and an EC domain (domains 3 and 4) later defined by Hohnester and colleagues to contain an amphipathic helix and two EF hands (high-affinity, Ca^{+2}-binding loops) similar to those in calmodulin (reviewed in References 3 and 4). With these discrete regions it became clear to us that this protein had several functions, some of which might be mimicked by short, synthetic peptides or, more importantly, by peptides released from the native protein by protease cleavage.

An early and persistent clue to the function of SPARC was the high levels at which it was secreted by cultured cells. Was the biosynthesis of SPARC associated with cellular injury or survival? That it was produced at sites of injury in vivo, in cancer, or wherever there was a response to injury became clear in subsequent studies. The identification of SPARC as a survival factor for cells in vitro was only recently confirmed (5). Luisa Iruela-Arispe and Timothy Lane, two excellent graduate students, contributed substantially to our early understanding of the function(s) of SPARC. Luisa showed that SPARC was present in most tissues of the embryonic and adult mouse, but in the latter its expression was limited to remodeling or regenerating cells, e.g., bone, gut, epithelia, steroid-producing organs (adrenal cortex, testicular Leydig cells), and hair follicles (6). Why these locations? Another compelling clue was provided by Bob Vernon, a postdoctoral fellow, who added purified SPARC to cells and saw deadhesion and changes in shape. We later published these findings in the *Journal of Cell Biology* (*JCB*) and showed, for the first time, that certain proteins (for example, SPARC) could inhibit cell spreading and modulate cellular interaction with the ECM (7). Consistent with this finding, Goldblum et al. reported that SPARC also regulates endothelial barrier function in vitro by its retraction of intracellular junctional associations (8). A consequence of this activity was an inhibition of cell proliferation, as shown by Sarah Funk, a research technologist who lived with SPARC in our laboratory for more than twenty years (9). The block occurred in the mid-G1 phase of the cell cycle and was reversible upon removal of SPARC. That part of the antiproliferative effect was due to sequestration of certain growth factors and/or inhibition of their cognate receptors as later shown by Raines et al. (10)—SPARC bound with high affinity to platelet-derived growth factor (PDGF) and inhibited its interaction with the PDGF receptor—and by Kupprion et al. (11), who reported similar binding to vascular endothelial growth factor (VEGF) and inhibition of the phosphorylation of VEGF receptor 1, a result with major implications in vivo (see section 17.7).

There were two other discoveries in the mid-1990s that were important for our own understanding of the role of SPARC in cellular function. Joanne Murphy-Ullrich was the first to show that SPARC reduced the number of focal adhesions after its addition in micromolar quantities to cultured endothelial cells (12), a result consistent with the diminishment of barrier function reported earlier (8). A second observation made by Luisa Iruela-Arispe was the striking absence of SPARC in the connective tissue ECM of *mov 13* mice, a strain deficient in type I collagen, despite the normal production of SPARC by *mov 13* cells in vitro (13). These two insights—loss of focal adhesions and binding to type I collagen-containing ECM—were to set the stage for further experiments and hypotheses concerning the function(s) of SPARC in both normal and pathologic settings.

17.4   EMERGENCE OF MATRICELLULAR PROTEINS AS A FUNCTIONAL CLASS

It was clear to me that counter/antiadhesion was a major function of SPARC, and that this property would affect cell spreading and shape, differentiation, proliferation, migration, and production or communication with the ECM. That the counteradhesive effect of SPARC was operative in vivo was nicely shown by Maurice Ringuette and colleagues in *Xenopus* during gastrulation (14); moreover, injection of SPARC into rats with glomerulonephritis was associated with robust production of a collagen I and fibronectin-rich ECM (15). Thus, the effects we were observing in vitro appeared relevant to processes in vivo and justified our continuance of the work.

The provocative studies of Joanne Murphy-Ullrich gave me the idea about a distinct class of counteradhesive proteins—she had shown that SPARC, thrombospondin 1, and tenascin C all were capable of diminishing endothelial cell focal adhesions, albeit through separate molecular pathways (reviewed in 16). In a *JBC* minireview in 1991, Paul Bornstein and I proposed a group of "extracellular proteins that modulate cell-matrix interactions: SPARC, tenascin, and thrombospondin" (17). Later, more proteins were added to this list (e.g., osteopontin, tenascin X, thrombospondin 2, hevin/SC1 (a SPARC homolog), CCN-1, and PD1α/autotaxin), and the term "matricellular" was coined by Paul to denote their unique functional properties (18). Our ideas were developed further, as more data became available concerning receptors and mechanisms of action for some of these proteins, in a review in *Current Opinions in Cell Biology* in 2002 (19). Validation of this genetically unrelated "functional family" was forthcoming when the phenotypes of mice with disruptions of matricellular genes became apparent (see section 17.6).

17.5   PROCESSING OF SPARC INTO BIOACTIVE PEPTIDES

The modular structure of SPARC, exons 2–9 encoding discrete subdomains, was indicative of potential proteolysis in the extracellular space, accompanied by the release of unique peptides that might have biological activity. Earlier work with Patrice Tremble and Zena Werb had shown an induction of matrix metalloproteinases (MMPs) in fibroblasts exposed to SPARC, via a novel but uncharacterized ECM-dependent pathway (20). Might one or more of these proteinases recognize SPARC as a substrate? Luisa Iruela-Arispe, now a postdoctoral fellow, next published data substantiating a regulated proteolysis of SPARC during the development of the chicken chorioallantoic membrane; peptides from the Cu^{+2}-binding domain 2 were identified, but we were limited by the reagents on hand and by the quantities of peptides that could be extracted from this tissue (21). Concomitantly, Tim Lane published an exceptional paper showing the release of Cu^{+2}-binding peptides (containing the sequence [k]GHK) from SPARC by serine proteases in vitro and their angiogenic activity on endothelial cells in vitro and in vivo (22). Importantly, a second Cu^{+2}-binding sequence, KHGK, located N-terminal to KGHK, was not angiogenic (and in fact was shown to be an inhibitor of proliferation [Reference 4, and section 17.6]). These data indicated that the activities of these peptides were not dependent on their high-affinity binding of Cu^{+2}. It is interesting that the sequence KGHK is highly conserved among SPARC proteins from different vertebrates, and that the tripeptide GHK, which circulates in the bloodstream, has been marketed commercially for years as an additive ("growth-stimulator") to cell culture media. We calculated that cleavage of SPARC, deposited for example by platelet degranulation at sites of tissue injury, would produce GHK-containing peptides in the micromolar amounts that were shown to be effective in proliferation and angiogenesis assays. Subsequently we identified MMP-3 (stromelysin 1) as one of the proteinases in vivo that cleaved SPARC into discrete peptides with differential activities in the chick chorioallantoic membrane angiogenesis assay: angiogenesis, cell migration, and proliferation (23). In 1997 I proposed, in an article in *Trends in Cell Biology*, that bioactive fragments of extracellular proteins were in vivo regulators of angiogenesis (24). The title of the paper, "Pieces of Eight," referred to eight proteins, fragments of which regulated angiogenesis with effects different from those of the native, parent protein (SPARC, platelet factor 4, plasminogen, high-molecular-weight kininogen, prolactin, osteopontin, and types IV and XVIII collagen). Although a complete list in 1997, there are now many more examples, and the concept of endogenous proteolysis of cryptic sites that exhibit new activities is well accepted (see Chapter 14).

17.6   LEARNING FROM THE SPARC-NULL MOUSE

The SPARC gene had been inactivated in mice by Evans's group in the United Kingdom, and these animals had languished for want of a phenotype. By a fortuitous turn of events, we were later offered another SPARC-null mouse by Chin Howe, who had made the knockout but could ascertain no phenotype. Of course, we were disappointed with the purported lack of phenotype but reckoned that the SPARC-null cells could be of use as controls in our in vitro experiments. When the animals arrived, several people in our laboratory noticed that their eyes were "cloudy." Again, we were fortunate in having a colleague, John Clark, a member of our department (Biological Structure) at the University of Washington and an internationally recognized expert in cataract and protein chaperone biochemistry. John immediately identified the "cloudy" eyes as advanced cataracts, and in collaboration with his laboratory we began a systematic investigation of this phenotype (25). Qi Yan, a postdoctoral fellow and ophthalmologist, readily identified the cataractogenesis in these animals as the result of a poorly assembled lens capsular basement membrane, which was leaky and led to the precipitation of crystallins characteristic of this disease (26). Unfortunately for us, a colleague at the University of Washington who knew about these results communicated them to the group in the United Kingdom, whereupon the English mice were re-examined and found to have developed cataracts. These investigators published their findings in *EMBO J* (27), and our work followed shortly thereafter, in *Invest Ophthal Vis Sci* (25). The latter journal was quite rigorous in demanding precise characterization of the cataract and its development in the SPARC-null mice, data that we were able to supply with the assistance of the Clark laboratory. An additional important observation, which had been overlooked in the other publication, was the alteration in the ECM/basement membrane surrounding the lens, with laminin appearing in punctuate deposits suggestive of denaturation. These data, and several other studies cited below, led to the identification of a major role for SPARC in the secretion and assembly of ECM.

An interesting finding that developed from the cataract work was the capacity of SPARC to act as a molecular chaperone (28). Clark and colleagues showed that SPARC, associated with partially unfolded proteins (the unfolded protein response [UPR] is involved in the formation of cataracts, among other responses to cellular injury) and could re-fold model substrates. Perhaps this finding explains the high levels of SPARC present in the endoplasmic reticulum of many cells in vivo and in vitro? That SPARC was produced upon cell injury was appreciated years ago (29) and has provided, in recent work by a postdoctoral fellow Matt Weaver, an explanation for the induction of SPARC by cultured lens cells. Thus SPARC acts as an anti-apoptotic or survival factor for these cells after exposure to tunicamycin or serum withdrawal, both of which lead to a UPR (5). This capacity for maintenance of cell survival becomes important in cancer biology, although it is likely to be restricted to certain types of tumor cells (30).

Other than the early-onset cataractogenesis, was the SPARC-null mouse really "normal"? Through the efforts of many talented students, postdocs, and technologists, a list of the most compelling characteristics of this animal includes (a) severe osteopenia (31), (b) intervertebral disc degeneration (32), (c) increased adiposity (33), (d) attenuated dermis, (e) decreased connective tissue and collagen, and (f) a hooked or curled tail (reviewed in 4 and 34). The last characteristic was initially overlooked due to "tailing" of the mice for DNA analysis, a procedure now unnecessary, as the phenotype is 100% penetrant in SPARC-null offspring. This list reflects alterations in tissues during the course of development, but many of the changes were exacerbated upon injury or in various pathologies (see below).

Amy Bradshaw, then a postdoctoral fellow in the laboratory, was the first to show the striking decrease in the diameter of collagen type I fibrils in the dermis of SPARC-null mice, and a concomitant reduction in dermal tensile strength (35). Subsequently, Amy showed that in dermal fibroblasts, SPARC regulated both the processing of procollagen I and collagen fibrillogenesis (36). The reduced capacity for collagen assembly was manifested in several disease models in these mice. The pulmonary fibrosis associated with bleomycin treatment was markedly attenuated in animals lacking SPARC (37). Moreover, SPARC-null mice exhibited a reduced foreign body response to implanted biomaterials, especially apparent in the compromised encapsulation of the implant (38). SPARC-null mice were also characterized by an accelerated wound closure, in comparison to wild-type counterparts, that was attributed in part to the reduction in dermal connective tissue and to an enhanced, migratory capacity of dermal fibroblasts (39). As I was beginning my postdoctoral studies, I recall a quote from a famous collagen biochemist: "You can't go wrong working on connective tissue—it is everywhere and likely to play a role in most types of disease." This certainly was prophetic in the case of SPARC, as mice lacking this prominent matricellular protein exhibited so many abnormalities (some obvious, others quite subtle, and clearly not all identified) that could be attributed to problems with fibrillar collagen (I) or basement membrane (collagen IV and laminin 1) production and/or assembly (40). As an example, Amy Bradshaw was attempting to extract collagen from mouse skins, and noted that those from SPARC-null mice floated in aqueous buffers. A quick recollection of college biochemistry told us that these skins were extremely fatty—in fact, SPARC-null versus wild-type mice showed a greater accumulation of dermal and inter-organ fat, with larger and more numerous adipocytes in their fat depots, and a significant reduction of collagen type I (33). Despite the accumulation of excessive adipose tissue, the SPARC-null mice were not obese—rather, their body weights were nearly equivalent to those of wild-type animals, data reflecting the loss of connective tissue and bone that occurs in the absence of SPARC. In the same year, Anne Delany published that SPARC-null bone marrow cells exhibited an enhanced propensity for differentiation into adipocytes instead of osteoblasts, and that the absence of SPARC compromised osteoblast maturation and survival (41).

These exciting studies were continued by Jing Nie, a postdoctoral fellow, who isolated preadipocytes from wild-type and SPARC-null fat and showed, definitively, that SPARC inhibited their differentiation into mature adipocytes (42). Two mechanisms, acting at different times during the differentiation process, were identified: (a) SPARC, signaling through integrin-linked kinase (ILK, see below) and GSK-3β, enhanced the accumulation and translocation of β-catenin into the nucleus of the preadipocyte, and the subsequent interaction of β-catenin with the transcription factor TCF/LEF. As the wnt/β-catenin pathway has been shown to inhibit adipogenesis and enhance osteoblastogenesis, we now have at least a partial explanation for the adiposity and osteopenia that characterize SPARC-null mice. (b) SPARC enhanced the deposition of fibronectin and expression of its α5 integrin receptor, and inhibited that of laminin and its α6 integrin, in preadipocytes. During differentiation, the fusiform preadipocyte associated with a fibrillar ECM undergoes substantial morphological changes to become a spherical adipocyte surrounded by a basal lamina—SPARC favors the less differentiated state by its promotion of ECM production that retards adipocyte morphogenesis.

A second example of biological regulation by ECM is the growth and metastasis of solid tumors. There is an increasing appreciation of tumor-stromal cell interactions, and of the role of angiogenesis and cells of the immune system, in the progression of many cancers. Herein lies a propitious future for SPARC and, indeed, other matricellular proteins. A landmark paper in *Nature Medicine* by Ledda et al. reported the enhancement of melanoma progression by SPARC and set the stage for subsequent studies designed to uncover mechanisms by which SPARC, generally associated with poor prognosis, mediates tumor growth (reviewed in 43). Whereas the melanoma as well as other studies focused on the deadhesive properties of SPARC associated with increased migration, other investigators concentrated on the inhibition of angiogenesis as a mechanism to thwart tumor growth. Recently, Chlenski et al. have shown that a SPARC peptide containing the sequence KHGK was an effective inhibitor of neuroblastoma (44). Our group, however, was the first to show that the growth of solid tumors was enhanced in SPARC-null mice, in part due to a compromised ECM, poor encapsulation of the tumor, and reduced infiltration of macrophages (45,46). There have been many publications on the role of SPARC in cancer, some of which present apparently conflicting results that arise from differences in the tumor models used in mice; the tumor types themselves; the criteria for assessment of tumor take, growth, and metastasis; and the particular areas of expertise of the investigators (reviewed in 47–49). Given time, it is likely that most, if not all, of these numerous studies will be proven correct for the experimental paradigm that was described. That SPARC plays a significant role in cancers has been given additional credence by gene array analyses identifying SPARC, as well as its hevin homologue, as part of an invasion-specific cluster, especially in juxta-tumoral stromal cells (47).

17.7 RECONCILIATION OF DATA AND FUTURE PROSPECTS

One of the highlights of working on SPARC was the International Hermelin Brain Tumor Symposium, organized by Sandra Rempel in 2004, on matricellular proteins in normal and cancer cell–ECM interactions (50). The meeting was centered around SPARC, and for the first time many of the investigators made contact with one another, agreed to exchange reagents, and established productive collaborations. But, what were the new directions? What major questions had we not answered?

Clearly there was the question of how SPARC communicated with cells. As a high-affinity cell-surface receptor for SPARC had not been identified (despite years of affinity chromatography, cross-linking experiments, and immunological approaches), I reasoned that, indeed, like many of the integrins, the K_D for SPARC and its cognate, cell-surface receptor must be low, i.e., $<10^{-5} – 10^{-6}$ M. Such affinity would explain the deadhesive, promigratory, and antiproliferative properties of SPARC attributed to intermediate states of adhesion (16); at the same time it provided challenges for the isolation of a specific receptor(s). After reading the papers published by the Arap and Pasqualini laboratory, in which phage display technology was used to identify cell-surface receptors and their ligands, I surmised that this was the technique our group had been looking for—one by which low-affinity receptors could be identified, and the corresponding, limited recognition sequence on the ligand validated. Wadih Arap and Renata Pasqualini graciously hosted a mini-sabbatical and introduced me to Marina Cardó-Vila, at that time a postdoctoral fellow, who screened a random phage display library on SPARC. The resulting data were both exciting and definitive, and several receptor and binding partner candidates were identified and subsequently confirmed. The first of these was ILK, a Ser/Thr kinase that binds to the cytoplasmic tails of β1 and β3 integrins. Tom Barker, a postdoctoral fellow in our laboratory, showed that SPARC, by virtue of its activation of ILK, stimulated the assembly of fibronectin by cultured fibroblasts (51). One of the targets of ILK is myosin light chain phosphatase, a regulator of stress fiber assembly and cell migration. Although we believe that an association between SPARC and ILK occurs, it has not yet been proven. A link between the two proteins, however, was recently provided by Mikhail Kolonin, who identified α5β1 integrin as a receptor for SPARC on adipose stromal cells (52). Similarly, Matt Weaver identified β1 integrin as a signaling receptor for SPARC, via ILK, that enabled the survival function of SPARC on lens cells (5). The interaction of SPARC with β1-containing integrins, with the subsequent activation of ILK, explains many of the properties previously attributed to SPARC, such as ECM assembly, induction of intermediate adhesion, and changes in cell shape. Moreover, the effect of SPARC on endothelial barrier function can now be attributed to its interaction with VCAM-1, a cell-surface receptor also identified by phage display, that mediates the transmigration of leukocytes (53).

A third receptor for SPARC that was discovered through phage display was a surprise, but one that has stimulated new hypotheses concerning the role of SPARC in responses to injury that involve inflammation. With Julia Kzhyshkowska (University of Heidelberg) we showed that a third candidate binding partner discerned from the phage display, stabilin 1, was in fact a receptor for SPARC (54). Expressed principally by alternatively activated macrophages, this scavenger receptor (with no previously identified ligand) cleared SPARC from the extracellular milieu and trafficked it to the lysosomal compartment, presumably for degradation. This population of macrophages might therefore be responsible in part for the titration of SPARC at sites of injury, where excessive levels of the protein could be counter-productive to resolution of the pathology. It now becomes important to identify the types of macrophages present in tumor stroma, adipose tissue, and healing wounds, and to determine whether via stabilin 1 SPARC recruits these alternatively activated cells and/or is consumed by them in situ.

SPARC belongs to a protein family, the signature of which is the extracellular Ca^{+2}-binding (EC) domain (reviewed in 4). A member of this family that was initially discovered in the central nervous system (synaptic cleft [SC]-1) and in *h*igh *e*ndothelial *v*enules of lymphatic tissue (hevin) (reviewed in 55) has become a focus in cancer research due to its high levels of expression in certain tumors, tumor stroma, and their neovasculature (reviewed in 47). Among other functions, hevin (also known in the cancer literature as SPARC-like 1) was shown to be deadhesive and to regulate radial glia-guided migration in the cerebral cortex (56). And then there was the challenge of the reported absence of phenotype in the hevin-null mouse (reviewed in 55). Working from a hypothesis that domains or peptides released by endogenous proteolysis at evolutionarily conserved sites in hevin and SPARC might compensate for each other in different tissues, we cloned murine hevin, produced the recombinant protein and corresponding antibodies, and generated a colony of 129/SVe mice (wild-type, SPARC-null, hevin-null, and SPARC/hevin double-null), along with the reagents necessary to address structure–function relationships between these two matricellular proteins.

Millicent Sullivan, a recent postdoctoral fellow in the laboratory, noted that the hevin-null dermis was unusually stiff, with a high tensile modulus, and contained tightly packed collagen fibrils of aberrant morphology (57). Further investigation revealed that hevin-null fibroblasts secreted substantially reduced levels of the proteoglycan decorin, an accessory molecule that promotes the formation of collagen type I fibrils. Furthermore, recombinant hevin was shown to enhance the rate of collagen I fibrillogenesis and to affect the adhesion of fibroblasts to this substrate by its regulation of decorin production. Thus, we had found yet another mechanism by which a matricellular homologue of SPARC influenced cell-ECM interaction.

Comparison of SPARC-null and hevin-null mice revealed a few similarities and some interesting differences: (a) both developed cataracts, although at different ages,

(b) both exhibited enhanced growth of solid tumors, and (c) both closed dermal wounds faster than wild-type counterparts. In contrast, hevin-null mice were (a) not fat, (b) did not have kinked tails, (c) did not exhibit intervertebral disc degeneration, and (d) presented with a stiff dermis, as opposed to the extremely lax counterpart seen in SPARC-null animals. Tom Barker was at the same time performing experiments on the entire colony with respect to the foreign body reaction, and his results were both provocative and confirmatory of earlier data. Each protein exhibited a unique function: hevin was associated with the recruitment of macrophage-like cells, whereas SPARC dictated the formation and morphology of the capsule surrounding the implant. However, both proteins exhibited compensatory functions in their collective inhibition of angiogenesis (58). Clearly, in this injury model, SPARC dominated the connective tissue response, while hevin regulated the influx of inflammatory cells. Given the similarity between SPARC and hevin in the Cu^{+2}-binding and EC domains, it was not difficult to predict how each might enhance the activity of the other in the angiogenic response, especially at the peptide level. Indeed, Matt Weaver and Gail Workman have sequenced specific MMP-3 cleavage products of hevin and established their near identities to those of SPARC (unpublished experiments). Moreover, evidence for endogenous proteolysis of hevin is readily apparent in certain tissues, e.g., brain and lung (unpublished experiments). Continuation of this work should answer our questions of compensation at the peptide level and might contribute to therapeutic strategies for resolution of injury or inhibition of tumor growth by receptor mimicry or interference.

There remain open questions concerning some functions of SPARC and their implications for regulation of cell behavior and differentiation. For example, why does SPARC traffic to the nucleus under certain conditions (59)? What is the common function that SPARC appears to mediate between tumor growth and wound healing (60)? Does the relegation of stem cells to certain developmental compartments reflect a fundamental role for SPARC (41) via its influence on ECM and signaling through ILK? More time, and more funding …

A fascinating and largely unexplored area relates to the role of SPARC as a molecular chaperone. Maurice Ringuette and coworkers have written an interesting review on the evolution of the SPARC gene family and have posited a chaperone-like function of SPARC in *Drosophila* (61). As this organism lacks a major chaperone of secreted (ECM) proteins, heat shock protein (HSP) 47, SPARC might act as a substitute molecule directing the correct folding and export of proteins such as collagen type IV. Correlative data supporting this hypothesis was recently provided in collaboration with Jackie Hecht, in which we showed that chondrocytes from patients with pseudoachondroplasia, which exhibit impaired secretion of specific ECM components, retained SPARC in the endoplasmic reticulum (62). It would be important to identify the sequence(s) of SPARC that mediate association with specific proteins in the endoplasmic reticulum, and whether SPARC is actually secreted with its re-folded partner. Would SPARC from *Caenorhabditis elegans*, which lacks several domains found in vertebrate SPARC (63), exhibit chaperone-like activity in different animal systems? Throughout this speculation it is important to remember that, in the adult mammal, SPARC is found chiefly in cells and not extracellularly. We have always suspected that part of its function lies inside the cell, whereas the

TABLE 17.1

Functions of SPARC *In Vitro* and *In Vivo*: Reconciliation

In Vitro	*In Vivo*
Addition of SPARC to cells:	Platelet release of stored SPARC:
Loss of focal adhesions	Change in cell shape
Loss of endothelial barrier function	↑ Leukocyte transmigration *via* VCAM-1
Inhibition of VEGFR-1 phosphorylation	↓ Ocular angiogenesis by VEGF
↑ ECM synthesis	↑ ECM in glomerulonephritis
Targeted disruption of SPARC:	Developmental and challenge phenotypes:
↑ Proliferation/migration	↑ Wound closure
↓ Collagen synthesis	Defective lens capsule
	↓ Collagen in bone and skin
	↓ Tumor capsule/stroma
	↓ Encapsulation of foreign body

secreted form interacts with receptors or ECM components and is rapidly degraded, perhaps to bioactive peptides under conditions of injury or repair.

Table 17.1 provides a reconciliation of data published over nearly 30 years, with an emphasis on the agreement among experiments in vitro (mostly earlier work) and their counterparts in vivo (allowing for limitations in experimental design and murine models). Most of these points have been covered in the text; a particularly satisfying correlation was afforded by J. Ambati and colleagues, who showed that SPARC suppressed VEGF receptor 1 activation in an ocular model of choroidal neovascularization (64). Ten years previously Christine Kupprion had published similar data from cultured endothelial cells (11).

One of the most rewarding aspects of my rather extended life with SPARC has been the society of talented and successful graduate students and postdoctoral fellows who chose to work with me on what was often a difficult, confusing, and frustrating protein that was not particularly generous with its secrets. It is gratifying to see that several of these fellows are continuing, in their own laboratories, to explore SPARC and its functions in a variety of developmental and pathologic settings. Their contributions will be important to us all.

> We shall not cease from exploration
> And the end of all our exploring
> Will be to arrive where we started
> And know the place for the first time.

—T. S. Eliot, "Four Quartets"

REFERENCES

1. H. Sage, C. Johnson, and P. Bornstein "Characterization of a Novel Serum Albumin-Binding Glycoprotein Secreted by Endothelial Cells in Culture." *J Biol Chem* 259 (1984): 3993–4007.

2. I. J. Mason, A. Taylor, J. G. Williams, H. Sage, and B. L. M. Hogan. "Evidence from Molecular Cloning That SPARC, a Major Product of Mouse Embryo Parietal Endoderm, Is Related to an Endothelial Cell 'Culture Shock' Glycoprotein of Mr=43,000." *EMBO J* 5 (1986): 1465–72.

3. T. F. Lane and E. H. Sage. "The Biology of SPARC, a Protein That Modulates Cell-Matrix Interactions." *FASEB J* 8 (1994): 163–73 (PMID: 8119487).

4. R. A. Brekken and E. H. Sage. "SPARC, A Matricellular Protein: At the Crossroads of Cell-Matrix Communication." *Matrix Biol* 19 (2001): 569–80 (PMID: 11223341).

5. M. S. Weaver, G. Workman, and E. H. Sage. "The Copper-Binding Domain of SPARC Mediates Cell Survival In Vitro via Interaction with Integrin Beta 1 and Activation of Integrin-Linked Kinase." *J Biol Chem* 283 (2008): 22826–37.

6. H. Sage, R. B. Vernon, J. Decker, S. Funk, and M. L. Iruela-Arispe. "Distribution of the Calcium-Binding Protein SPARC in Tissues of Embryonic and Adult Mice." *J Histochem Cytochem* 37 (1989): 819–29.

7. H. Sage, R. B. Vernon, S. E. Funk, E. A. Everitt, and J. Angello. "SPARC, a Secreted Protein Associated with Cellular Proliferation, Inhibits Cell Spreading In Vitro and Exhibits Ca+2-Dependent Binding to the Extracellular Matrix." *J Cell Biol* 109 (1989): 341–56.

8. S. E. Goldblum, X. Ding, S. E. Funk, and E. H. Sage. "SPARC (Secreted Protein Acidic and Rich in Cysteine) Regulates Endothelial Cell Shape and Barrier Function." *Proc Natl Acad Sci USA* 91 (1994): 3448–52 (PMID: 8159767).

9. S. E. Funk and E. H. Sage. "The Ca^{2+}-Binding Glycoprotein SPARC Modulates Cell Cycle Progression in Bovine Aortic Endothelial Cells." *Proc Natl Acad Sci USA* 88 (1991): 2648–52 (PMID: 2011576).

10. E. W. Raines, T. F. Lane, M. L. Iruela-Arispe, R. Ross, and E. H. Sage. "The Extracellular Glycoprotein SPARC Interacts with Platelet-Derived Growth Factor (PDGF)-AB and -BB and Inhibits the Binding of PDGF to Its Receptors." *Proc Natl Acad Sci USA* 89 (1992): 1281–85 (PMID: 1311092).

11. C. Kupprion, K. Motamed, and E. H. Sage. "SPARC (BM-40, Osteonectin) Inhibits the Mitogenic Effect of Vascular Endothelial Growth Factor on Microvascular Endothelial Cells." *J Biol Chem* 273 (1998): 29635–40 (PMID: 9792673).

12. J. E. Murphy-Ullrich, T. F. Lane, M. A. Pallero, and E. H. Sage. "SPARC Mediates Focal Adhesion Disassembly in Endothelial Cells through a Follistatin-Like Region and the Ca^{2+}-Binding EF-Hand." *J Cell Biochem* 57 (1995): 341–50 (PMID: 7539008).

13. M. L. Iruela-Arispe, R. B. Vernon, H. Wu, R. Jaenisch, and E. H. Sage. "Type I Collagen-Deficient Mov-13 Mice Do Not Retain SPARC in the Extracellular Matrix: Implications for Fibroblast Function." *Dev Dynamics* 207 (1996): 171–83 (PMID: 8906420).

14. M. H. Huynh, E. H. Sage, and M. Ringuette. "A Calcium-Binding Motif in SPARC/ Osteonectin Inhibits Chordomesoderm Cell Migration during *Xenopus laevis* Gastrulation: Evidence of Counteradhesive Activity In Vivo." *Develop Growth Differ* 41 (1999): 407–18 (PMID: 10466928).

15. J. A. Bassuk, R. Pichler, J. D. Rothmier, J. Pippen, K. Gordon, R. L. Meek, A. D. Bradshaw, et al. "Induction of TGF-β1 by the Matricellular Protein SPARC in a Rat Model of Glomerulonephritis." *Kidney Int* 57 (2000): 117–28 (PMID: 10620193).

16. J. E. Murphy-Ullrich. "The De-adhesive Activity of Matricellular Proteins: Is Intermediate Cell Adhesion an Adaptive State?" *J Clin Invest* 107 (2001): 785–90 (PMID: 11285293).

17. E. H. Sage and P. Bornstein. "Minireview: Extracellular Proteins That Modulate Cell-Matrix Interactions: SPARC, Tenascin, and Thrombospondin." *J Biol Chem* 266 (1991): 14831-34 (PMID: 1714444).

18. P. Bornstein. "Diversity of Function Is Inherent in Matricellular Proteins: An Appraisal of Thrombospondin-1." *J Cell Biol* 130 (1995): 503–6 (PMID: 7542656).
19. P. Bornstein and E. H. Sage. "Matricellular Proteins: Extracellular Modulators of Cell Function." *Curr Opin Cell Biol* 14 (2002): 608–16.
20. P. M. Tremble, T. F. Lane, E. H. Sage, and Z. Werb. "SPARC, a Secreted Protein Associated with Morphogenesis and Tissue Remodeling, Induces Expression of Metalloproteinases in Fibroblasts through a Novel Extracellular Matrix-Dependent Pathway." *J Cell Biol* 121 (1993): 1433–44 (PMID: 8509459).
21. M. L. Iruela-Arispe, T. F. Lane, D. Redmond, M. Reilly, R. Bolender, T. J. Kavanagh, and E. H. Sage. "Expression of SPARC during Development of the Chicken Chorioallantoic Membrane: Evidence for Regulated Proteolysis In Vivo." *Molec Biol Cell* 6 (1995): 327–43 (PMID: 7612967).
22. T. F. Lane, M. L. Iruela-Arispe, R. S. Johnson, and E. H. Sage. "SPARC Is a Source of Copper-Binding Peptides That Stimulate Angiogenesis." *J Cell Biol* 125 (1994): 929–43 (PMID: 7514608).
23. E. H. Sage, M. Reed, S. E. Funk, T. Truong, M. Steadele, P. Puolakkainen, D. H. Maurice, and J. Bassuk. "Cleavage of the Matricellular Protein SPARC by Matrix Metalloproteinase 3 Produces Polypeptides That Influence Angiogenesis." *J Biol Chem* 278 (2003): 37849–57 (PMID: 12867428).
24. E. H. Sage. "Pieces of Eight: Bioactive Fragments of Extracellular Proteins as Regulators of Angiogenesis." *Trends Cell Biol* 7 (1997): 182–86.
25. K. Norose, J. I. Clark, N. A. Syed, A. Basu, E. Heber-Katz, E. H. Sage, and C. C. Howe. "SPARC Deficiency Leads to Early Onset Cataractogenesis." *Invest Ophthal Vis Sci* 39 (1998): 2674–80 (PMID: 9856777).
26. Q. Yan, J. I. Clark, T. Wight, and E. H. Sage. "Alterations in the Lens Capsule Contribute to Cataractogenesis in SPARC-Null Mice." *J Cell Sci* 115 (2002): 2747–56 (PMID: 12077365).
27. D. T. Gilmour, G. J. Lyon, M. B. Carlton, J. R. Sanes, J. M. Cunningham, J. R. Anderson, B. L. Hogan, M. J. Evans, and W. H. Colledge. "Mice Deficient for the Secreted Glycoprotein SPARC/Osteonectin/BM40 Develop Normally but Show Severe Age-Onset Cataract Formation and Disruption of the Lens." *EMBO J* 17 (1998): 1860–70 (PMID: 9524110).
28. R. O. Emerson, E. H. Sage, J. G. Ghosh, and J. I. Clark. "Chaperone-Like Activity Revealed in the Matricellular Protein SPARC." *J Cell Biochem* 98 (2006): 701–5 (PMID: 16598771).
29. H. Sage, J. Tupper, and R. Bramson. "Endothelial Cell Injury In Vitro Is Associated with Increased Secretion of an Mr 43,000 Glycoprotein Ligand." *J Cell Physiol* 127 (1986): 373–87 (PMID: 2423540).
30. Q. Shi, S. Bao, L. Song, Q. Wu, D. D. Bigner, A. B. Hjelmeland, and J. N. Rich. "Targeting SPARC Expression Decreases Glioma Cellular Survival and Invasion Associated with Reduced Activities of FAK and ILK Kinases." *Oncogene* 26 (2007): 4084–94 (PMID: 17213807).
31. A. M. Delany, M. Amling, M. Priemel, C. Howe, R. Baron, and E. Canalis. "Osteopenia and Decreased Bone Formation in Osteonectin-Deficient Mice." *J Clin Invest* 105 (2000): 915–23 (PMID: 10749571).
32. H. E. Gruber, E. H. Sage, H. J. Norton, S. Funk, J. Ingram, E. N. Hanley. "Targeted Deletion of the SPARC Gene Accelerates Disc Degeneration in the Aging Mouse." *J Histochem Cytochem* 53 (2005): 1131–38 (PMID: 15879573).
33. A. D. Bradshaw, D. C. Graves, K. Motamed, and E. H. Sage. "SPARC-Null Mice Exhibit Increased Adiposity Without Significant Differences in Overall Body Weight." *Proc Natl Acad Sci USA* 100 (2003): 6045–50 (PMID: 12721366).

34. A. D. Bradshaw and E. H. Sage. "SPARC, a Matricellular Protein That Functions in Cellular Differentiation and Tissue Tesponse to Injury." *J Clin Invest* 107 (2001): 1049–54 (PMID: 11342565).

35. A. D. Bradshaw, P. Puolakkainen, J. Dasgupta, J. M. Davidson, T. N. Wight, E. H. Sage. "SPARC-Null Mice Display Abnormalities in the Dermis Characterized by Decreased Collagen Fibril Diameter and Reduced Tensile Strength." *J Invest Dermatol* 120 (2003): 949–55 (PMID: 12787119).

36. T. J. Rentz, F. Poobalarahi, M. Collins, P. Bornstein, E. H. Sage, and A. D. Bradshaw. "SPARC Regulates Processing of Procollagen I and Collagen Fibrillogenesis in Dermal Fibroblasts." *J Biol Chem* 282 (2007): 22062–71 (PMID: 17522057).

37. T. P. Strandjord, D. K. Madtes, D. J. Weiss, and E. H. Sage. "Collagen Accumulation Is Decreased in SPARC-Null Mice with Bleomycin-Induced Pulmonary Fibrosis." *Am J Physiol* 277 (1999): L628–35 (PMID: 10484471).

38. P. Puolakkainen, A. D. Bradshaw, T. R. Kyriakides, M. Reed, R. Brekken, T. Wight, P. Bornstein, B. Ratner, and E. H. Sage. "Compromised Production of Extracellular Matrix in Mice Lacking Secreted Protein, Acidic and Rich in Cysteine (SPARC), Leads to a Reduced Foreign Body Reaction to Implanted Biomaterials." *Am J Pathol* 162 (2003): 627–35 (PMID: 12547720).

39. A. D. Bradshaw, M. J. Reed, and E. H. Sage. "SPARC-Null Mice Exhibit Accelerated Cutaneous Wound Closure." *J Histochem Cytochem* 50 (2002): 1–10 (PMID: 11748289).

40. A. Francki and E. H. Sage. "SPARC and the Kidney Glomerulus: Matricellular Proteins Exhibit Diverse Functions under Normal and Pathological Conditions." *Trends Cardiovasc Med* 11 (2001): 32–37 (PMID: 11413050).

41. A. M. Delany, I. Kalajzic, A. D. Bradshaw, E. H. Sage, and E. Canalis. "Osteonectin-Null Mutation Compromises Osteoblast Formation, Maturation, and Survival." *Endocrinology* 144 (2003): 2588–96.

42. J. Nie and E. H. Sage. "SPARC Inhibits Adipogenesis by Its Enhancement of β-Catenin Signaling." *J Biol Chem* 284 (2009): 1279–90.

43. E. H. Sage. "Terms of Attachment: SPARC and Tumorigenesis—News & Views." *Nat Med* 3 (1997): 144–46 (PMID: 9018225).

44. A. Chlenski, S. Liu, L. J. Baker, Q. Yang, Y. Tian, H. R. Salwen, S. L. Cohn. "Neuroblastoma Angiogenesis Is Inhibited with a Folded Synthetic Molecule Corresponding to the Epidermal Growth Factor-Like Module of the Follistatin Domain of SPARC." *Cancer Res* 64 (2004): 7420–25 (PMID: 15492265).

45. R. A. Brekken, P. Puolakkainen, D. C. Graves, G. Workman, S. R. Lubkin, and E. H. Sage. "Enhanced Growth of Tumors in SPARC-Null Mice Is Associated with Changes in the ECM." *J Clin Invest* 111 (2003): 487–95 (PMID: 12588887).

46. P. A. Puolakkainen, R. A. Brekken, S. Muneer, and E. H. Sage. "Enhanced Growth of Pancreatic Tumors in SPARC-Null Mice Is Associated with Decreased Deposition of Extracellular Matrix and Reduced Tumor Cell Apoptosis." *Mol Cancer Res* 2 (2004): 215–24 (PMID: 15140943).

47. P. E. Framson and E. H. Sage. "Prospects: SPARC and Tumor Growth: Where the Seed Meets the Soil? *J Cell Biochem* 92 (2004): 679–90 (PMID: 15211566).

48. C. J. Clark and E. H. Sage. "A Prototypic Matricellular Protein in the Tumor Microenvironment—Where There's SPARC, There's Fire." *J Cell Biochem* 104 (2008): 721–32.

49. O. L. Podhajcer, L. Benedetti, M. R. Girotti, F. Prada, E. Salvatierra, and A. S. Llera. "The Role of the Matricellular Protein SPARC in the Dynamic Interaction between the Tumor and the Host." *Cancer Metastasis Rev* 27 (2008): 523-37 (PMID: 18459035).

50. T. J. Bos, S. L. Cohn, H. K. Kleinman, J. E. Murphy-Ullrich, O. L. Podhajcer, S. A. Rempel, J. N. Rich, J. T. Rutka, E. H. Sage, and E. W. Thompson. "International Hermelin Brain Tumor Symposium on Matricellular Proteins in Normal and Cancer Cell-Matrix Interactions." *Matrix Biol* 23 (2004): 63–69 (PMID: 15230275).

51. T. H. Barker, G. Baneyx, M. Cardó-Vila, G. A. Workman, M. Weaver, P. M. Menon, S. Dedhar, et al. "SPARC Regulates Extracellular Matrix Organization through Its Modulation of Integrin-Linked Kinase Activity." *J Biol Chem* 43 (2005): 36483–93 (PMID: 16115889).

52. J. Nie, B. Chang, D. O. Traktuev, J. Sun, K. March, L. Chan, E. H. Sage, R. Pasqualini, W. Arap, and M. G. Kolonin. "IFATS Collection: Combinatorial Peptides Identify $\alpha 5 \beta 1$ Integrin as a Receptor for the Matricellular Protein SPARC on Adipose Stromal Cells." *Stem Cells* 26 (2008): 2735–45.

53. K. A. Kelly, J. R. Allport, A. M. Yu, E. H. Sage, R. E. Gerszten, and R. Weissleder. "SPARC Is a VCAM-1 Counter-Ligand That Mediates Leukocyte Transmigration." *J Leukocyte Biol* 81 (2007): 748–56 (PMID: 17178915).

54. J. Kzhyshkowska, G. Workman, M. Cardó-Vila, W. Arap, R. Pasqualini, A. Gratchev, L. Krusell, S. Goerdt, and E. H. Sage. "Novel Function of Alternatively Activated Macrophages: Stabilin-1-Mediated Clearance of SPARC." *J Immunol* 176 (2006): 5825–32 (PMID: 16670288).

55. M. Sullivan and E. H. Sage. "Molecules in Focus: hevin/SC1, a Matricellular Glycoprotein and Potential Tumor-Suppressor of the SPARC/BM-40/Osteonectin Family." *Int J Biochem Cell Biol* 36 (2004): 991–96 (PMID: 15094114).

56. V. Gongidi, C. Ring, M. Moody, R. Brekken, E. H. Sage, R. Rakic, and E. S. Anton. "SPARC-Like 1 Regulates the Terminal Phase of Radial Glia-Guided Migration in the Cerebral Cortex." *Neuron* 41 (2004): 57–69 (PMID: 14715135).

57. M. M. Sullivan, T. H. Barker, S. E. Funk, A. Karchin, N. S. Seo, M. Höök, J. Sanders, et al. "Matricellular hevin Regulates Decorin Production and Collagen Assembly." *J Biol Chem* 281 (2006): 27621–32 (PMID: 16844696).

58. T. H. Barker, P. Framson, P. Puolakkainen, M. Reed, S. E. Funk, and E. H. Sage. "Matricellular Homologs in the Foreign Body Response: Hevin Suppresses Inflammation, but Hevin and SPARC Together Diminish Angiogenesis." *Am J Pathol* 166 (2005): 923–33 (PMID: 15743803).

59. M. D. Gooden, R. B. Vernon, J. A. Bassuk, and E. H. Sage. "Cell Cycle-Dependent Nuclear Location of the Matricellular Protein SPARC: Association with the Nuclear Matrix." *J Cell Biochem* 74 (1999): 152–67 (PMID: 10404386).

60. M. J. Reed and E. H. Sage. "SPARC and the Extracellular Matrix: Implications for Cancer and Wound Repair," in U. Günthert and W. Birchmeier, eds., *Attempts to Understand Metastasis Formation I. Curr. Top. Microbiol. Immunol.*, vol. 231, 81–94 (Berlin: Springer-Verlag, 1996).

61. N. Martinek, J. Shahab, J. Sodek, and M. Ringuette. "Is SPARC an Evolutionarily Conserved Collagen Chaperone?" *J Dent Res* 86 (2007): 296–305 (PMID: 17384023).

62. J. T. Hecht and E. H. Sage. "Retention of the Matricellular Protein SPARC in the Endoplasmic Reticulum of Chondrocytes from Patients with Pseudoachondroplasia." *J Histochem Cytochem* 54 (2006): 269–74 (PMID: 16286662).

63. J. E. Schwarzbauer and C. S. Spencer. "The *Caenorhabditis elegans* Homologue of the Extracellular Calcium Binding Protein SPARC/Osteonectin Affects Nematode Body Morphology and Mobility." *Mol Biol Cell* 4 (1993): 941–52 (PMID: 8257796).

64. M. Nozaki, E. Sakurai, B. J. Raisler, J. Z. Baffi, J. Witta, Y. Ogura, R. A. Brekken, E. H. Sage, B. K. Ambati, and J. Ambati. "Loss of SPARC-Mediated VEGFR-1 Suppression Post Injury Reveals a Novel Antiangiogenic Activity of VEGF-A." *J Clin Invest* 116 (2006): 422–29 (PMID: 16453023).

Index